1524722-1
8/11/03

C0-ALP-197

Friedrich G. Barth
Joseph A. C. Humphrey
Timothy W. Secomb
(eds.)

Sensors and Sensing
in Biology and Engineering

SpringerWienNewYork

Friedrich G. Barth
Biocenter, Institute of Zoology – Neurobiology, University of Vienna, Austria

Joseph A.C. Humphrey
Department of Mechanical and Aerospace Engineering, University of Virginia,
Charlottesville, VA, USA

Timothy W. Secomb
Department of Physiology, University of Arizona, Tucson, AZ, USA

This work is subject to copyright.
All rights are reserved, whether the whole or part of the material is concerned, specifically those of translation, reprinting, re-use of illustrations, broad-casting, reproduction by photocopying machines or similar means, and storage in data banks.

© 2003 Springer-Verlag Wien
Printed in Austria

Product Liability: The publisher can give no guarantee for all the information contained in this book. This does also refer to information about drug dosage and application thereof. In every individual case the respective user must check its accuracy by consulting other pharmaceutical literature. The use of registered names, trademarks, etc. in this publication does not imply, even in the absence of a specific statement, that such names are exempt from the relevant protective laws and regulations and therefore free for general use.

This book is based upon work supported by the National Science Foundation under Grant No. 0196067. Any opinions, findings, and conclusions or recommendations expressed in this material are those of the authors and do not necessarily reflect the views of the National Science Foundation.

Cover illustration: © Nicolas Franceschini
Head of a fly in the center of a complex routing diagram used to integrate visual information about obstacles and target onboard a fly-inspired visually-guided robot (cf. chapter 16)
Typesetting: Thomson Press (India) Ltd., Chennai
Printing and binding: Druckerei Theiss GmbH, A-9431 St. Stefan im Lavanttal
Printed on acid-free and chlorine-free bleached paper
SPIN: 10847682
CIP data applied for

With 169 partly coloured Figures

ISBN 3-211-83771-X Springer-Verlag Wien New York

R
857
.B54
B37
2002

Preface

All living beings, ranging from bacteria to highly developed plants and animals and man possess sensors and sensing processes, by means of which they perceive and respond to the environments within and outside them. For all of them, sensing is critical for the guidance of their particular behavior, and for regulating metabolic, reproductive and other processes. The sensors of living organisms have been perfected by evolutionary advances and permanent "quality control" through natural selective pressures over hundreds of millions of years. This evolutionary process has led to a fascinating number and variety of sensory systems, rich in terms of design variations. Biological sensors are often remarkably selective and sensitive, and, in addition, exquisitely small. Their "designs" exhibit impressive efficiency, successfully adjusted by variation and natural selection to fulfill the needs of a particular species in its particular habitat.

Both the miniaturization and the integration of many functions are typical features of natural sensory systems. During the last few decades, technology has made possible the design and fabrication of corresponding artificial sensors for scientific, industrial, and commercial purposes. Increasingly, the trend is towards miniaturization with Micro- and Nano-Electro-Mechanical Systems (MEMS, NEMS) being the focus of attention. Likewise there is a trend towards systems integrating functions lying beyond the mere generation of the desired measuring signal such as signal storage, permanent recalibration and malfunction monitoring. Artificial microsensors have the potential to revolutionize the way science and engineering are practiced, over a wide spectrum of professions and industries ranging from medicine and bioengineering to robotics and manufacturing and intra- and extraterrestrial exploration.

Curiously, there has to date been little organized and serious cross-disciplinary discussion and information exchange on the topic of sensors and sensing, among biologists, engineers, mathematicians, and physical scientists. In an effort to redress this situation, under the auspices of the National Science Foundation and the United Engineering Foundation we organized an international symposium on the subject of *Sensors and Sensing in the Natural and Fabricated Worlds.* The symposium was held from June 11–16, 2000 in Il Ciocco, Italy, with over sixty internationally recognized participants attending. The chapters in this book are expanded versions of several of the many noteworthy lectures given at the symposium.

Our premise is that many of the complex biophysical and biochemical processes underpinning sensors and sensing in living organisms are governed by the same fundamental physical-chemical principles and laws that describe artificial sensors. With its focus on *Sensors and Sensing in Biology and Engineering,* this book seeks to uncover some biological, physical-chemical and engineering fundamentals that will allow a fuller understanding of the nature and performance of natural sensors and, as a result, the improved and enriched conceptualization, design, fabrication and range of applicability of artificial microsensors. The book contains 26 chapters that in an exemplary way serve this purpose. Its first two chapters are essays introducing the reader to the world of sensors and sensing as seen from the viewpoints of a biologist and an engineer, respectively. 22 of the remaining chapters deal with different forms of stimulus and are grouped under the headings *Mechanical Sensors, Visual Sensors and Vision*, and *Chemosensors and Chemosensing*. The concluding chapters are devoted to *embedded sensors* and *active dressware*, as examples of the development of novel types of sensors.

We would like to thank the National Science Foundation (Grant No. 0196067) for financial support of both our Conference in Il Ciocco and the production of this book, and Springer Verlag, Wien New York, for a constructive and pleasant collaboration.

July 2002   *Friedrich G Barth*, Wien
Joseph AC Humphrey, Charlottesville
Timothy W Secomb, Tucson

Contents

Visual Sensors and Vision

Chemosensors and Chemosensing

The Embedding of Sensors

List of Contributors

Friedrich G. Barth
Biocenter, Institute of Zoology
Vienna University, Althanstr. 14
A-1090 Vienna, Austria
friedrich.g.barth@univie.ac.at

J. Mark Blanchard
Institute of Neuroinformatics
University/ETH Zürich
Winterthurer Str. 190
CH-8075 Zürich, Switzerland
jmb@ini.phys.ethz.ch

William E. Brownell
Bobby R. Alford Dept. of Otorhinolaryngology
and Communicative Sci.
NA 505 – Baylor College of Medicine
Houston, Texas 77030, USA
brownell@bcm.tmc.edu

Paul Calvert
Dept. of Materials Science and Engineering
University of Arizona
Tucson, AZ 85721, USA
calvert@engr.arizona.edu

Maria Chiara Carrozza
ARTS & CRIM Labs
Scuola Superiore Sant'Anna
Piazza Martiri della Libertà, 33
I-56127 Pisa, Italy
Chiara@mail-arts.sssup.it

Paolo Dario
ARTS & CRIM Labs
Scuola Superiore Sant'Anna
Piazza Martiri della Libertà, 33
I-56127 Pisa, Italy
dario@mail-arts.sssup.it

Danilo De Rossi
Centro E. Piaggio
Faculty of Engineering
University of Pisa
Via Diotisalvi 2

I-56126 Pisa, Italy
derossi@piaggio.ccii.unipi.it

Lori J. Dodson-Dreibelbis
Gas Dynamics Laboratory, Mechanical
and Nuclear Engineering Dept.
301D Reber Bldg.
University Park,
PA 16802, USA
ljd3@psu.edu

David B. Dusenbery
School of Biology
Georgia Institute of Technology
310 Ferst Drive
Atlanta, GA 30332-0230, USA
david.dusenbery@biology.gatech.edu

Tony Farquhar
Dept. of Mechanical Engineering
Room 234, Engineering & Computer
Science (ECS) Building
University of Maryland
Baltimore County 21250, USA
farquhar@research.umbc.edu

Nicolas Franceschini
Dept. of Biorobotics
Laboratory "Motion and Perception"
CNRS/Univ. de la Méditerranée
31, Chemin J. Aiguier
F-13402 Marseille
Cedex 20, France
franceschini@laps.univ-mrs.fr

Henry W. Haslach, Jr.
Department of Mechanical Engineering
University of Maryland, College Park
College Park, MD 20742, USA
haslach@eng.umd.edu

Travis W. Hein
Department of Medical Physiology
Cardiovascular Research Institute
College of Medicine

Texas A + M University System Health
Science College Station
TX 77843-1114, USA
thein@tamu.edu

Joseph A.C. Humphrey
Department of Mechanical and
Aerospace Engineering
University of Virginia
Charlottesville, Virginia 22904, USA
jach@virginia.edu

John S. Kauer
Dept. of Neuroscience
Tufts University School of Medicine
136 Harrison Av.
Boston, MA 02111, USA
john.kauer@tufts.edu

Douglas A. Kester
c/o York International, 1419 Monroe St.,
York, PA 17404, USA
douglas.kester@york.com

Walter G. Kropatsch
Vienna University of Technology
Institute of Computer Aided
Automation, 183/2
Pattern Recognition+Image
Processing Group
Favoritenstraße 9
A-1040 Wien, Austria
krw@prip.tuwien.ac.at

Tsuneko Kumagai
Neuro Cybernetics Laboratory
Research Institute for Electronic Science
Hokkaido University
Sapporo 060-0812, Japan
kmg@ncp8.es.hokudai.ac.jp

Lih Kuo
Department of Medical Physiology
Cardiovascular Research Institute
College of Medicine
Texas A&M University System Health
Science Center, College Station
Texas 77843-1114, USA
lkuo@tamu.edu

Cecilia Laschi
ARTS & CRIM Labs
Scuola Superiore Sant'Anna
Piazza Martiri della Libertà, 33
I-56127 Pisa, Italy
cecilia@mail-arts.sssup.it

Federico Lorussi
Centro E. Piaggio
Faculty of Engineering
University of Pisa
Via Diotisalvi 2
I-56126 Pisa, Italy
lorussi@piaggio.ccii.unipi.it

Arun Majumdar
Department of Mechanical Engineering
University of California
Berkeley, CA 94720-1740, USA
majumdar@me.berkeley.edu

Barbara Mazzolai
ARTS & CRIM Labs
Scuola Superiore Sant'Anna
Piazza Martiri della Libertà, 33
I-56127 Pisa, Italy
barbara@mail-arts.sssup.it

Alberto Mazzoldi
Centro E. Piaggio
Faculty of Engineering
University of Pisa
Via Diotisalvi 2
I-56126 Pisa, Italy
alberto@piaggio.ccii.unipi.it

Hans Meixner
Corporate Technology Siemens AG
Otto-Hahn-Ring 6
D-81730 Munich, Germany
hans.meixner@siemens.com

Arianna Menciassi
ARTS & CRIM Labs
Scuola Superiore Sant'Anna
Piazza Martiri della Libertà, 33
I-56127 Pisa, Italy
arianna@sssup.it

Giorgio Metta
LIRA-Lab
DIST – University of Genova
Viale Causa 13
I-16145 Genova, Italy
metta@dist.unige.it

Silvestro Micera
ARTS & CRIM Labs
Scuola Superiore Sant'Anna
Piazza Martiri della Libertà, 33
I-56127 Pisa, Italy
micera@sssup1.sssup.it

Axel Michelsen
Centre for Sound Communication
Institute of Biology
University of Southern Denmark
DK-5230 Odense M, Denmark
a.michelsen@biology.sdu.dk

Joachim Mogdans
Institut für Zoologie
Universität Bonn
Poppelsdorfer Schloss
D-53115 Bonn, Germany
mogdans@uni-bonn.de

Jun Murakami
Neuro Cybernetics Laboratory
Research Institute for Electronic Science
Hokkaido University
Sapporo 060-0812, Japan
murakami@ncp8.es.hokudai.ac.jp

Piero Orsini
Azienda Ospedaliera Pisana
U.O. Neurologia S.A.
Recupero e Rieducazione Funzionale Presidio
Ospedaliero Cisanello
Via Paradisa 2
I-56100 Pisa, Italy
orsini@piaggio.ccii.unipi.it

Axel R. Pries
Department of Physiology
Freie Universität Berlin
Arnimallee 22, D-14195
Berlin, and German Heart Center Berlin
Augustenburger Platz 1
D-13353 Berlin, Germany
pries@zedat.fu-berlin.de

Per Rasmussen
G.R.A.S. Sound & Vibration
Staktoften 22D
DK-2950 Vedbaek, Denmark
pr@gras.dk

Michael Reed
Department of Electrical and
Computer Engineering
University of Virginia
Charlottesville, VA 22904, USA
reed@virginia.edu

F. Claire Rind
School of Neurosciences and Psychiatry
Henry Wellcome Building
University of Newcastle

Framlington Place
Newcastle upon Tyne NE2 4HH, UK
claire.rind@ncl.ac.uk

John J. Rosowski
Eaton-Peabody Laboratory
Massachusetts Eye and Ear Infirmary
243 Charles Street
Boston, MA 02114, USA
jjr@epl.meei.harvard.edu

Frederick Sachs
Hughes Center for Single Molecule
Biophysics Physiology and Biophysical
Sciences
SUNY Buffalo
320 Cary, Buffalo, NY 14214, USA
sachs@buffalo.edu

Giulio Sandini
LIRA-Lab
DIST – University of Genova
Viale Causa 13
I-16145 Genova, Italy
sandini@dist.unige.it

Roger D. Santer
School of Biology
Henry Wellcome Building
University of Newcastle
Framlington Place
Newcastle upon Tyne NE2 4HH, UK
r.d.santer@ncl.ac.uk

Rahul Sarpeshkar
Massachusetts Institute of Technology
Room 38-294
77 Mass. Avenue
Cambridge, MA 02139, USA
rahuls@mit.edu

Enzo Pasquale Scilingo
Centro E. Piaggio
Faculty of Engineering
University of Pisa
Via Diotisalvi 2
I-56126 Pisa, Italy
pasquale@piaggio.ccii.unipi.it

Timothy W. Secomb
Department of Physiology
University of Arizona
Tucson, AZ 85724-5051, USA
secomb@u.arizona.edu

Gary Settles
Dynamics Laboratory, Mechanical
and Nuclear Engineering Dept.
301D Reber Bldg.
University Park, PA 16802, USA
gss2@psu.edu

Tateo Shimozawa
Neuro Cybernetics Laboratory
Research Institute for Electronic Science
Hokkaido University
Sapporo 060-0812, Japan
tateo@ncp8.es.hokudai.ac.jp

Kenneth V. Snyder
Hughes Center for Single Molecule Biophysics
Physiology and Biophysical Sciences
SUNY Buffalo
320 Cary, Buffalo, NY 14214, USA
ksnyder@acsu.buffalo.edu

Aaron Spak
Northrop Grumman Corporation
1212 Winterson Road
Linthicum, Maryland 21090, USA
aaron_j_spak@md.northgrum.com

Thomas Thundat
Life Sciences Division
Oak Ridge National Laboratory
Mail Stop 6123, Rm. G148, 4500S
Oak Ridge, TN 37831-6123, USA
thundattg@ornl.gov

Fabrizio Vecchi
ARTS & CRIM Labs
Scuola Superiore Sant'Anna
Piazza Martiri della Libertà, 33
I-56127 Pisa, Italy
fabrizio@mail-arts.sssup.it

Paul F.M.J. Verschure
Institute of Neuroinformatics
University/ETH Zürich
Winterthurer Str. 190
CH-8075 Zürich, Switzerland
pfmjv@ini.phys.ethz.ch

Massimiliano Zecca
ARTS & CRIM Labs
Scuola Superiore Sant'Anna
Piazza Martiri della Libertà, 33
I-56127 Pisa, Italy
max@sssup.it

Jiang Zhou
Department of Mechanical Engineering
University of Maryland, Baltimore County
Baltimore, MD 21250, USA
jizhou@umbc.edu

Steven W. Zucker
Dept. of Computer Science and
Electrical Engineering, Yale University
New Haven, CT 06520-8285, USA
steven.zucker@yale.edu

Introductory Remarks

1. Sensors and Sensing: A Biologist's View

Friedrich G. Barth

Abstract

Sensors and sensing are essential for all forms of
life. Correspondingly, there is a fascinating richness and diversity of sensory systems throughout
the animal kingdom. This essay was mainly written for the non-biologist who is introduced to a
few selected aspects of sensory biology. Foremost
among these are some working principles common to biological sensors, whatever the form of
energy is they respond to. The physiology of
membrane channels and the outstanding sensitivity of many bio-sensors are highlighted. It is
stressed that besides the cellular and molecular
details the refinement of a sense organ's "engineering" reflects the needs of an entire behaving
animal. Accordingly, the relevant and often complex natural stimuli, which the sensors evolved to
process, have to be found and applied, in order to
fully understand any sensory system. This is a
particular challenge in cases where animals show
sensory capabilities alien to our human perceptions. The two main links between biology and
engineering are "technical biology" and "bionics".
The application of computational skills and mathematical quantification are the most valuable inputs from engineering to sensory biology. The
biologist on the other hand will not provide ready
made instructions of how to build a device but
offer familiarity with a treasure house of inspiring
solutions to a wealth of sensory problems.

I. Introduction

Life without sensors and sensing would be
like an opera without singers or a violin
without strings. Such life does not exist.
Sensors and sensing, on the contrary, are
basic properties of life and as characteristic
of it as are metabolism and reproduction.

From bacteria to vertebrates organisms
critically depend on information about
their outside and inside worlds. They
need this information to behave in a way
that makes sense, that is to say, that guar-
antees survival and provides fitness in
terms of successful reproduction and the
passing on of their genes. In more practical terms, sensing is essential for orientation and communication, for finding the
mate, prey, or a host, for body and locomotion control, and other functions. In
more technical terms, one may say that
sensors and sensing are responsible for
the closed-loop real time control of what
is going on inside the organism and how
it reacts to the outside situation.

Animals have developed a stunning diversity of sensory systems to gain the information they need. In all cases, however, basically the same thing happens.
All sensors absorb some kind of energy

(an exception being cold receptors responding to loss of thermal energy), typically in minute quantity, and generate electrical signals. In all but the phylogenetically most primitive animals these signals carry the information to a central nervous system for further processing and filtering, for planning, initiating, or modifying motor actions, or just for being stored in the memory for future nervous activity and for behavior. This short introductory essay mainly intends to address engineers and physical scientists not too familiar with biology. It will try to make clear a few aspects threaded through sensory neuroscience which attract the interest and curiosity of a biologist. The chapter will start with a short explanation of the basic events taking place in a sense organ and it will end with a personal view (though based on many years of practical experience) on the interaction between biology and engineering.

II. Basics

The sequence of events that lies between the uptake of a particular stimulus and the generation of electrochemical signals such as action potentials is the following in all cases of biological sensors:

A. Stimulus Uptake and Transformation

Non-nervous auxiliary structures "designed" to absorb energy of a particular form take up a stimulus and pass it on to the sensory cells proper. While doing this they transform the stimulus in a way that makes it suitable for being taken up by the sensory cells. The term *stimulus transformation* refers to the fact that the form of energy of the input stimulus does not change. Thus, a visual stimulus still is a visual one after having passed through the cornea and lens of our eye, an acoustic stimulus still is a mechanical stimulus after its transformation by the eardrum and middle ear ossicles, and a tactile stimulus still is a mechanical event after its transformation by the leverage of an arthropod cuticular hair. It is fair to say that the most fascinating diversity among biological sensors is due to the diversity of their non-nervous auxiliary structures. It is here where the nicest tricks are found and where a multitude of solutions to technically complex problems is to be expected (see for instance chapters 3, 5, 9, 10 and 11, this volume).

B. Stimulus Transduction

Contrasting stimulus transformation the processes of stimulus transduction are much more uniform across the spectrum of biological sensors. Take a photoreceptor cell, a hair cell in the inner ear, or an olfactory cell from an animal's nose and basically you will find the same phenomena, independent of the type of energy characterizing the stimulus. Sensory physiologists call these processes stimulus transduction because the energy form of the already-transformed (modified) stimulus is transduced into the electro-chemical form of a nervous signal. The structure where this happens is the membrane of the sensory cell. An essential feature of membranes of all living cells is that they represent a condenser (with a capacitance of about $1 \mu F$ per cm^2) which, due to an active process (the electrogenic $Na+/K+$ pump) using up metabolic energy, separates ions and thus charges in the cell interior from those in the extracellular medium. The results are concentration gradients e.g. for $Na+$, $K+$, and $Cl-$ across the cell membrane which, combined with a selective permeability for $K+$ (in its resting state the membrane is more permeable to $K+$ than e.g. to $Na+$

by a factor of about 25) leads to a so called resting membrane potential of roughly 40 to 90 mV, the inside of the membrane being negative relative to the outside. The specialization of sensory cells is that their respective adequate stimulus modulates a resistance (conductivity, respectively) in the membrane and in a circuit which is driven by the batteries across the membrane. Importantly, the processes of transduction and signal generation are not driven by the energy contained in the stimulus. Instead, the stimulus mainly serves to modulate a resistance r in a gating process similar to the turning of a water hose (valve) which is not the driving force behind the actual water flow out of the tap. Sometimes the electrical current is not only driven by the batteries across the sensory cell membrane alone but in addition by specialized auxiliary cells and tissues as for example in the inner ear or various cuticular sensors of insects. A conductance $(1/r)$ change of the sensory cell membrane is a common feature of sensors taking up such different types of energy as that contained in a visual, a tactile, or an olfactory stimulus. Its consequence is a so-called receptor potential of some 20 mV which is nothing but a change in the resting potential (in most cases moving it towards a more positive value). The receptor potential is a graded phenomenon. It changes amplitude with stimulus strength and also with the temporal and spatial pattern of the incoming stimuli. The stronger the stimulus and the smaller the spatial and temporal distances of two stimuli are the larger the receptor potential will be.

C. Encoding

What happens next is referred to as encoding. In most cases of sensors, though not in all, information about the stimulus is carried by so called action potentials. As soon as the receptor potential (then called a generator potential) has reached a threshold value (threshold depolarization of the cell membrane) the Na+ conductance (or in some cases that of another ion) of the membrane suddenly (due to a positive feedback loop) increases about 1000-fold. This sudden change is due to voltage sensitive Na+ – channels and leads to an explosive complete depolarization and even a reversal of the resting membrane potential towards the positive Na+ equilibrium potential. This event is the action potential. It lasts for not much more than one millisecond. After about two milliseconds the resting potential returns to its normal "unstimulated" value. Action potentials differ from generator potentials in that they are all – or – nothing phenomena with essentially always the same size and thus assuming the properties of digital instead of analog signals.

The information about the stimulus is encoded in the temporal pattern of successive action potentials referred to as action potential frequency and its modulation. Again, this is independent of the type of stimulus energy taken up by the sensory organ. It is the universal language in the nervous systems of animals as different as a hydrozoan, an earthworm or man. A crucial question then is how the central nervous system and the brain distinguish stimulus quality? For instance, is it a visual or an acoustic stimulus? The answer to this fundamental question lies in the specificity of connections. A message coming through the optic nerve is always interpreted as visual even if it is elicited by artificial electrical stimulation by the experimenter. The main advantage of using the frequency modulation code and "digital" action potentials is the resulting insensitivity towards background noise, at the expense of temporal resolution at low frequencies.

Whereas the formal relation between the size of the stimulus and the generator potential typically is linear in the case of proprioreceptors, it is logarithmic in case of long distance sensors such as our eyes and ears. The "logic" behind this difference is that, typically, proprioreceptors such as a crayfish abdominal stretch receptor or a mammalian muscle spindle are part of a negative feedback loop informing about the smallest deviation from the desired state of the system which does not require a large working range but rather a high difference sensitivity. An eye or a nose, on the other hand, covers an enormous range of stimulus intensity. Sometimes this range stretches over 12 or even more powers of 10 which requires a logarithmic relationship between stimulus strength and nervous response. Another rather general property of most biological sensors is that much more weight is given to the dynamic properties of the stimulus than to its static ones. In other words: It is in particular the change in environmental properties which the sensor responds to and which the organism needs to know.

III. Membrane Channels

Like action potentials, the receptor potential and thus transduction is closely linked to membrane channels which upon stimulation of the sensor change their conductance allowing currents in the range of picoamperes while open. In case of mechanoreceptors like the hair cells of our inner ear or the tactile hairs of insects the stimulating forces are believed to gate the channels directly within a few microseconds by pulling them open. At least in some cases the mechanosensory stretch – activated channels are cation-channels and a steep $K+$ or $Na+$ gradient across the membrane drives the depolarizing current. Subunits of the channel protein are believed to be linked to the extracellular matrix and/or the cytoskeleton so that membrane tension

modulates tension across the channel and increases its open-probability. Single – channel conductances in mechanosensory neurons are estimated to be between about 5 and 10 pS. In a spider slit sensillum it was recently estimated to be ca. 7.5 pS (Höger and French 1999), whereas a value of 50–70 pS is given for the crayfish stretch – receptor neuron (Erxleben 1989), and 12 pS for vestibular hair cells of the bullfrog (Holton and Hudspeth 1986). The estimated number of channels involved in mechanotransduction is ca. 250 for the spider sensor and 7–280 for the bullfrog hair cells. (lit. s. Sackin 1995, Hamill and McBride 1996, Höger et al. 1997, Höger and French 1999). Obviously, mechano – gated ion channels are a class of their own and distinct from voltage and ligand-gated channels.

In contrast, transduction channels involved in the reception of other forms of energy such as in chemoreception and vision are modified indirectly by a cascade of chemical reactions coupled to G-protein, and involving so-called second messengers (like cyclic-AMP, adenosine monophosphate, and IP3, inosittriphosphate) (Firestein 2001, Hardie and Raghu 2001). This process takes at least 100 times as long as direct gating. It is found in similar form for the action of hormones and the action of certain neurotransmitters in synaptic transmission. In all these cases one prime advantage of the gated device is a high rate of amplification. A single molecular interaction may result in a flux of up to a million ions per second (Cornell et al. 1997). In the vertebrate photoreceptor cell the transduction of stimulus energy goes along with an amplification of about 100.000 times.

IV. Sensitivity

A striking feature of many biological sensors is their outstanding sensitivity. This

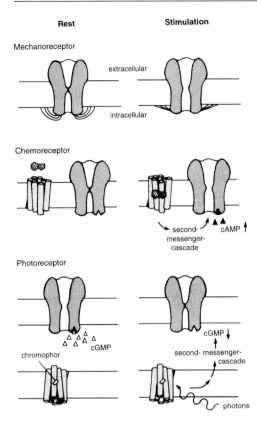

Fig. 1. *Sensory transduction mechanisms.* The conductivity of ion channels in the membrane of the sensory cell is modified by different molecular mechanisms when different forms of stimulus energy are transduced into nervous signals. Panels on the left and right depict non-stimulated and stimulated situation, respectively. *Mechanoreceptor.* The transduction channel is assumed to be anchored to the cytoskeleton intracellularly and exposed to tension which results from a mechanical stimulus applied from outside. *Chemoreceptor.* Particular membrane proteins serve as receptors which bind the stimulating chemical such as a tastant. This binding in turn triggers a cascade of molecular events leading to an increase (↑) of the concentration of a so-called second messenger (like cyclic adenosine monophospate, cAMP). The second messenger may bind directly to the inside of the ion channel, increase its open probability and thus its conductivity. *Photoreceptor.* In photoreceptor cells light (photons) is absorbed by a photopigment (rhodopsin). Phototransduction then starts with the 11-cis to all-trans photoisome-

applies to both quantum – detecting systems as in photoreception and in olfaction and to noise limited systems such as in hearing and electroreception. Photoreceptor cells throughout the animal kingdom have been known for a long time to respond to single quanta of light (e.g. Baylor et al. 1979). Chemoreceptor cells such as those associated with the olfactory hairs of the male silkworm moth and related species respond to a single or a few molecules of the female pheromone which is secreted in order to attract the sexual partner for reproduction (e.g. Kaissling and Priesner 1970). The stereovilli of vertebrate hair cells (like those in our ear) need to be deflected on the scale of atomic dimensions only (Pickles and Corey 1992) in order to elicit a receptor potential. The threshold dendritic displacement of insect tactile hairs due to hair deflection is in the range of 0.05 nm only (Thurm 1982). Vibration thresholds of spiders and other arthropods reach values down to a few Ångström units (Barth 1998). Weakly electric fish, which actively probe their immediate environment by sending out short electrical pulses, have electrosensors which respond to changes in the electrical field due to objects as small as a few nV/cm and detect capacitances in the order of a few nF (Bullock and Heiligenberg 1986, von der Emde 2001). Skates which are able to sense low-frequency electric fields of biotic and abiotic origin respond to voltage gradients less than 0.01 µV/cm which

risation of the chromophore (retinal). Light-activated rhodopsin activates a second messenger cascade, which modifies the open probability of ion channels. In the vertebrate case shown the reduction (↓) of the cGMP (cyclic guanosine monophosphate) level reduces the conductivity of ligand dependent ion channels and thus the membrane current already flowing in the dark (modified from Kandel et al. 1996)

corresponds to a voltage of 1V over 1.000 km (von der Emde 2001).

These are just a few examples taken from a plethora of fascinating cases which are commonly interpreted as showing perfect "design" and even true optimization. Sensors of such high sensitivity, sometimes at the limit of the physically possible, are particularly welcome to a sensory physiologist because the basic principles underlying sensory functions are much more likely to be uncovered when effectivity is carried to an extreme as opposed to a more sloppy just good enough solution. It should be pointed out that the existence of the latter is all but a non-biological thought. Many sensory systems are just as good as they need to be.

A particularly interesting case of extreme sensitivity is that of an arthropod filiform hair, a structure analogous to the spider trichobothria (see chapters 9 and 10, this volume). It has been known for a while (Thurm 1982) that stimuli containing fractions of the energy contained in a single quantum of green light suffice to elicit a response of the sensory cell. The dendrite changes its diameter only by 0.05 nm when the hair is exposed to threshold deflection. These hairs are movement detectors par excellence and are deflected by the viscous forces exerted on the hair shaft by the slightest movement of the surrounding air. It is well understood now, how intriguing the complex physics of the air-hair interaction is (Shimozawa and Kanou 1984, Barth et al. 1993, Humphrey et al. 1993, 2001; chapter 9, this volume). Recent calculations by Shimozawa et al. (1998, chapter 10, this volume) suggest that the cricket filiform hairs are the most sensitive sensors so far known with a stimulus energy (better: mechanical work) at threshold of only 10^{-21} Ws. These hairs then are more sensitive by a factor of 100 than the most sensitive sensors so far known including human vision and

hearing (Autrum 1984, Gitter and Klinke 1989). No doubt, these hairs operate at the limit of the physically possible and fully exploit the non-quantum structure of their adequate stimulus. As a matter of fact the energy discussed is of the order of the thermal noise ($k_B T$ at 300°K, 4×10^{-21} Ws). By making use of stochastic resonance the filiform hair receptors are even able to exploit noise for the detection of extremely weak subthreshold signals. External noise can improve the signal-to-noise ratio instead of degrading the hair's performance (see chapter 10, this volume).

V. Selectivity and Meaning

One of the most biologically relevant features of animal sensors and indeed the main source of their striking diversity is the seemingly simple notion that there is no information without meaning. What we find as the astounding refinement of the "engineering" of sense organs not only reflects the universal laws of physics to which the processes of evolution have been subjected as much as any device designed by an engineer but, in addition, it reflects the needs of the animal. The biological sense of senses is not so much found in an abstract unlimited truth but rather in the behavioral relevance of the information provided. Instead of abstract and complete truth, survival and reproduction are the important issues. The implication is that in order to understand animal sensors we not only have to study physical and chemical aspects but also the sensory demands resulting from the particular behavior of a particular animal species in a particular and species specific environment. In order to fully understand a sensory system the stimuli have to be found and applied which the system evolved to process. A honey bee needs to see the world in a way drastically different from a pigeon or a deep-sea crustacean. And even what the same individual animal sees is not a constant but changing with the behavioral program, with time of day and year, developmental stage, hormonal influences and with what the brain wants and expects to see. Vision is just the example taken for animal senses in general. We depend on a considerable amount of integrative and organismic knowledge about an animal species if we strive to understand the structural and functional specialization of sense organs and sensory systems (which include the central nervous system and the processing of the input received from the sensory periphery by it).

The questions a biologist has to ask are like: How does the bee see and recognize the flowers where it collects its food? How does it find its way (often several kilometers) from the hive to a small patch of flowers? How does a male moth find its way to a female many kilometers away and how does a female spider recognize a conspecific sexual partner and distinguish it from prey and predator? How does this extraordinary beetle, which deposits its eggs in freshly burnt trees find a woodfire (see Fig. 2) and how does a weakly electric fish use its electric organs and electrosensors to find its way in lurky water and communicate with conspecifics? What does a bird do to overcome problems of signal attenuation when singing and communicating in a dense forest?

By trying to answer such questions one usually finds a close match between the properties of a sensory system and what the organism needs from it. To put it another way, we learn to understand senses as specialized filters which only very partially represent the physical world but rather a very limited species-specific reality. This is by no means trivial when it comes to mechanisms and a quantitative understanding of a biological sensor, or when we even try to understand the way

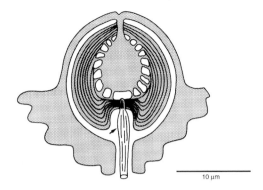

Fig. 2. *Infrared sensor.* Some beetles (here: *Melanophila acuminata*, Buprestidae) have so-called pit organs laterally in their thoracic exoskeleton which help them to detect forest fires. These organs are infrared (IR) detectors responding best to wavelengths of about $3\,\mu m$. Stimulus intensity eliciting a neural response may be as small as ca. $5\,mW/cm^2$. The graph shows the structure taking up the stimulus and serving as a photomechanical transducer. The cuticular spherule is made up of molecules with stretch resonances in the IR region best absorbed. Following stimulation the vibrational energy is converted into heat which in turn changes the spherule's volume and finally deforms and thereby stimulates the dendrite of a mechanosensory cell (arrow) innervating the spherule from below (modified from Schmitz and Bleckmann 1998)

in which evolution may have led to a particular match between the demands of an animal species and its senses (Wehner 1987, Barth and Schmid 2001, Humphrey et al. 2001).

In chapter 9, this volume, Humphrey et al. report on spider filiform hairs. We may take them to illustrate the match biologists are referring to. Trichobothria are hairs measuring only between $100\,\mu m$ and $1400\,\mu m$ in length and 5 to $15\,\mu m$ in base diameter. They have a tiny mass in the order of $10^{-9}\,g$ and their articulation with the general cuticular exoskeleton is very flexible, the damping constant R being of the order of $10^{-15}\,Nms/rad$ and the elastic restoring constant S (joint stiffness) being of the

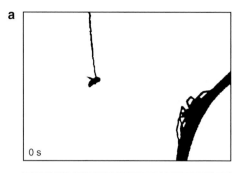

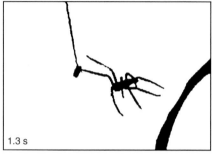

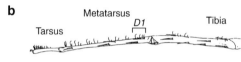

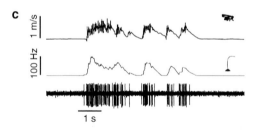

order of 10^{-12} Nm/rad (Barth et al. 1993, Barth 2000). Such hairs are perfect movement detectors, taken along and deflected by the surrounding air. From their high sensitivity it is clear that the air-hair interaction must be optimized. The underlying physics and fluid mechanics is quite well understood thanks to a very fruitful collaboration between engineering and biology (Barth et al. 1993, Humphrey et al. 1993, Barth 2000, 2002).

The biologist has learned from a combination of these data with behavioral and electrophysiological experiments why the trichobothria are so well suited to tell the spider that a "fly signal" is around which elicits prey capture behavior (Fig. 3). The spider indeed jumps towards a fly flying by, alerted and guided not by visual cues (it does it equally well after being blinded) but by the air movement caused by the fly.

Thrichobothria do not send any action potentials at rest, nor do they send any when deflected stationarily. They only respond to dynamic stimuli. Indeed, air movement sensitive interneurons were found to behave in such a strictly "phasic" way as well. The system is a *matched filter* because a most prominent feature of the fly signal indeed is its high degree of fluctuation of its velocity (Fig. 3) which nicely contrasts the background air flow at the time of day when the spider sits on a plant waiting for prey. Distinction between such background "noise" and the signal relevant for the animal is a problem to be solved by any sensory system. Like in the given case a match between the frequencies contained in the signal and the frequency sensitivity of the

Fig. 3. *Behavioral relevance.* **a** When a hungry wandering spider (*Cupiennius salei*, Ctenidae) is exposed to the airflow produced by a flying insect (here: an experiment with a tethered flying fly) it responds to it with prey capture behavior, jumping into the air towards a stimulus source as far away as 30 cm. The two video frames shown were taken 1.3 s apart. **b** The flow sensors involved are fine cuticular hairs (trichobothria) of which there are nearly 100 on each leg (see also chapters 9 and 10, this volume). **c** The response characteristics of the sensory cells attached to the basis of a hair are finely tuned to

the biologically relevant frequency range and to the fluctuations which are characteristic of the biologically relevant airflow but not of the background wind prevailing in the spider's habitat at the time of its nocturnal activity (modified from Barth 2002)

sensor is part of its solution. Sensor properties in this way reflect stimulus properties which therefore are an important guide when analyzing the "design" properties of a biological sensor. Ecology comes in here massively. Detailed knowledge of the natural signals occurring under the conditions experienced by the animal when showing the behavior under study is of obvious relevance (Dusenbery 1992, Barth and Schmid 2001).

VI. Richness and Diversity

As a biologist one often wonders how anthropocentric many humans are, all too often considering man as something set apart and completely distinct from all other creatures and not sharing a long stretch of evolutionary history with them. More than often such an attitude has a blinding effect and not only is the richness and diversity in the animal kingdom (the evertebrates and so called lower animals explicitly included) not adequately appreciated but also the fact that many animals can do many things much better than we can. This also applies to their sensors.

As zoologists we cannot follow the classical view dating back to Aristotle which classifies the senses according to human sensations. Animals do, of course, have vision, taste, hearing etc. as well. But often they are organized in a very different way. In addition one finds many sensory capabilities in animals which are completely alien to our own perceptions. Think for instance of the ability of bees and ants to see and make use of the polarization pattern of the sky for orientation (Wehner 2001), of hearing ultrasound in grasshoppers and bats (Neuweiler 1989, Römer 1998), of detecting a source of infrared radiation by some snakes and beetles (Bullock and Diecke 1956, Schmitz and Bleckmann 1998, Hammer et al. 2001), magnetic field detection by pigeons and

bees (Wiltschko and Wiltschko 2001), and the electric sense of weakly electric fish (Bullock and Heiligenberg 1986, Moller 1995, von der Emde 2001). For this and other reasons senses must not be classified according to human perceptions but according to the form of energy of the respective adequate stimulus. It is important to stress this because it opens the eyes for all the fascinating surprises and the diversity and inventiveness biologists should share with engineers.

VII. The Link Between Biology and Engineering

Without adequately working sense organs life would not be possible – from bacteria to man. It is also clear that living organisms are the most complex "machines" we know. Their machinery works incredibly smoothly and unfailingly, and in addition it is adaptive and to a large extent regenerative. For the insider there is a stunning wealth in the inventions made by millions of years of evolutionary history. All of this even applies when we restrict our considerations to the sensors per se as done in this introductory essay and neglect the complex processing of sensory information in the central nervous system. It does not need a particular sensibility or a large amount of specialized training to see "technical efficiency and perfection" in nature. The obvious consequence is to ask: What can engineers and biologists learn from each other?

1. Technical biology. What the biologist can do and indeed does is to apply approaches and methods otherwise used by physical scientists and engineers to biological objects and phenomena. The kind of biology resulting is referred to as *technical biology*. Regarding sense organs the application of fluid mechanics to the study of wind sensitive hairs of spiders and insects (chapters 9 and 10, this volume), of acoustics to the hearing organs of arthropods (chapter 3, this volume), or of micro-mechanical analysis to the middle ear and tactile hairs (chapters 5 and 11, this volume), are just a few examples. Information theory, computational modeling and finite element analysis are among the many methodological approaches already applied by many biologists. The benefit is obvious: quantification often becomes much more rigorous than it could ever be without such technical approaches and very often it allows predictions about the system and thus deepens our understanding in a way not otherwise possible. My personal view for the future is that biology increasingly needs as a major input of engineering the application of computational skills and mathematical quantification. This would allow us to increasingly tackle questions at a level of complexity (for instance arrays of large numbers of sensors distributed in specific ways over the body surface) which so far cannot be adequately treated in a quantitative manner. More biomathematics and computational biology are highly desirable in sensory biology.

2. *Bionics.* The idea to use nature as a model for human technology is an old one. Leonardo da Vinci's and Otto Lilienthal's efforts to build a flying machine spring to mind and a wealth of other examples (Nachtigall 1998, 2000). The motivation is highly justified because animals and plants and machines all are subject to the same laws of physics. As a rule the engineer will not find ready made "instructions" of how to build a device in nature. A camera is quite different from an eye, as much as an airplane differs from a bird, a ship from a fish and a car from a horse. Nature, however, offers a multitude of ideas of how to solve a problem. In many cases evolution seems to have found practically any solution prin-

cipally possible. As an example the reader is referred to chapter 3, this volume by Michelsen on arthropod hearing organs and to a similar diversity regarding the exploitation of the possibilities provided by the laws of optics in arthropod eyes (Land and Nilsson 2002).

The effort to explicitly transform a biological solution of a problem into a technical construction is the essence of *bionics*. Examples contained in the present book are provided by Sarpeshkar and Franceschini and Kauer and White (chapters 7, 16 and 22). The reader is also referred to the authorative and very stimulating books by Nachtigall (1998, 2000). With regard to recent developments in technology it should be added that biological sensors not only are exquisitely sensitive and highly specific but also operating at the micro- and nano-scale.

In engineering a "bio-sensor" is defined (Turner 1997, 2000) as a "compact analytical device incorporating a biological or biologically derived sensory element either integrated with or intimately associated with an physiochemical transducer. The usual aim of a bio-sensor is to produce either discrete or continuous digital electronic signals which are proportional to a single analyte or a related group of analytes".

From this it is also clear that biosensors in a common engineering language are molecular sensors with a biological recognition mechanism mainly associated with chemical sensing, the ideal being a stable lipid membrane forming an ion gate that opens and closes by the binding of single molecules (Cornell et al. 1997). In a recent report Bayley and Cremer (2001) describe pores in a lipid bilayer based on staphylococcal alpha-haemolysin (αHL). These pores can be protein engineered for a variety of analytes by placing the proper binding sites within the lumen of the pore. A transducer combined with it provides the electronic signal. The most well known bio-sensor and probably the most beneficial one is the hand-held blood glucose sensor used by many patients suffering from diabetes for self monitoring at home. This sensor uses an enzyme (glucose oxidase fixed onto an oxygen electrode). In other cases the receptor is an antibody, a nuclear acid, DNA and RNA (binding to nuclear acids with complementary sequence), or even a whole microorganism like genetically modified bacteria with the *lux* gene that glow in response to pollutants, recognizing and interacting with the analyte of interest (Turner 2000).

From a biologist's point of view the impression prevails that the most intriguing diversity in animal sense organs is not so much found in the cellular transduction processes proper (which are quite conservative over evolutionary times) but in stimulus transformation. It should be stressed that here in particular seem to be hidden many starting points for inventive technological developments. Examples may be beetle infrared sensors (Schmitz and Bleckmann 1998), spider trichobothria for the measurement of fluid flow at the microscale (chapter 9, this volume), spider micro-force sensors (Barth 2002), eyes (chapters 16, 17 and 18, this volume) and ears of arthropods (chapter 3, this volume) and other animals to name just a few.

3. Difficulties. There are a few difficulties in the interaction between biology and engineering which seem to be common and often in the way of long lasting productive cooperation. When an engineer has to design something he or she typically knows precisely what the goals are and the conditions under which they have to be reached. Biological research very often is a kind of "backward engineering" dealing with a complex ready-made structure whose detailed functions

and "design" principles very often are all but clear (which of course is the challenge driving research). Clearly, "design" is the wrong terminology in biology. Evolution does not "design". It works under a multitude of diverse selective pressures, with errors and variation being associated with the passing on of genetic information to the next generation, and in many small steps over long stretches of time. We call the sensors, that have resulted from this process *matched filters* because their properties serve the particular needs of a particular animal species performing a particular behavior in its particular species-specific habitat (see also Wehner 1987). The process leading to matched structures and functions is blind regarding its "goal", not teleological, and not design, but rather trial and error and tinkering. There is no brain which would think before a sensor is built. Among pre-existing variants, natural selection favors the better ones; one should therefore rather talk about the selection of adapted individuals than about adaptation.

Biologists quite often have an attitude towards uncertainty, under-determination and quantification different from a physical scientist for that very reason, that is the necessity of "backwards" analyses and the complexity of the objects they try to understand. Final perfection in a technical sense is not the rule in biology. A single parameter can be carried to its extreme only at the cost of other characteristics. Organisms are highly multifunctional and consequently "multi-optimal" and selection works on whole organisms. While wondering that so much "perfection", high-tech, and harmony is obviously at work in nature and in many instances clearly ahead of technology we should never forget that evolution has been tinkering since about 3 billion years and its products have been quality controlled for the same period of time.

References

Autrum H (1984) Leistungsgrenzen von Sinnesorganen. Verh Ges Dtsch Naturforsch Ärzte 113: 87–112

Barth FG (1998) The vibrational sense of spiders. In: Hoy RR, Popper AN, Fay RR (eds) Comparative Hearing: Insects. Springer Handbook of Auditory Research. Springer-Verlag, New York Berlin Heidelberg, pp 228–278

Barth FG (2000) How to catch the wind: spider hairs specialized for sensing the movement of air. Naturwissenschaften 87(2): 51–58

Barth FG (2002) A Spider's World: Senses and Behavior. Springer-Verlag, Berlin Heidelberg New York

Barth FG, Schmid A (eds) (2001) Ecology of Sensing. Springer-Verlag, Berlin Heidelberg New York Tokyo

Barth FG, Wastl U, Humphrey JAC, Devarakonda R (1993) Dynamics of arthropod filiform hairs. II. Mechanical properties of spider trichobothria (Cupiennius salei KEYS.). Phil Trans R Soc Lond B 340: 445–461

Bayley H, Cremer PS (2001) Stochastic sensors inspired by biology. Nature 413: 445–461

Baylor DA, Lamb TD, You KW (1979) Responses of retinal rods to single photons. J Physiol 288: 613–634

Bullock TH, Diecke FPJ (1956) Properties of an infrared receptor. J Physiol 134: 47–87

Bullock TH, Heiligenberg W (1986) Electroreception. John Wiley and Sons. New York

Cornell BA et al. (1997) A biosensor that uses ion-channel switches. Nature 387: 580–583

Dusenbery DB (1992) Sensory Ecology. WH Freeman and Company, New York

Erxleben C (1989) Stretch-activated current through single ion channels in the abdominal stretch receptor organ of the crayfish. J Gen Physiol 94: 1071–1083

Firestein S (2001) How the olfactory system makes sense of scents. Nature 413: 211–218

Gitter AH, Klinke R (1989) Die Energieschwellen von Auge und Ohr in heutiger Sicht. Naturwissenschaften 76: 160–164

Hamill OP, Mc Bride DW Jr (1996) The pharmacology of mechano-gated membrane ion channels. Pharmacol Rev 48: 231–252

Hammer DX, Schmitz H, Schmitz A, Rylander HG, Welch AJ (2001) Sensitivity threshold and response characteristics of infrared detection in the beetle Melanophila acuminata (Coleoptera: Buprestidae). Comp Biochem Physiol A 128: 805–819

Hardie RC, Raghu P (2001) Visual transduction in Drosophila. Nature 413: 186–193

Holton T, Hudspeth AJ (1986) The transduction channel of hair cells from the bullfrog characterized by noise analysis. J Physiol 375: 195–227

Höger U, French AS (1999) Estimated single-channel conductance of mechanically-activated channels in a spider mechanoreceptor. Brain Res 826: 230–235

Höger U, Torkkeli PH, Seyfarth E-A, French AS (1997) Ionic selectivity of mechanically activated channels in spider mechanoreceptor neurons. J Neurophysiol 78: 2079–2085

Humphrey JAC, Barth FG, Voss K (2001) The motion-sensing hairs of arthropods: using physics to understand sensory ecology and adaptive evolution. In: Barth FG, Schmid A (eds) The Ecology of Sensing. Springer-Verlag, Berlin Heidelberg New York, pp 105–125

Humphrey JAC, Devarakonda R, Iglesias I, Barth FG (1993) Dynamics of arthropod filiform hairs. I. Mathematical modelling of the hair and air motions. Phil Trans R Soc Lond B 340: 423–444 (see also Erratum Phil Trans R Soc London B 352:1995 (1997)

Kaissling K-E, Priesner E (1970) Die Riechschwelle des Seidenspinners. Naturwissenschaften 576: 25–28

Kandel ER, Schwartz JH, Jessell TM (eds) (1996) Neurowissenschaften. Eine Einführung. Spektrum Akad Verlag, Heidelberg

Land MF, Nilsson D-E (2002) Animal Eyes. Oxford University Press, Oxford

Moller P (1995) Electric Fishes. History and Behavior. Chapman and Hall, London.

Nachtigall W (1998) Bionik. Grundlagen und Beispiele für Ingenieure und Naturwissenschaftler. Springer-Verlag, Berlin Heidelberg New York

Nachtigall W, Blüchel KG (2000) Das große Buch der Bionik. Neue Technologien nach dem Vorbild der Natur. DVA, Stuttgart München

Neuweiler G (1989) Foraging ecology and audition in echolocating bats. Trends Ecol Evol 4: 1650–1666

Pickles JO, Corey DP (1992) Mechanotransduction by hair cells. Trends Neurosci 15: 254–259

Römer H (1998) The sensory ecology of hearing in insects. In: Hoy RR, Popper AM, Fay RR (eds) Comparative Hearing: Insects. Springer, New York Berlin Heidelberg, pp 63–96

Sackin H (1995) Stretch-activated ion channels. Kidney Int 48: 1134–1147

Schmitz H, Bleckmann H (1998) The photomechanic infrared receptor for the detection of forest fires of the beetle *Melanophila acuminata* (Coleoptera: Buprestidae). J Comp Physiol A 182: 647–657

Shimozawa T, Kanou M (1984) The aerodynamics and sensory physiology of range fractionation in the cercal filiform sensilla of the cricket *Gryllus bimaculatus*. J Comp Physiol A 155: 495–505

Shimozawa T, Kumagai T, Baba Y (1998) Structural scaling and functional design of the cercal wind-receptor hairs of a cricket. J Comp Physiol A 183: 171–186

Shimozawa T, Murakami J, Kumagai T (1998) Cricket wind receptor cell detects mechanical energy of the level of kT of thermal fluctuation. Abstract 112, International Soc of Neuroethol Conf San Diego

Thurm U (1982) Grundzüge der Transduktionsmechanismen in Sinneszellen. Mechanoelektrische Transduktion. In: Hoppe W, Lohmann W, Markl H, Ziegler H (eds) Biophysik. Springer, Berlin pp 681–696

Turner APF (1997) Biosensors: Realities and aspirations. Annali di Chimica 87: 255–260

Turner APF (2000) Biosensors – sense and sensitivity. Science 290: 1315–1317

Von der Emde G (2001) Electric fields and electroreception: How electrosensory fish perceive their environment. In: Barth FG, Schmid A (eds) Ecology of Sensing. Springer-Verlag, Berlin Heidelberg New York, pp 313–329

Wehner R (1987) "Matched filters" – neural models of the external world. J Comp Physiol A 161: 511–531

Wehner R (2001) Polarization vision-a uniform sensory capacity? J Exp Biol 204: 2589–2596

Wiltschko W, Wiltschko R (2001) The geomagnetic field and its role in directional orientation. In: Barth FG, Schmid A (eds) Ecology of Sensing. Springer, Berlin Heidelberg New York, pp. 289–312

2. Sensors and Sensing: An Engineer's View

Hans Meixner

Abstract

Concern about the future of industrial societies has stimulated intensive discussions around the world and has led to a number of recommendations. Leading the list is the necessity for a speedy and permanent turn towards resource-sparing and environmentally compatible technologies and innovations. Incentives for this development include the exponential growth of the world's population, the increasing environmental burdens and our diminishing primary resources. A solution to these problems presupposes ongoing improvements in our technical competence and its optimum utilization.

The following points are to be viewed as especially important fields of application for micro-technologies and nanotechnologies: communication, information/education/entertainment, transportation energy, the environment, buildings/housing/ industrial plants, production, safety/security, medicine and health.

For these fields of application the technological branches of new materials, sensor and actuator technology, microelectronics, optoelectronics, and information storage, electrooptical and electromechanical transducers, generation, distribution and storage of electrical energy, information engineering, intelligence functions and software have key effects that are both wide-ranging and enduring. In coming years, research and development must concentrate on these technological branches in particular. Industry is responsible first and foremost for innovations. It must above all: (i) increase innovative power by more effective cooperation between research, development production and marketing and (ii) improve the climate of innovation by working in cooperation with academia and government.

An improvement in the innovative climate requires a reorientation of the classical division of the research and development process and more effective forms of cooperation between government, national research establishment and industry. Therefore, in order to better convert our scientific competence into economically measurable results, nationally supported research and research in the private sector must move appreciably closer together. It is beyond human capabilities to anticipate technological developments over a relatively long period of time (e.g. a century). Such capabilities are adequate enough at best to obtain information for the next 10 to 20 years on the basis of currently existing knowledge and results, by the use of extrapolation and scenario techniques. The implementation probabilities decrease very rapidly with increasing period of consideration.

I. Micro- and Nanotechnologies for Sensors

Whereas in the 1970s microelectronics was one of the dominant strategic research and development goals, in the 1980s materials research and information engineering had priority. Then in the early 1990s, development work was started mainly in the field of miniaturization and integration of extremely small functional units within a system, in order to open up new technologies of the future, ranging from micro- and nanostructures down to molecular and atomic units, by utilizing also phenomena of quantum physics and quantum chemistry. Sensors are of essential importance for most products and systems and for their manufacture. To some extent the development of sensors has not been able to keep pace with the tumultuous developments in microelectronic components. For this reason sensorics is in a restructuring phase in the direction of achieving increased miniaturization and integration of sensors and signal processing within a total system. This is giving substantially more importance to technologies that permit low-cost manufacture of both the sensors and the related electronics (Muller 1991, Muller et al. 1991, de Raolj 1991, Wise and Najafi 1991, Benecke and Petzold 1992).

A. Microtechnologies – Microsystems Engineering

The basic philosophy of microsystems engineering can be described as using the smallest possible space to record data, process it, evaluate it and translate it into actions. The special feature of this engineering is its combining of a number of miniaturization techniques or basic techniques.

Technical developments in the fields of sensorics, actuators, ASICs (application specific integrated circuit) and micromechanics are growing together into a "system". Innovations in the area of field-bus engineering and mathematical tools (computer logic) can improve these systems and optimize communication between them. Thus a complex technology is available that autonomously processes information and directly translates it into actions in a decentralized fashion in peripheral equipment, without the need for large-scale central data processing. Microsystems engineering is thus not only an enhancement of microelectronics, it also represents a qualitative innovation. Microelectronics has entered into nearly all devices in which information is processed or processes are regulated or controlled, from the computer to the automobile and extending to self-sufficient robot systems. Why should not other components and technologies be miniaturized and integrated on a chip as well and the intelligence of the system be expanded, with a simultaneously greatly reduced energy consumption? The total sum of these future changes will bring a large number of new applications with great benefits for society.

By combining a number of miniaturization techniques a great deal of interdisciplinary knowledge concerning technical possibilities and technologies will come together in the design and realization of systems. In the ideal case, this "knowledge" should come from one entity, because otherwise a high degree of cooperation is required. This in turn can be successful only if the exchange of information and the logistics are good and it must be based on standardization and high quality. A number of technologies have already been developed separately in recent years. High-frequency circuits, power semiconductors and displays are still part of the field of electronic-elements and they expand the functional scope of the "classical" microelectronics. Micro-mechanics, integrated optics, electrooptics, chemosensors, biosensors, polymer sensors, sensors in thin-film and/or thick-film engineering and radio-readable passive surface acoustic wave (SAW) sensors are opening up new dimensions.

Up to now, these technologies have only been pursued separately from one another and in part they are also based on materials different from silicon, for example on gallium arsenide, ceramics, glass or even monocrystals such as quartz or lithium niobate. Nevertheless, today microsystems are frequently constructed in hybrid fashion from various different parts, from various technologies, with the goal of miniaturization and enhancing functionality. New procedures aim at combining chips directly with one another, either as a "chip in a chip" or as a "chip within a chip". Microsystems engineering will provide manufacturing machinery for much finer work than is possible today. The door to the submicro world has in any case already been opened. The former magical limits of micrometer dimensions are being considerably lowered by the present-day memory chips and SAW (surface acoustic wave) components. Microsystems engineering leads us into nanotechnology and micromachines will be able to produce systems in the molecular and atomic range.

Current research work is aimed first of all at bringing together sensors, actuators and logic components into self-contained units (Fig. 1). Here it is not a multitude of elements that is in great demand, but a multiplicity of functions. But to integrate this multiplicity means to bring together different production processes and technologies. Difficulties are necessarily associated with this task, since frequently many processes are "not compatible" with one another (Howe et al. 1990, Kroy et al. 1992, Stix 1992, Bundesministerium für Forschung und Technologie, Bonn 1994).

B. Systems-Development: Methods and Tools

Closely linked to these process-engineering difficulties are the problems of designing systems. Here, microelectronics has attained an important position: nowadays logic chips can be designed right on the computer and all the important basic elements can be retrieved from libraries. There are design tools also for other technologies. Microsystems engineering, however, still lacks general design tools. This is true especially for the simulation of systems, which is much more difficult than in the electronics field. Aspects which in the macroscopic range scarcely play any role or can be easily compensated for may have a tremendous effect in the microscopic range. Simulation programs must take into account all of these "cross-sensitivities". They must model electrical, thermal, and mechanical behavior in three dimensions, which necessitates a complex mathematical description. Whether there will ever be development tools with control over all aspects of micro-systems, from design to the generation of masks and extending to simulation and testing, cannot be predicted from our present viewpoint. Technological leadership will have to depend more and more on powerful subsystem tools.

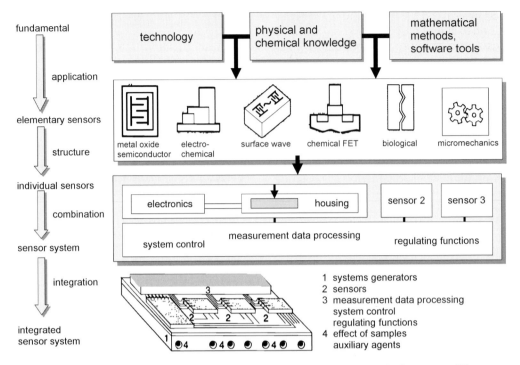

Fig. 1. Sensors in microsystems engineering: from the elementary sensor to the bus-compatible sensor system

C. Signal Processing

In order to make use of the growing power of microelectronic systems, corresponding improvements in peripheral areas are required. Information storage devices such as cartridges, diskettes or hard disks must make more information available along with ever smaller space requirements; sensors must be able to provide more information and printers or display units must be able to pass on more detailed information. Just as the spatial extent of the storage device or the fineness of the printed or displayed picture have to be improved, so also the space needed by a single informational element must decrease. The geometrical dimensions of the components of the peripheral devices must follow.

Signal processing stands as a necessary connecting link between sensors and ac-

tuators and the superordinate levels of information systems. This processing treats and modifies an input signal of the sensor in a way that makes the desired network information available. This is then converted by other processing measures into an output signal, which causes the desired action (by triggering an actuator) or is forwarded to a superordinate system for further processing.

The future development of signal processing within the next 10 years will primarily bring together sensor and signal processing at the site of the measurement and also permit the processing of measured values by intelligent devices. Problems that still need to be solved here concern manufacturing processes providing these integration steps and the mastering of increased reciprocal effects as a consequence of miniaturization. The availability of suitable signal-processing concepts and design tools will be important. In addition to signal processing on an electrical basis, implementations of a non-electrical type will be increasingly needed; for example, optical and acoustic signal processing and transmission, ultrasonic and surface applications and biocybernetic systems. Also high-temperature electronics will become increasingly more important. Complete, intelligent microsystems will be available. They will function as independent, teachable, or adaptable systems, will have geometries extending down into the nano range, will include non-electrical signal processing components, and will be based partly on materials from high-temperature electronics.

D. Testing and Diagnosis of Sensor Microsystems

Suitable self-testing and diagnostic components and the testing of supportive signal-processing algorithms can be anticipated even simultaneously with the design of sensor microsystems and these will go beyond previous procedures used in the development of microelectronic circuits. For physical and chemical functions to be adequately tested in complex systems additional "test sensors" may have to be implemented. Via suitable diagnostic interfaces these test components are to be coupled with external diagnostic facilities in order to thereby test the functionality of the systems under realistic conditions. Therefore, in the future, we must progress towards unified and system-overarching strategies and signal processing concepts for the testing of microsystems.

E. Malfunction Susceptibility, Reliability

One significant deficiency of microsystems is paradoxically their malfunction susceptibility. The significant advantage of an increased reliability should not obscure the fact that in case of a breakdown,

for example, of an interconnection or of a component, a microsystem can at best only detect its non-functionality. This basic malfunction susceptibility can possibly be removed by a massive parallelization of the processing chain. Materials with inherently intelligent properties are under discussion for this purpose (see section II G). In the extreme case, these materials combine sensor, actuator and signal processing. Two classes should be distinguished: materials that when exchanged for unintelligent materials increase operational reliability in familiar electronic systems and materials that perform their function independently, without electronics. The latter have the advantage of being able to work without an electrical power supply.

F. Constructive and Connective Techniques

Constructive and connective techniques (CCT) encompasses the totality of process engineering and design tools that are needed for the implementation of microsystems in an extremely confined space. It therefore forms the bridge between micro- and optoelectronic as well as micromechanical components for the complete system. It is a significant factor in determining the functionality, quality and economic efficiency of microsensors and microsystems. In recent years, constructive and connective engineering has made decisive advances with respect to miniaturizability, fail-safe reliability and ease of handling. Major bottlenecks still exist for non-electronic components of microsystems engineering. Examples are optical, mechanical, chemical, biological microcomponents. Therefore, in addition to progress in the electronics field, in the next few years it will be above all important to provide adequate technology in this sector as well.

In detail the following is needed: (i) provision of new materials with definite macroscopic and microscopic properties (alloys, polymers, material systems and film systems); (ii) investigation of phenomena that result from the combination of different materials, and the controllability of such phenomena (e.g. interdiffusion processes, adhesive strength, thermo-mechanical correlation); (iii) non-material-dependent removable and permanent connection procedures (memory alloys); (iv) provision of simulation tools; (v) provision of highly integrated fabrication technologies, availability of devices, reproducibility under fabrication conditions and combinability of different technology levels; (vi) system integration based on biological models; (vii) development of block-integration techniques (stacked chips, stacked modules); (viii) attendant studies on reliability and on the degradation behavior of the constructed sensor systems.

G. Housing Techniques

Housings in microsensorics and microsystems engineering have to protect components or systems and at the same time to establish the connection to the physical or chemical quantities to be measured. In sensorics and in microsystems engineering the "housing" cannot be regarded as independent. It is an indispensable component of the overall system.

There is a demand for multifunctional materials that on the basis of their properties also allow wide-ranging designer freedom in configuring the sensor system. Sensor housings with sensor windows are needed in particular. These multifunctional materials include, among others: (i) shape-memory alloys/polymers; (ii) piezoelectric ceramics, piezopolymers; (iii) diamond and diamond-like films; (iv) magnetoresistive materials; (v) optically activatable gels; (vi) electro-rheological, magneto-rheological liquids; (vii) optomechanical (especially electrochrome) mate-

rials; (viii) materials with self-diagnostic and alarm functions; (ix) membranes with self-regulating permeability behavior.

Shape-memory alloys, piezoceramics, or magnetostrictive alloys, embedded in a filamentous or layer-like way in a matrix material, can affect in a well defined manner the state of stress and state of strain of a housing. Structures with a variable rigidity or with adjustable bending and deflection properties might be realizable. Composite structures that are provided with piezoelectric foils/fibers or photoconductive fibers can perform sensor functions in addition to their structural-mechanics tasks, possibly even for the purpose of self-monitoring their material condition. This in situ damage recognition would be desirable precisely where there are reliability-relevant structural units. There is still developmental potential in connection with electro-rheological and magneto-rheological liquids, which react upon electrical or magnetic fields with a rapid and pronounced change in their viscosity and thereby permit, for example, the coupling of forces. These materials are self-supplying, work without electric power and are fail-safe, since even when substantial parts are destroyed the remaining undamaged parts can continue to function. Moreover, many such materials are self-adjusting, can survive overloading and are self-repairing.

Special requirements are placed on housing engineering that involves sensitive windows that are exposed directly to caustic and corrosive chemical and biological media under measurement.

H. Microtechnologies

Micro- and nanotechnologies provide the prerequisites needed for miniaturization and for adjusting the individual functions or components of the system. Most microtechnologies have their origin in technologies of microelectronics or conventional sensors/actuators and these are being further developed in a way that is geared to the interests of micro- and nanosystem engineering.

1. Micromechanics

The fabrication methods developed for microelectronics are not, of course, restricted to the production of electronic components. As a child of microelectronics, micromechanics could rely from the beginning on a large reservoir of fabrication techniques that permitted the design of extremely fine, complex structures in the "batch" procedure. The previously mentioned great application potential for micromechanics arises from the utilizability of five technical dimensions: (i) materials; (ii) microstructures; (iii) complex arrays; (iv) combination (system-capability) with microelectronics; (v) utilization of the entire scope of physical phenomena.

Among the available materials, silicon plays a predominant role. The procedures familiar from microelectronics permit the production of three-dimensional forms down to the size of a few microns. Other materials beside silicon can likewise be used when appropriate shaping-treatment possibilities are available. For example, a large number of electrical non-conducting materials are of interest for shaping via heavy-ion bombardment.

Starting with lithographic procedures for transferring structures, three-dimensional shaping can be achieved by means of various micromechanical processes, also used in anisotropic etching of silicon, micro-metal plating, and laser processes, or by way of ion-tracking technology, which is still under basic development. With the LIGA procedure (lithography, electroforming, molding), microstructures can be implemented from a wide range of materials, such as metals, plastics, and

ceramics. Silicon clearly dominates the materials used at present, but metals, plastics and glasses are gaining in importance. Special advantages result from the possibility of combining these with microelectronics. Thereby local "intelligence" can be imparted to the micromechanical components. Sensors will be equipped with their own microprocessors. Of particular importance is the possibility of combining micromechanics with a multitude of physical phenomena. In connection with magnetism, micromechanical magnetic heads detecting the data of magnetic plates can be used as non-contact potentiometers, or when applied as superconducting SQUIDs (Superconducting quantum interference device) can accept information from brain waves (Peterson 1978, Kroy et al. 1988, Heuberger 1989, McEntee 1993, Bley and Menz 1993, Menz and Bley 1993).

2. Microoptics

In addition to microelectronics and micromechanics, microoptics also is a basic technology. It permits the implementation of communication channels of very high transmission capacity in a very small space. Its goal is to make miniaturized optical components, integrated in wave guide structures, with the aid of planar technologies.

Microengineering methods play a crucial role in the production of microoptics components: from the production of semiconductor lasers to the adjustment of glass fibers at the output window of the laser or the low-loss splicing of glass fibers, diffraction gratings, or holograms. Factors of great influence on the quality of the products are micron-precise adjustments and also the creation of new structures and the inexpensive replication of the structures. In this context, certain material systems (e.g. III–V semiconductors) (GaAs, InP) and production processes (e.g. molecular-beam epitaxy) are being used that go far beyond those employed in silicon technology. Other basic elements of microoptics are wave guides, which for example can be made by ion exchange in special glasses, oxynitrides on silicon, by means of polyimides on glass substrates, or with special structures on the basis of III–V semiconductors (GaAs, InP). With wave guide structures such as branches and mirrors, complex beam-guiding elements can be achieved on a substrate.

3. Fiber Optics

Fiber optics utilizes the guidance of optical and infrared radiation in fiber-optic cables. Accordingly, advantageous features familiar from optical transmission techniques can be utilized: immunity to electromagnetic effects, extremely low attenuation values and considerable miniaturization possibilities. The primary application areas of fiber optics in microsystems engineering are fiber-optic sensors and optical power supplies for microsystems. The fiber-optic solutions are at present characterized by a steady but somewhat gradual penetration into microsystems engineering. They are relatively expensive and therefore used only where an above-average advantage can be expected or where other solutions are impossible.

4. Film-Coating Techniques

By using methods of epitaxy, it is possible to deposit various materials on each other with atomic precision. By this means, novel concepts for microelectronic components can be realized (see nanocomponents). From microelectronics, powerful technologies for miniaturization tech-

niques are available, which in modified form are suitable for microsystems engineering. Thick-film techniques are widespread in microsystems engineering. Thin-film technology is likewise used to a great extent, although the process engineering involved is more difficult. At present, developments are being pursued with the following objectives: (i) double-sided lithography and assimilation of lithography engineering for three-dimensional micromechanical structures; (ii) coating and microstructurization of new materials (e.g. biomaterials); (iii) combining various materials in film systems (thin-film stacking techniques, additive techniques); (iv) modifying applied films; (v) ultra-thin impervious insulating films.

5. Surface Acoustic Waves

Acoustic waves that propagate on the surface of a solid similar to the waves on the water surface of a pond, are called surface acoustic waves (SAW). Surface waves arise through transient or periodic distortions on the surface. Electrical signals can excite surface waves on piezoelectric solids (quartz, lithium niobate $LiNbO_3$, piezoceramics) or on materials provided with thin piezoelectric films. To that end use is made of thin (about $0.1\,\mu m$) strip-shaped metal electrodes, which are applied to the surface by vapor-deposition and photolithographic processes. When the signals are applied, electric fields are built up between the electrodes, and these fields are coupled to mechanical distortions by the piezoelectric effect.

Given a propagation velocity of about $3500\,m/s$, in the typically used frequency range of about $10\,MHz$ to a few GHz for surface waves, microscopically small wavelengths and penetration depths are produced (for example, $35\,\mu m$ at $100\,MHz$). The frequency range is limited upwards by the technologically feasible structural fineness of the electrodes. The unrestricted accessibility of the surface in conjunction with the flexibility of the photolithographic process allows diverse electrode structures to be applied with high precision. The surface waves are affected by the selection of the crystal and the crystal cut and also by the shaping of the conducting structures on the surface. With this approach the transmission behavior for electrical signals that are transmitted via surface waves is determined in a highly reproducible fashion. A multitude of different surface-wave components of high quality with tailor-made electrical transmission properties that are stable over a long time is thereby made possible. In addition to the filters,

resonators, correlators, etc. now being manufactured – in some cases in enormous unit numbers – sensor applications of surface-wave components are becoming more and more important.

By means of surface-wave resonators, oscillators can be built that are tuned with the introduction of other effects on the resonator, such as temperature or tensile stress (Fig. 2). In this way the oscillator becomes a sensitive frequency-to-analog sensor. For example, the introduction of the appropriate force causes it to become a torque sensor that measures the torque on a rotating shaft, without the need for slip rings. Surface-wave components can be used as passive identification flags, so-called SAW-ID tags, for sensor systems

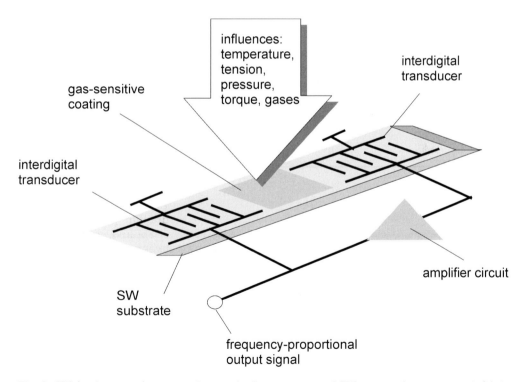

Fig. 2. SW (surface-wave) sensors: changes in the parameters of SW propagation are converted into frequency changes of an oscillator

designed for non-conducting identification. Radio-pulses are intercepted by a simple antenna and converted into surface waves. Electrodes configured as reflectors are positioned in a characteristic sequence (similar to a bar code) on the surface of the crystal, where they reflect these waves. The reflected waves are re-converted into radio signals and received by the evaluation device. The arrangement of the reflectors is manifested in the response signal as a characteristic pattern in a chain of partial pulses, which for example embody 32 bits, giving 4 billion different distinct possibilities.

The advantages of this process are the following: (i) remote inquiry, even at inaccessible locations, such as in hazardous areas, through windows and walls, without the expense of an additional installation; (ii) passive functioning, no need for its own electric power at the ID tag (no battery needed, cost, maintenance, and safety aspects); (iii) small, lightweight, simple to install; (iv) well defined inquiry range, interference effects from echoes of other objects limited because of a relatively short transit time; (v) high reliability, "crystal-stable".

These advantages will open up a multitude of additional applications, both technical and economic, for the SAW ID tag. Interesting examples of applications in process control engineering would be (i) identification of roadway and rail vehicles that are bringing in raw materials or are shipping out products; (ii) marking of containers, receptacles, semi-finished products, and workpieces, tracking them within the process, in hazardous areas; (iii) checking the access authorization of persons, recording of the access of persons in hazardous areas; (iv) travel flags and orientation aids for autonomously navigating driverless vehicles, speed measurement, position determination, distance measurement; (v) selective reflex

gates, danger recognition, room security devices; (vi) malfunction indicators created by the setting of a switchable reflector ("read-me flag") by passive means, e.g. upon damage from an impact or excess temperature, thereby preventing more extensive damage (Bulst 1994).

6. Techniques of Chemical Sensors

Chemical sensors are like measuring probes when they identify a specific component in an unknown medium and determine its concentration. Such a component can be, for example, nitric oxide or oxygen within the exhaust gas of a motor vehicle or carbon monoxide in room air. Such sensors must not have any cross-sensitivities, i.e. the measured value must not change due to the presence of other substances. They also have to respond rapidly, in order, for example, to regulate optimally the combustion process or to indicate malfunctions. Important areas of application of chemical sensors are: (i) medical diagnostics; (ii) workplace monitoring; (iii) chemical process engineering; (iv) environmental monitoring of water and air.

The chemical microsensors that have been in use up to now do not completely satisfy the demands. Only by means of system approaches satisfactory results can be achieved. Thus, the chemical composition of the sensitive film of a sensor element is very crucial. The physical processes occurring on the surface or within the sensitive film likewise influence the result. In the case of semiconducting metal oxides, parameter changes can be obtained, for example, by doping the sensitive film or changing the working temperature. The structural architecture and the thickness of the sensitive film can significantly influence the result and especially the sensor's long-term stability.

It is often necessary to use a number of sensor elements of differing cross-sensitivity, in order to draw definitive conclusions about the concentration of the components from the different signals. So far there have been only insular solutions and these are still not very perfect. Signal-processing concepts based on neural networks or fuzzy logic are still in their infancy. Developmental approaches to complex chemical analysis systems are at the initial stage of research.

7. Biocomponents

Many people see a contradiction between what has naturally evolved and what has been created by man. Yet when capabilities are analyzed that in the course of evolution have been impressed by nature alone, it becomes clear that even the living world makes use of certain "technologies". From combining certain biological building blocks used by organism to perceiving signals with components of mensuration and analysis engineering, a number of interesting applications are arising. Biosensors can be regarded as a bridge between biology and engineering (especially electrical engineering and electronics). They lead to measuring methods that can bring about important advances in medicine, in environmental protection, in food technology and in connection with process control in other fields (Bundesministerium für Forschung und Entwicklung Bonn 1992).

When considering the development of biosensors it can be seen that the trend of microsystems engineering – smaller and smaller, more and more precise – is being repeated here. Overall, for biosensors the following lines of application are emerging: (i) medical diagnostics, including patient monitoring in intensive care and bed-side analysis; (ii) environmental analysis, above all waste-water and water-

pollution control; (iii) food quality control, for example, for purity and freshness; (iv) process control substance testing and other utilizations in the chemical-pharmaceutical industry; (v) drug detection.

These applications become possible because biological probes are characterized by a highly sensitive and selective recognition of specific chemical compounds. Analyses can be provided considerably more rapidly and also more reliably than before and almost continuously. Another advantage is that these analyses can be made directly at the site of events instead of in the laboratory. Thereby not only a distinct reduction in costs per measurement, but also enormous savings in connection with the disposal of the most extremely poisonous analytes are possible even today.

In using biological probes, the capability of biological materials is exploited to find certain substances, so-called analytes, among a multitude of other materials, in accordance with the lock-and-key principle of molecular biology and to react with them. The number of possible analytes (keys) is virtually unlimited and so are the biological recognition materials (locks). Among the analytes there can be both organic and inorganic substances, as they are found, for example, in the living organism and in food. The chemical reaction of the biological material with the analyte affects physical parameters such as the electrical potential connected with extremely small changes in mass, fluorescence, or temperature. These changes are converted into measurable signals and are amplified (Fig. 3). In this way it is feasible to conduct analyses, which have so far been possible only by painstaking separation and measuring processes, and with the need to use experimental animals. One disadvantage is the limited stability of the biological measuring probe. Regardless of whether these are enzymes, antibodies, nucleic acids,

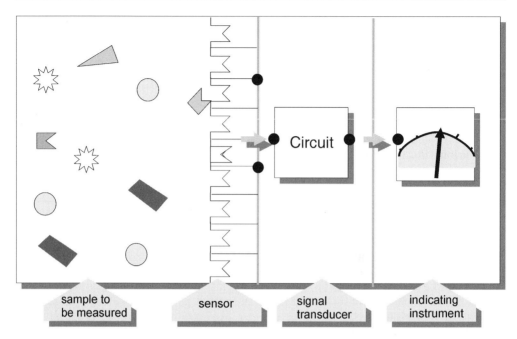

Fig. 3. Functional principle of biosensors. According to the so-called lock-and-key principle, the biological measuring device/probe ("lock") reacts with a specific substance to be analyzed, the analyte ("key"). The physical parameters that change due to the reaction are converted by the signal transducer into measurable signals, then amplified and read out (Illustration: UNIVERS/Max Ley)

cells, or even complete microorganisms, they exist in an aqueous milieu. As a rule, the biological components are confined (immobilized) between membranes. The analytes can diffuse out of the sample via one of the membranes and react with the biological component. A signal thereby generated is transformed by the signal transducer and is passed on. At present biosensors have been described for about 120 different analytes. Among these are sensors for low-molecular-mass substances, sensors for enzyme components, and sensors for macromolecules, for viruses, and for microorganisms. However, most of these developments are still at the level of general solutions.

I. Fields of Application

For greater functionality, modern technical systems require more information. Information is provided by communication-capable (teachable, radio-readable) sensors. The availability of low-cost sensors plays an important role in ensuring that the implementation of these systems will be marketable, especially when large numbers of units are involved. This goal is usually achieved only via "batch-capable" microtechniques. Some of these technologies for miniaturization are already available.

At this point, the great potential of sensors now at the market must also be addressed, because here also the evolution of sensor techniques is going in the direction of miniaturization, multi-functionality, integration, and intelligence. The fields of application that have a high demand for sensors have already been cited in the Introduction. Here the following fields of application are relevant: (i) transportation; (ii) safety; (iii) modern conveniences; (iv) vehicle-drive management; (v) vehicle running gear; (vi) traffic guidance; (vii) intelligent rail systems; (viii) navigation; (ix) roadway-route and obstacle recognition; (x) health and usage monitoring (Scheller and Schubert 1989, Scheller et al. 1992, Bundesministerum für Forschung und Entwicklung, Bonn 1994). Other applications are in building engineering, illumination management, safety, property protection, energy management and air-conditioning management, emission and immission monitoring, intelligent household appliances, water and waste-water management and security (building access, ID tags, property protection, theft-proofing of motor vehicles, monitoring of national borders, fire detection and fire fighting).

Innovations in the field of measurement, control and regulation engineering have advanced far enough so that sensors are available for the technologies in their various areas of application. But at the same time it must be noted that there is a shortage of efficient sensors at moderate prices that could promote rapid innovations of product and processes with the aid of microtechniques and microelectronics. Whereas the automobile, the environment and residential property are fields of application that call for inexpensive and rugged sensors, the fields of application of production and automation, security, health and traffic-control engineering pay particular attention to the precision of sensors.

From the multitude of possible sensor applications, the following fields of application seem to be attractive in terms of demand, numbers of units and complexity: *Traffic engineering*: automobiles, trains, aircraft, and ships. *Health*: affordable health care, monitoring of TLVs (industrial threshold limit values), non-invasive analysis, minimally invasive surgery, exercise monitoring and maintaining physical fitness, drug dosing, intensive-care medicine (Buck 1990,

Renneberg 1992, Scheper 1992). *Environment*: monitoring of soil, water bodies, air, agricultural areas, dumps, leakage detection, material recognition and byproduct separation, emission and immission monitoring. *Production techniques*: process control and regulation, process analysis, quality monitoring, testing, operational and environmental safety.

II. Nanotechnology

In nanotechnological development work, it is the integrated application of various sciences and engineering disciplines, such as physics, chemistry, biology, process engineering, and communication and information science which is innovative. However, this development rests on successful basic research. At present, a corresponding market expectation cannot yet be seen. Moreover, all of the innovative production, characterization, and modification methods suitable for nanotechnology are oriented consistently towards the idea of "engineering on the atomic and molecular level". In this regard, not only the creation of complex molecular electronics components or biochips using neural networking are discussed at present, but also, for example, the implementation of nanomachines, which by means of physical, chemical, and biological methods could be built molecule by molecule or atom by atom into molecular systems with submicroscopic dimensions (Fig. 4).

Nanotechnology is gaining importance in all fields of application where smaller and smaller structures and greater and greater precision are becoming decisive for the success on the market. Examples of such fields are optics, information engineering and nano-electronics, solar and environmental engineering, biotechnology, sensorics, robotics, and medicine (Corcoran 1991, Bachmann 1994).

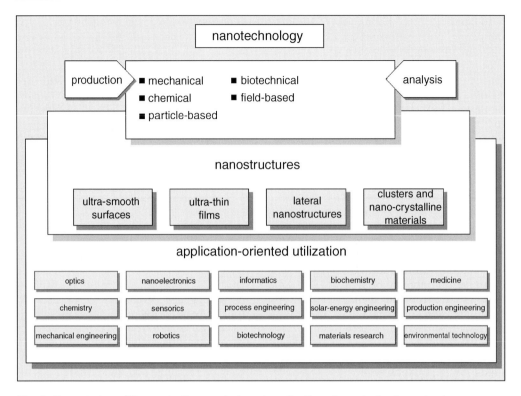

Fig. 4. General view of the production, analysis and application of nanotechnology structures

A. Product Prospects – Application Trends

The reduction of microscopic structural dimensions down to the nanometer scale, in order to utilize dimension-dependent properties and nanoscopic phenomena, is called a "top-down strategy". The most important advance here is from microelectronics to nanoelectronics which aims to utilize new physical interactive mechanisms whose causes are accounted for in terms of the transition from continuum physics to quantum physics. In order to make such complicated components, it is also important to press on with the basic technologies needed for this development (Reed 1993).

In quantum electronics, the making of layer arrangements for three-dimensional electron gases plays a major role. Chain arrangements or lattice arrangements of quantum dots – the building blocks of the individual electronics – permit, for example, under the influence of the Coulomb barrier, the correlated and controllable tunneling of individual electrons (single electron logic). Since quantum dots can be viewed as artificial atoms with discrete energy levels, pronounced optical non-linearities can furthermore arise from external excitation by means of photons. Therefore, such structures are also suitable in optoelectronics. Furthermore, quantum dots can also be used as a storage medium.

For special applications (e.g. high temperatures, high radiation exposure, high frequency), vacuum electronics can contribute to the opening up of new markets. One significant area of application is in display technology. In the area of information storage, nanotechnology will be the basis for future developments. The limit of storage density is technologically reached only when atomic dimensions are employed for structurization. Moreover, future storage technologies will take into consideration the actuating of three-dimensional storage mechanisms. Another conceivable application is the interlinking of zeolite crystals with polymer wires and the subsequent local storing of chemical information. Optical storage in bacteriorhodopsin, holographic storage technology and electron-hole burning are other potential options for three-dimensional storage (Engineering a small world 1991).

For optics it is possible to open wide fields of application from the development of structurization and super-polishing procedures for ultraprecision treatment, to nanometer-precise positioning and control methods necessary for this. In chemistry and biology, the applications of nanotechnology go beyond analysis (biosensorics) or the physical manipulation of molecules (tailor-made molecules). We could mention, for example, catalysis (by clusters or nanoporous materials) and filtration engineering (e.g. biological ultrafilters with pores in the micrometer diameter range). Visionary fields of application for nanotechnology can be seen in medical engineering, for example, in connection with the physical coupling of biological systems with microelectrodes, the use of nanometer-sized capillary tubes for micro-dosing devices, the use of micro-nanomachines in surgery and the development of a silicon-retina chip. In the field of mechanical engineering there is the possibility of minimizing wear and tear on workpieces by giving them a surface treatment of the highest quality. In solar-energy engineering the development of granular films with tailor-made optical properties is conceivable.

Bottom-up strategy: another approach in nanotechnology consists in the selective building up of atomic or molecular aggregates into larger systems. Here the main distinctions made are those of principles of self-organization, organic/inorganic interface phenomena and the selective chemical or physical joining of molecular systems to prepared surfaces. The Langmuir-Blodgett technique represents a multi-disciplinary field in this regard. Material combinations of such monomolecular layers permit the selective preparation of materials with highly anisotropic electrical properties, good insulation properties, non-linear optical properties or pyroelectrical activity.

In molecular electronics, through inspection and control, molecular processes and system concepts are to be developed that show a strong parallelism with information handling in their general tendency. Serving as a model here are properties of molecular and biological systems such as efficiency, packing density, self-organization, fault tolerance, reproducibility, and low energy consumption. With regard to the implementation of three-dimensional integrated circuits on a molecular level, technological upheavals are expected similar to those that came about in the course of supplanting the electron tube by the transistor.

Information being gained from cluster research also shows that there is a great range of applications here, for example, with respect to heterogeneous catalysis, microstructure physics, development of nanotechnological structure units, medicine, sensorics, and photovoltaics. Clusters of intermediate size show new properties between molecules and the solid state, whereby electrical and optical characteristics can be adjusted in a well defined way.

Analytics: both in the field of material and bio-sciences and also in the sector of micro- and nanoelectronics/sensorics, important scientific and technological advances will depend on whether, for example, molecular structures can be imaged in a non-destructive way and whether time changes on the atomic scale can be directly made visible. The currently available maximum lateral resolution in scanning electron microscopy amounts to about 1–5 nm. With transmission electron microscopes, local resolutions of up to 0.1 nm can be achieved. An especially important role is played by scanning-probe methods. Their most important representative is the scanning tunnel microscope (STM). The STM is based on the distance-dependent electron tunnel current between a peak and a surface. With it, spatial resolution in the range of sub-atomic diameters is possible. With the development of the scanning force microscopy (AFM), the particular advantages of STMs could be extended to non-conducting samples. A disadvantage of this special scanning technique is that chemical elements cannot be unequivocally specified.

B. Procedures and Techniques

The listings offered below give a general idea of the most important methods and techniques used in nanotechnology.

1. The Making of Ultrathin Films

By the construction and connecting of various ultra-thin films in the nm-range, complex combinations of mechanical, optical, electrical or chemical properties can be created in a three-dimensional config-uration with a function-oriented character and with an extremely high integration density. This is true not only for inorganic and organic substrate films but also for biological molecular structures, which can be used for making switches, storage devices, sensors, processors, actuators, membranes, catalysts, etc. and also for multivalent applications.

The requirements placed on the making of films in nanotechnology are characterized by the demand for sharp atomic interfaces and for controlling depositions in atomically thin layers. These are mostly vacuum methods, which are based either on molecular beam epitaxy (MBE) or precipitation from the gaseous phase (e.g. atomic layer epitaxy, ALE; chemical beam epitaxy, CBE). For organic films the Langmuir-Blodgett method can be mentioned as an important preparation technique.

2. Creation of Lateral Nanostructures

In addition to the creation of extremely thin films, another central problem of nanotechnology is to make lateral structures and localized doping profiles. The motivation for making smaller and smaller structures has come mainly from microelectronics. For the making of masks, in addition to optical lithography also direct-writing methods are being used, where focused laser, ion, or electron beams are employed. The resolutions are roughly 200 nm, 50 nm and about 10 nm, respectively. Alternative structurization methods are holographic structurization and STM lithography. Nanolithography per se involves structural dimensions < 100 nm.

One disadvantage of lithography that uses scanning-probe methods is its low writing speed and its small visual field. An array of structurization probes would provide a remedy for this.

3. Creation of Clusters and Nanocrystalline Materials

Clusters are agglomerates consisting of about 10–1000 atoms. With these, the dependence of their physical and chemical properties on cluster size is directly correlated with their major surface-to-volume ratio. There are many different techniques that can be used to make clusters in the condensed phase, in the gaseous phase, or in vacuum. The scientific and technical interest in clusters is in essence based on their physical and chemical properties, which specifically for clusters consisting of up to about 100 atoms may depend greatly on their size. Thus, for example, structure, stability, electronic state density, optical absorption behavior, ionization potential, melting-point lowering and also catalytic, magnetic and chemical properties of cluster materials depend significantly on the number of particles in the cluster.

4. Principles of Self-Organization

With the use of atomic and molecular architectures, new material structures are being provided target-selectively for the execution of well defined chemical, physical and biological functions. These systems have new capabilities. They are among those intelligent products that record and evaluate data and execute actions resulting from this. For example, molecules can become electric switches or can be used for information reception, storage and transmission.

Molecular architectures can also be used for pattern recognition, self-structurization, selforganization, self-reproduction, for the selective linking of atoms or for the creation or separation of molecular groups (clusters). By the fabrication of new structurally defined molecular species, metals can be transformed into semicon-ductors or into optical or "switchable" industrial materials. Filters or membranes become controllable, since they close their pores upon contact with certain molecules. It can be expected that on this basis, the development of new sensor generations, biosensors, information storage devices and information conductors, biochips, molecular transistors, switches and storage devices, structural and functional materials, materials for artificial intelligence systems, biocompatible materials, optical components, medical applications, pharmaceuticals, etc., will become feasible.

The general goal of molecular/atomic architecture is to see a certain property being realized in a molecule or material (Engineering a small world 1991, Weisburch and Vinter 1991, Kirk and Reed 1992).

5. Analysis of Ultrathin Films

Highly developed procedures and devices that work with beams of electrons, ions, neutrons, neutral particles and photons, with field-emission and tunnel effects and also on the basis of acoustic, electrical, magnetic and optical principles are used in nanoanalytics. The detection of lateral element distribution is done with electron-beam or ion-beam techniques. Depth-dependent element analysis is done with ion sputtering. A relatively new analytical procedure for nanotechnology can clearly be seen in the use of scanning tunnel microscopy/scanning force microscopy.

References

Bachmann DG (1994) Nanotechnologie. VDI

Benecke W, Petzold HC (1991) Proceedings IEEE: Micro Electro Mechanical Systems. Travemünde

Bley P, Menz W (1993) Aufbruch in die Mikrowelt. Phys Bl 49: 179–184

Buck P (1990) Biosensor Technology. Marcel Dekker: New York

Bulst WE (1994) Developments to Watch in Surface Acoustic Wave Technology. Siemens Rev: 2–6

Bundesministerium für Forschung und Entwicklung (1992) Biosensorik. Bonn

Bundesministerium für Forschung und Technologie (1994) Mikrosystemtechnik. Bonn

Bundesministerium für Forschung und Entwicklung (1994) Produktionsintegrierter Umweltschutz. Bonn

Corcoran E (1991) Nanotechnik. Spektrum Wiss Jan 1991: 76–86

de Raolj NF (1991) Current status and future trends of silicon microsensors. Proceedings of Transducers. San Francisco, 1991, pp 79–86

Engineering a Small World; From Atomic Manipulation to Microfabrication Science (1991) 254: 1300–1342

Heuberger A (1989) Mikromechanik. Springer: Berlin

Howe RT, Muller RS, Gabriel KJ, Trimmer WSN (1990) Silicon Micromechanics: Sensors and Actuators on Chip. IEEE- Spektrum, July 1990: 29–37

Kirk WP, Reed MA (1992) Nanostructures and Mesoscopic Systems. Academic Press: New York

Kroy W, Fuhr G, Heuberger A (1992) Mikrosystemtechnik. Spektrum Wiss May 1992: 98–115

Kroy W, Heuberger A, Sehrfeld W (1988) Nerven und Sinne für die Chips. High Tech 4: 22–36

McEntee J (1993) Start Making Microsensors. Phys World Dec 1993: 33–37

Menz W, Bley P (1993) Mikrosystemtechnik. VCH: Weinheim

Muller RS, Howe RT, Senturia StD, Smith RL, White RM (1991) Microsensors. IEEE Press: New York

Peterson K (1978) Dynamic Micromechanics of Silicon. IEEE Trans Electron Devices ED-25: 1241–1250

Reed MA (1993) Quantenpunkte. Spektrum Wiss März 1993: 52–57

Renneberg R (1992) Molekulare Erkennung mittels Immuno- und Rezeptorsensoren. Spektrum Wiss Sept 1992: 103–112

Scheller F, Schubert F (1989) Biosensoren. Birkhäuser: Basel

Scheller F, Schubert F, Pfeiffer D (1992) Biosensoren. Enzym- und Zellsensoren: Anwendungen, Trends und Perspektiven. Spektrum Wiss Sept 1992: 99–103

Scheper Th (1992) Wärme- und Lichtsignale aus dem Bioreaktor. Spektrum Wiss Sept 1992: 103–144

Stix G (1991) Trends in Micromechanics "Micron Machinations". Sci Am Nov 1991: 73–82

Weisburch C, Vinter B (1991) Quantum Semiconductor Structures: Fundamentals and Applications. Academic Press: New York

Wise KD, Najafi N (1991) The coming opportunities in microsensor systems. Proceedings of Transducers. San Francisco, pp 2–7

Mechanical Sensors
A. Waves, Sound and Vibrations

3. How Nature Designs Ears

Axel Michelsen

Abstract

Most ears respond to the pressure component of sound, but some ears detect the oscillatory flows of the medium associated with sound. Ears generally receive sound pressure by means of an ear drum, the vibrations of which may either be guided to an inner ear (e.g., via middle ear ossicles in terrestrial vertebrates) or detected by receptor cells attaching to the ear drum (e.g., moths, grasshoppers). Most hearing animals can determine the direction to the sound source, but the cues exploited depend on whether the animal is sufficiently large to use differences in diffraction and/or in the time-of-arrival of sound at the ears. In small animals, the sound can often reach both surfaces of the ear drum, and the ears may then have directivity properties similar to those of pressure gradient microphones. While some animals such as the insect prey of echolocating bats do not need to analyse sound frequency, most animals using sounds for social communication have a capacity for frequency analysis. The mechanism of frequency analysis generally involves mechanical filters, but evidence for additional active mechanisms (mechanisms depending on metabolic energy) has been found in both vertebrates and insects.

I. Introduction

A sense of hearing is found only in vertebrates and arthropods, but within these groups it evolved independently more than twenty times (Webster 1992, Yager 1999). The ears again evolved into a large number of different types. In different groups of animals, the ears are located on the head, thorax, abdomen, legs or wings. In all cases, the ears seem to be derived from mechanoreceptors at or near the body surface (e.g., Lakes-Harlan et al. 1999). Many ears evolved for the detection of sounds made by predators. Echolocating bats evolved 50 million years ago. Apparently they caused much panic among nocturnal insects, many of which evolved ears and suitable behavioural responses to ultrasound. Other ears may have evolved or been adapted for social communication. These ears are often adapted to certain types of sound signals, which are particularly well transmitted in the animals' habitats.

A. The Context of Hearing

The frequency range heard by various animals extends from about 10 Hz to more than 200 kHz, but no single animal covers the entire range. The use of sounds for communication is limited by the physical properties of both the animals and their environments (Michelsen

1992), and this is often reflected in the hearing capabilities of the animals. Most insects and small birds cannot emit sounds efficiently below 1 kHz, and insects smaller than about 1 cm are restricted to emitting ultrasound. Ultrasound is heavily absorbed by vegetation, however, and small animals living in dense vegetation are left with a very limited frequency range for communication (Michelsen and Larsen 1983). Ultrasound can be used in free air spaces, both for communication (e.g., bush crickets) and for locating prey (echo-locating bats). Sound emission depends on the properties of the medium. For example, it requires less energy to emit sounds at low frequencies in water than in air. Several small fishes use low frequency sounds in their social communication.

Sound is generally described as tiny fluctuations of pressure, but sound also involves small oscillatory flows of air (or water) "particles" (defined as small volumes of the medium, which move as a unit). Many animals can detect the pressure component of sound and are thus able to hear in the same sense as we humans do. Other animals detect only the oscillating flows of the medium associated with the sound wave, and it has been debated whether this ability should also be named hearing (see chapters 9 and 10, this volume). The hearing of fishes provides an example of both modalities. The otolith organs of the inner ear respond to a variety of mechanical stimuli including oscillatory vibrations at auditory frequencies, but not fluctuations of pressure. Sound causes movements of the otolith relative to the sensory macula, which contains groups of receptor (hair-) cells with different preferred directions. Some fishes are also sensitive to pressure oscillations, because the inner ear is mechanically coupled to a gas-filled structure like a swim bladder or a prootic auditory bulla (Blaxter 1981).

The information carried by sound is often briefly summarized by "Who, what, and where?" The identity of the sound emitter (*who*) and the content of the message (*what*) can be revealed by an analysis of the spectral characteristics and rhythm of the sound. Spectral analysis is generally performed by means of some sort of frequency analyser in the ear, whereas the temporal characteristics are analysed in the periphery by receptor cells and in the brain by neural networks or even by single neurons. The localization of the sound emitter (*where*) involves a determination of both direction and distance. Solutions for the latter (ranging) are not well understood (Naguib and Wiley 2001), whereas substantial amounts of information exist about the mechanisms for directional hearing (e.g., Yost and Gourevitch 1987, Michelsen 1998).

B. Hearing Organs

Most ears responding to the pressure component of sound receive sound by means of an ear drum. Ear drums are generally membranes (in physical terms), but some contain plates surrounded by membrane (e.g., bush crickets). The vibrations of the ear drum may either be guided to an inner ear via middle ear ossicles (terrestrial vertebrates) or be detected by receptor cells attaching to the ear drum (e.g., moths, grasshoppers). However, in some cases there is no obvious connection between the ear drum and the inner ear, and the transmission path is not known (e.g., crickets, bush crickets). Ear drums tend to be fairly linear systems with a wide dynamic range, but the receptor cells respond to the vibrations in a non-linear manner and have only limited dynamic ranges. Groups of receptor cells with different sensitivities may be used for increasing the entire dynamic range of the ear (range fractionation).

At a symposium for comparing sensors and sensing in the natural and fabricated worlds, it is tempting to compare the mechanics of microphones and ears. This is difficult, however, because both the structure (anatomy) and material properties vary considerably. Physical data for ear drums are available only for a few species, and often the response of the ear drum to sound is determined mainly by other components like the load of structures attached to the drum (e.g., the ossicles transmitting vibrations to an inner ear in vertebrates). It is easier to compare the simple ears of some insects to condenser microphones.

Condenser microphones are generally made to be used at frequencies below or including the first, heavily damped resonance of the membrane (see chapter 4, this volume). In contrast, higher modes of vibration are quite common in ear drums. For example, the locust ear (Fig. 1) has no less than four modes of lightly damped vibrations (similar to circularly symmetrical modes in circular membranes) between 1 and 20 kHz (Michelsen 1971). In the human ear, the higher modes are located at parts of the ear drum such that these vibrations are not transmitted to the inner ear. In contrast, in the locust ear the higher modes are exploited for frequency analysis (see below).

The theory of membrane vibrations is fairly complicated, but up to the first resonance the average behavior of a circular membrane is not far from that of a simple second order system such as a mass on a spring. Ignoring friction, the resonance frequency is proportional to

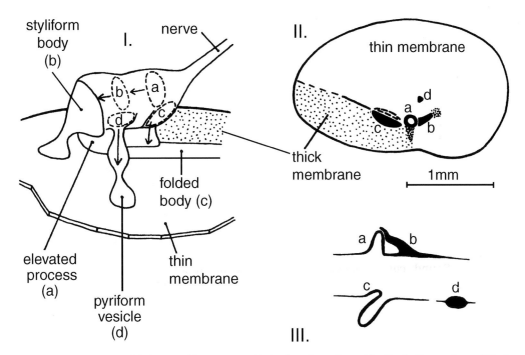

Fig. 1. The locust ear. I Right ear with receptor cells (a–d) and attachment bodies in the ear drum. II Left ear seen from outside; the attachment parts are shown in black. III Schematic sections through the attachment bodies (From Schwabe 1906 and Michelsen 1971)

the square root of the membrane tension and inversely proportional to the radius and the square root of the mass per unit area (Morse 1948). The tension is inversely proportional to the compliance. In the following, we shall consider the order of magnitude of these factors in ear drums and microphones.

The membranes of condenser microphones are made of steel or nickel and are typically 2–5 µm thick. A 5 µm thick membrane has a mass per square mm of about 40 µg. In comparison, the thin (3–5 µm) part of the ear drum of locusts has a mass of only about 4 µg per square mm. The effective mass of vibrating membranes/drums is somewhat larger, because a boundary layer of air is moved with the membrane. In the locust ear drum vibrating in its basic mode, this radiation mass is about 1 µg per square mm, i.e. about 20% of the total mass and far from negligible (Michelsen 1971).

At least two components contribute to the stiffness of microphone membranes and ear drums. The stiffness of the membrane/drum itself, the stiffness of the air trapped behind it, and – in ears – the stiffness of structures attaching to the drum. For microphones the total stiffness is generally given as an equivalent air volume, which is smaller than the actual volume of air trapped behind the membrane. Instead of stiffness we shall consider the acoustical compliance, which is the volume of the air displaced by a membrane acted upon by a certain (sound) pressure (in units of m^3 per $N/m^2 = m^5/N$). The compliance of condenser microphone membranes varies with their size. For $\frac{1}{2}''$ (area 110 mm²), $\frac{1}{4}''$ (area 28 mm²), and $\frac{1}{8}''$ (area 7 mm²) microphone membranes, the acoustical compliances are about 3×10^{-13}, 5×10^{-15}, and $10^{-16}\,m^5/N$, respectively. For comparison, the compliances of the ear drums of humans (area 65 mm²) and locusts (area 3 mm²) are

about $7 \times 10^{-12}\,m^5/N$ and $10^{-12}\,m^5/N$, respectively (Rosowski 1994; unpublished data, see also chapter 5, this volume). In both these ears the drums are heavily loaded, and yet they are more compliant than microphone membranes of similar size. The compliance value for the locust ear drum was calculated from its vibration at low frequencies, where the ear mechanics is controlled by stiffness (this can be seen from the phase angle between the sound and the vibration). The same approach provided the value $10^{-14}\,m^5/N$ in bush crickets with an ear drum area of about 0.5 mm² (Michelsen and Larsen 1978). In these and many other insects, a tracheal tube or a series of tracheal sacs connects the space behind the ear drum with other air spaces or with the outer world. The surprisingly large compliances probably reflect the fact that the air behind the ear drum is not trapped.

As expected from the relationships discussed above, the resonance frequencies of condenser microphones vary inversely with their diameter (8–10 kHz and 70 kHz in $1''$ and $\frac{1}{4}''$ microphones, respectively). This rule also applies to ear drums. However, compared to a condenser microphone membrane of similar size, the locust ear drum has approximately a 10 times smaller mass and a 10^5–10^6 times larger compliance. A smaller mass should cause a higher resonance frequency, whereas a larger compliance should cause a lower resonance frequency. The actual basic resonance frequencies are 3.5 kHz in the locust ear drum and above 100 kHz in the microphone membrane.

What do insects do in order to detect ultrasound? The solution of the locust is to exploit higher modes of ear drum vibration, but moths have somehow managed to increase the basic resonance frequency of their ear drum (Surlykke et al. 1999). How did they do this?

One possibility is to decrease the mass, i.e., the ear drum can be made thinner. In moth ears used for detecting the ultrasonic cries of hunting bats, the thickness of the ear drum is indeed well below $1\,\mu m$. The total mass does not decrease in proportion to the thickness of the drum, however, since the radiation mass remains constant (it is now larger than the mass of the drum). One may speculate that backing the ear drum with solid walls might cause a decrease of the compliance. The only parameter left is size, and it comes as no surprise that moth ears have 6–15 times smaller diameters than a $\frac{1}{8}''$ condenser microphone. It is interesting that the radius of moth ears increases less with the size of the body than other body parts, which "suggests that some auditory constraint (e.g., frequency tuning) limits the maximum size" (Surlykke et al. 1999). Although these trends do make some sense, it should be stressed that no experimental data are available for the physical properties of moth ear drums.

II. Directional Hearing

In theory, in order to estimate the direction of sound in three-dimensional space one needs four sound receivers, widely spaced, but not placed in the same plane. Preferentially, the receivers should be small and placed far away from large (sound reflecting) bodies. They could then make exact records of the temporal and spectral characteristics of sound. So far, evolution has not produced animals carrying small ears at the tips of four tentacles. Most animals have two ears on or close to the main part of the body. With such an arrangement, a satisfactory directional hearing has been obtained at the expense of the ability to determine the exact frequency spectrum of the sounds. Apparently, the need to determine

direction has been so urgent that the animals have paid this price.

A. Pressure Receivers

Man and most large animals can use two mechanisms for detecting the direction of sound incidence: the inter-aural difference in sound pressure caused by diffraction of sound by the body, and the difference in time of arrival of the sound at the two ears. In addition, the frequency spectrum at each ear varies with the angle of sound incidence, and the spectra at the two ears are different for most directions. It is thus possible for the brain to estimate the direction to the sound source by comparing the sound spectra at the two ears. This task is easier with broad-band sounds than with pure tones or narrow-band sounds.

The barn owl has evolved very precise directional hearing (Konishi 1983), making use of the two mechanisms also available to man: the inter-aural difference in sound pressure caused by diffraction of sound by the body, and the difference in time of arrival of the sound at the two ears. Humans are very good at using these cues for localising the horizontal direction to a sound source, but we have only a modest ability to judge the elevation of the source (Yost and Gourevitch 1987). In contrast, barn owls have approximately the same localization error in both directions when they are listening to noises like the rustling sounds from a mouse. The main reason for this ability is that the facial ruff (tightly packed feathers forming the face) is asymmetrical. Troughs in the ruff collect sounds and funnel them into the ear canals openings. The trough is lower at the right side and tilted up, whereas the opposite is true at the left side. At 5–8 kHz (typical for rustling), the right ear is more sensitive upward, and the left ear downward, and the owl can now separate the cues, using the binaural sound pressure differences and the time differences for the estimation of elevation and azimuth, respectively.

The possible use of diffraction as a cue is not determined by the absolute size of the animal, but by its size relative to the wavelength of the important sounds. For example, in moths exposed to the ultrasonic cries of bats, the two ears may experience a pressure difference of 10–20 dB, which is more than sufficient for the purpose of determining the direction away from the bat (Payne et al. 1966).

In many hearing animals (e.g., insects, frogs, birds) the part of the body carrying the ears is so small that a difference in the sound pressure at the two ears is available only at very high frequencies. For example, the wavelength of the calling song of many crickets and frogs is 5–10 times larger than the animal's body, and only little difference in pressure is found at the ears during the calling song. In addition, the time differences are often too small to be useful as directional cues, if they are to be processed by the nervous system.

Recently, it has been found that minute time (phase) differences can indeed be exploited for directional hearing, if they are processed directly by the hearing organ. Parasitic tachinid flies locate their insect hosts by means of an unorthodox directional hearing. The flies have a pair of adjacent and essentially fused ears in the ventral midline of the thorax, just behind the head. The ear drums are roughly triangular. The two ear drums span the same tracheal space and are mechanically coupled by an intertympanal bridge, which comprises two cuticular bars (attached at their ends to the ear drums) connected at a midline pivot point, which is not rigid. Vibrations of one ear drum are transmitted to the other with some damping and time delay. As a result, differences in time of arrival of sound to the

two ears of a few μs result in large differences in vibration amplitude. In the species *Ormia ochracea*, the optimal frequency for this mechanism (5 kHz) matches the calling song of its male cricket hosts (Robert and Hoy 1998). As pointed out by Miles et al. (1997), the ears of *Ormia* employ a mechanical realization of a concept devised by Blumlein (1931) to electronically process signals from two closely spaced microphones in order to record a stereophonic signal. Like the *Ormia* ear, Blumlein's circuit converts very small phase differences at two sensors into amplitude differences at the outputs.

B. Pressure Difference Receivers

In a *pressure difference receiver*, the sound waves reach both surfaces of the ear drum. In theory, this is possible in many animals, since the ears are often connected by an air-filled passage. Alternatively, the sound waves may enter the body and reach the inner surface of the ear drum through some other route (e.g., a tracheal tube). Such potentially sound-transmitting pathways are known in several insect groups and in some frogs, reptiles, birds and even mammals (Rosowski and Saunders 1980). The existence of an anatomical air space leading to the inner surface of the ear drum is a necessary prerequisite, but it does not automatically create a pressure difference receiver with proper directionality. The sound has to arrive at the inner surface of the membrane with a proper amplitude and phase. In addition, the sound inside the animal should be affected in a suitable manner by the direction from which the sound reaches the outer surface of the animal (Michelsen 1998).

The transmission of the sound to the inner surface of the tympanum is often a function of frequency. In locusts, low-frequency sound is transmitted from one ear to the other with little attenuation (Michelsen and Rohrseitz 1995). High-frequency sounds are attenuated, however, and the ear then almost becomes a pressure receiver. In most animals the ears are located at the surface of the head or body, and the difference in the sound pressure at the ears caused by diffraction is sufficient to provide a reliable directional cue at high frequencies.

Crickets have one of the most complicated hearing organs known (Fig. 2). The ears are located in the front legs, just below the "knees". Sound acts on the outer surface of the ear drum, and an *acoustic trachea* connects the inner surface of the ear drum to the lateral surface of the body. In addition, a *transverse trachea* connects the acoustic tracheae at the two sides of the body. The cricket ear is thus an acoustic four-input device (Michelsen et al. 1994). The velocity of sound propagation in the tracheae (264 m/s) is rather close to that expected for isothermal wave propagation in air (245 m/s), probably because an exchange

of heat occurs at the tracheal walls. The acoustic trachea behaves much like a tube open at one end, whereas the transverse trachea acts like an 8-pole bandpass filter. The physical mechanism behind this filter is not known, but a *central membrane* in the transverse trachea is important for its function (Michelsen and Löhe 1995). A hole of 10–25% of its area (made with the tip of a human hair) is sufficient to change the transmission properties to those observed in the acoustic trachea.

For a physical analysis of the directional hearing, one also needs to measure how the amplitude and phase of sound at the four sound inputs vary with the direction of sound incidence. From these data and the transmission gains one can calculate the amplitude and phase of the sounds arriving at the ear drum (Fig. 3, where the weak sound from the other ear (CT) has been ignored). The cardioid directionality pattern of one ear in Fig. 3 combines a good sensitivity in the forward (0°) direction with a much lower sensitivity to sounds from the other side of the body (270°). This is an ideal pattern for animals walking towards a sound source. The main contributor to this pattern is the changing phase of the sound from the other side of the body (CS). However, the phase variation of the sound CS would not lead to such a pattern, if the phase relationships of the three sounds had not been right. The crucial importance of the phase relationships can be confirmed by simple mathematical modelling.

A proper phase relationship occurs only within narrow ranges of frequency. At the calling song frequency (4.6–4.7 kHz in *Gryllus bimaculatus*) the differences between the ear drum vibrations at the sound directions 30° and 330° is about 10 dB. This gradient in the forward direction is tuned with a $Q_{(3\,dB)}$ of 14. The calls are emitted by

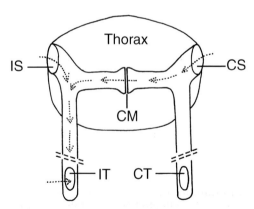

Fig. 2. The cricket ear has four acoustic inputs, one of which is the external surface of the ear drum (IT: ipsilateral tympanum). The other inputs are the ipsilateral spiracle (IS), contralateral spiracle (CS), and contralateral ear (CT). Sound propagating across the midline passes a central membrane (CM). Dotted arrows: paths of sounds affecting the left ear (From Michelsen and Löhe 1995)

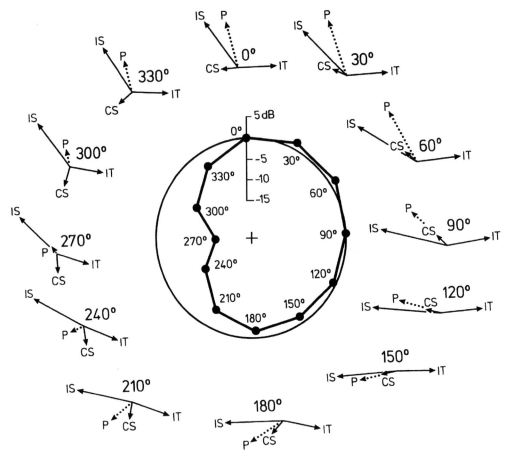

Fig. 3. Calculated directional pattern of the force acting on the ear drum (tympanum) of the right ear of a cricket at 4.5 kHz. The force driving the drum is proportional to a sound pressure, P, which is the sum of three vectors: the sound at the outer surface of the tympanum (IT) and the sounds transmitted to the inner surface from the ipsi- (IS) and contralateral (CS) spiracles (cf. Fig. 2). 180° has been added to the IS and CS components (From Michelsen et al. 1994)

the harp, a part of the wing, which is tuned with a $Q_{(3\,dB)}$ of 25. This communication system is thus based on two mechanical systems tuned to the same frequency. This allows the field cricket to engage in "long distance" (metres) sound communication close to ground, where the attenuation and degradation of sound are much larger than in most habitats (Michelsen and Rohrseitz 1997).

III. Frequency Analysis

Some animals do not need the ability to analyse sound frequencies (e.g., it is not interesting for the potential prey of echo-locating bats to find out, which kind of bat is looking for its dinner). The crucial information is the presence of ultrasound, the sound pressure level (a measure of the distance to the bat), and the sound direction. In contrast, frequency analysis greatly facilitates the analysis of social sound signals, and most animals using sounds for social communication have a capacity for frequency analysis (vertebrates, insects like grasshoppers, bushcrickets, crickets, and cicadas). The number of frequency bands in the frequency analysis vary from two to more than twenty. However, it is not at all obvious that this variation in capacity is correlated with the needs of the animals for information about sound frequency.

In both vertebrates and insects, the main mechanism of frequency analysis is based on the place principle: Receptor cells with different anatomical locations differ as to characteristic frequency. In some cases the receptor cells are known to attach to structures with frequency-dependent vibrations (e.g., mammals, locusts). A very simple frequency discrimination is found in the water boatman, in which the two ears are mechanically tuned to different frequency bands (Prager 1976). The water boatman does not seem to make use of the different tuning of the ears for frequency analysis, but rather for obtaining a sensitive hearing within two frequency bands that are characteristic for the resonances of two different sizes of the air bubbles carried with the diving water boatman.

In locusts and mole crickets the ear drum is used both for receiving sound and for analysing its frequency content. In locusts, the ear drum is bean-shaped and about 2.5×1.5 mm at its widest (Fig. 1). Four cuticular bodies are situated between the middle of the ear drum and its anterior edge. The sensory units, a total of 60–80 chordotonal sensilla, are attached in four groups (named a to d) to the cuticular bodies. The frequency responses of the four groups of receptor cells differ. The frequency analysis in the locust ear is partly due to the presence in the ear drum of both basic and higher modes of vibration, which are picked up by receptor cells attached to the cuticular bodies (Michelsen 1971). In addition, the cuticular bodies and the masses of receptor cells vibrate in a complicated manner, and this contributes to the frequency analysis (Breckow and Sippel 1985). A similar principle appears to be used by cicadas.

In other cases, it is not known why the receptor cells differ in frequency preference. Most frequency analysers are located in "inner ears" located some distance away from the ear drum (bush crickets, crickets, vertebrates). Despite intense research efforts we are still fairly ignorant about the fine mechanisms leading to frequency analysis in most inner ears. For example, the inner ear of bush crickets consists of a row of about 25 receptor cells, each of which have a preferred frequency that is about 2 kHz different from those of its neighbours (Fig. 4).

In addition to mechanical filters, active processes at the cellular level may sharpen the passive resonant responses of ears. At frequencies below about 1 kHz, an electrical tuning of the cell membrane may contribute to the frequency preference of receptor cells, e.g. in turtles and frogs (Fettiplace et al. 2001). The role of contractile elements in hair cells in the cochlea is discussed by Brownell (see chapter 6, this volume). Although originally discovered in mammals, such active mechanisms may also play a role in insect ears. Otoacoustic emissions (ear drum vibrations generated by active processes in receptor cells) have been described in grasshoppers and moths (Kössl and Boyan 1998, Coro and Kössl 1998), and active mechanisms in mosquito hearing have been found by Göpfert and Robert (2001).

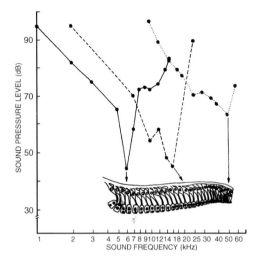

Fig. 4. Threshold curves for three receptor cells in the ear of a bush cricket (redrawn from Zhantiev and Korsunovskaya 1978)

IV. Discussion and Conclusions

In this paper, I have attempted (with much hand-waving) a biophysical explanation for the fact that most insect ears used to detect ultrasound are very small (typical diameter 200–500 µm). Obviously, a real understanding of the problems involved requires that the basic physical properties of the ears become properly investigated.

The use of sounds for communication is limited by the physical properties of the environment, but the physical processes attenuating and degrading sounds propagating outdoors are only known in principle, not in sufficient detail to explain the animals' specific adaptations. For example, it has been found that degradation of sound close to ground is more destructive to directional amplitude cues than to the phase cues (Michelsen and Rohrseitz 1997), and model calculations show that pressure difference receivers are especially sensitive to phase cues. It remains to be learned, however, whether pressure

difference receiver mechanisms may be advantageous to large animals living in dense vegetation.

It is easy to predict that future studies of hearing will focus on active mechanisms, but a much better understanding of the passive mechanics will be needed in order to reveal the contributions of the active processes.

Acknowledgements

Original research for this chapter was supported by The Centre for Sound Communication, which is financed by the Danish National Research Foundation. The author is most grateful to Anne-marie Surlykke and Ole Næsbye Larsen for their comments on the manuscript.

References

Blaxter JHS (1981) The swimbladder and hearing. In: Tavolga WN et al. (eds) Hearing and Sound Communication in Fishes. Springer, New York, pp 61–71

Blumlein AD (1931) British patent 394,325. Directional effect in sound systems

Breckow J, Sippel M (1985) Mechanics of the transduction of sound in the tympanal organ of adult and larvae of locusts. J Comp Physiol A 157: 619–629

Coro F, Kössl M (1998) Distortion-product otoacoustic emissions from the tympanal organ in two noctuid moths. J Comp Physiol A 183: 525–531

Fettiplace R, Ricci AJ, Hackney CM (2001) Clues to the cochlear amplifier from the turtle ear. Trends in Neurosci 24: 169–175

Göpfert MC, Robert D (2001) Active auditory mechanics in mosquitoes. Proc R Soc Lond B 268: 333–339

Konishi M (1983) Neuroethology of acoustic prey localization in the barn owl. In: Huber F, Markl H (eds) Neuroethology and Behavioral Physiology. Springer, Berlin, pp 303–317

Kössl M, Boyan GS (1998) Otoacoustic emissions from a non-vertebrate ear. Naturwissenschaften 85: 124–126

Lakes-Harlan R, Sölting H, Stumpner A (1999) Convergent evolution of insect hearing organs from a preadaptive structure. Proc R Soc Lond B 266: 1161–1167

Michelsen A (1971) The physiology of the locust ear. Z Vergl Physiol 71: 49–128

Michelsen A (1992) Hearing and sound communication in small animals: Evolutionary adaptations to the laws of physics. In: Webster DB et al. (eds) The Evolutionary Biology of Hearing. Springer, New York, pp 61–77

Michelsen A (1998) Biophysics of sound localization in insects. In: Hoy RR et al. (eds) Comparative Hearing: Insects. Springer Handbook of Auditory Research 10. pp 18–62

Michelsen A, Larsen ON (1978) Biophysics of the ensiferan ear. I. Tympanal vibrations in bush crickets (Tettigoniidae) studied with laser vibrometry. J Comp Physiol 123: 193–203

Michelsen A, Larsen ON (1983) Strategies for acoustic communication in complex environments. In: Huber F, Markl H (eds) Neuroethology and Behavioral Physiology. Springer, Berlin, pp 321–331

Michelsen A, Löhe G (1995) Tuned directionality in cricket ears. Nature 375: 639

Michelsen A, Rohrseitz K (1995) Directional sound processing and interaural sound transmission in a small and a large grasshopper. J Exp Biol 198: 1817–1827

Michelsen A, Rohrseitz K (1997) Sound localisation in a habitat: an analytical approach to quantifying the degradation of directional cues. Bioacoustics 7: 291–313

Michelsen A, Popov AV, Lewis B (1994) Physics of directional hearing in the cricket *Gryllus bimaculatus*. J Comp Physiol A 175: 153–164

Miles RN, Tieu TD, Robert D, Hoy RR (1997) A mechanical analysis of the novel ear of the parasitoid fly *Ormia ochracea*. In: Lewis ER et al. (eds) Diversity in Auditory Mechanics. World Scientific, Singapore, pp 18–24

Morse PM (1948) Vibration and Sound, 2. ed. McGraw-Hill, New York

Naguib M, Wiley RH (2001) Estimating the distance to a source of sound: mechanisms and adaptations for long-range communication. Animal Behaviour 62: 825–837

Payne R, Roeder KD, Wallman J (1966) Directional sensitivity of the ears of noctuid moths. J Exp Biol 44: 17–31

Prager J (1976) Das mesothoracale Tympanalorgan von *Corixa punctata* Ill. Heteroptera, Corixidae. J Comp Physiol 110: 33–50

Robert D, Hoy RR (1998) The evolutionary innovation of tympanal hearing in Diptera. In: Hoy RR et al. (eds) Comparative Hearing: Insects. Springer Handbook of Auditory Research Vol. 10, pp 197–227

Rosowski JJ (1994) Outer and middle ears. In: Fay RR, Popper AN (eds) Comparative Hearing: Mammals. Springer Handbook of Auditory Research Vol. 4, pp 172–247

Rosowski JJ, Saunders JC (1980) Sound transmission through the avian interaural pathways. J Comp Physiol A 136: 183–190

Surlykke A, Filskov M, Fullard JH, Forrest E (1999) Auditory relationships to size in noctuid moths: bigger is better. Naturwissenschaften 86: 238–241

Schwabe J (1906) Beiträge zur Morphologie und Histologie der tympanalen Sinnesapparate der Orthopteren. Zoologica 20: 1–154

Webster DB (1992) Epilogue to the conference on the evolutionary biology of hearing. In: Webster DB et al. (eds) The Evolutionary Biology of Hearing. Springer, New York, pp 61–77

Yager D (1999) Structure, development, and evolution of insect auditory systems. Microsc Res Tech 47: 380–400

Yost WA, Gourevitch G (1987) (eds) Directional Hearing. Springer, New York

Zhantiev R, Korsunovskaya O (1978) Morphological organization of the tympanal organs in *Tettigonia cantans* (Orthoptera, Tettigoniidae) (in Russian) Zool J 57: 1012–1016

4. How to Build a Microphone

Per Rasmussen

Abstract

Microphones are transducers for converting acoustic pressure fluctuations into analogous electrical output signals. Various types of microphones are available, each optimised to meet particular requirements regarding performance, price, reliability and stability. Measurement microphones are optimised for performance, stability and reliability, but are influenced by the same design parameters as are other types of microphones. For measurement purposes, the condenser microphone has proved to be the best choice. It combines accuracy with stability and can be constructed to meet the requirements for measuring sound in free-fields, diffuse fields and in the couplers of equipment for testing and adjusting hearing aids. The design parameters include such factors as size, diaphragm stiffness, internal damping and diaphragm mass. By varying these design parameters, measurement microphones can be designed to have the desired combination of frequency response characteristics and dynamic range. The important design characteristics to consider are sensitivity, dynamic range and directionality. These are looked at in some detail. In the case of a $\frac{1}{2}$" condenser microphone (a) limiting the frequency response to lie within 20 kHz to 40 kHz, (b) limiting diaphragm deflection to about 1/10 of the distance between itself
and the back plate and (c) correctly adjusting the acoustic damping behind the diaphragm, will produce the best results.

I. Introduction

Microphones are transducers converting acoustic pressure fluctuations into an analogous electrical output signal. In the past, a number of different conversion principles have been used with the help of piezo-electric materials, magnetic coils and carbon materials (Beranek 1949). However, these days the dominating principle makes use of the capacity changes caused by sound on a condenser type microphone. A condenser microphone is based on the movement of a diaphragm in relation to a fixed backplate. A change in the distance between the diaphragm and the backplate results in a change in capacity between these two components, and, by applying a polarization voltage across them, this change in capacity produces a corresponding change in voltage. Generally, microphones can be divided into three broad categories: (i) communication microphones, (ii) studio microphones and (iii) measurements microphones. While the basic operating principles and design rules are similar for all three, design and optimisation criteria are very different. Communication microphones are normally designed with price and reliability as the determining factors. They are used in large numbers in telephones, mobile telephones, walkie-talkies, intercoms etc. They generally have a limited frequency range and dynamic range and can be quite sensitive to environmental changes. It is not uncommon to find a change of sensitivity of 1–3 dB for a temperature change of 10°C. This, however, is easily compensated for either by talking louder into a telephone or by amplifying the microphone signal in the electronic signal-conditioning circuits. Typically, these microphones cost around USD 1–5 in large quantities. Studio microphones are normally optimised for specific characteristics and reliability but visual design may also be very important for on-stage or TV use. Furthermore, the ability to withstand rough handling, vibrations and dropping could be important factors. Prices can range from USD 100 or less to USD 1000 for a high performance recording microphone with special characteristics. The third type is measuring microphones. These are primarily designed to be insensitive to variations in environmental factors such as temperature, humidity, barometric pressure (and altitude) and magnetic fields. Moreover, the use of corrosion resistant materials gives excellent long-term stability in sensitivity which typically varies by less than 0.001 dB/year. Measurement microphones are standardized by the international organisation IEC[1] and are widely used in all areas of noise measurements, building acoustics, product-noise labelling etc.

II. Microphone Principle

While the criteria for optimum design are different for the different types of microphones, the design parameters and basic elements are nearly all the same. Therefore, we will base the following on the principles for designing measurement mi-

crophones. Almost all microphones used today are based on the principle of the condenser microphone (Fig. 1) as described by Rasmussen (1959). A conducting diaphragm close to a parallel backplate forms a capacitor with capacity C_m determined by the distance h_0 between the diaphragm and the backplate and the diaphragm area given by the effective diameter d of the diaphragm, i.e.:

$$C_m = \frac{\varepsilon_0 \pi d^2}{h_0} \qquad (1)$$

A

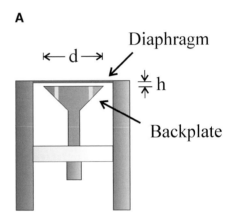

B

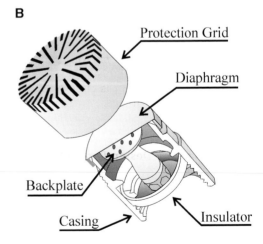

Fig. 1. Basic elements of a measurement microphone

---

[1] International Electrotechnical Commission.

where ε_0 is the dielectric constant of the air between the diaphragm and backplate. A voltage V_s from a regulated, stable voltage source then charges the capacitor. This is normally called the polarization voltage. This gives a charge Q on the capacitor given by:

$$Q = C \cdot V_s \qquad (2)$$

When a sound pressure deflects the diaphragm, a change of capacity takes place because the distance h_0 in Eq. (1) between the diaphragm and backplate changes. If the charge Q is kept constant, the output voltage V_o from the microphone is given by:

$$V_o = V_s \frac{\Delta h}{h_0} \qquad (3)$$

where Δh is the change in the distance between the backplate and diaphragm caused by the sound pressure.

The polarization voltage establishing the constant charge Q on the capacitor can come from either an external voltage supply (Fig. 2A) or an internally injected charge (polarized or electret type Fig. 2B). In externally polarized microphones, the polarization voltage comes from an external, very stable voltage source and is supplied via a very high resistor, typically in the range of 20 GΩ in order to keep the charge on the microphone constant. In polarized microphones, a thin layer of highly insulating material, usually PTFE[2], is charged statically by injecting it with an electrical charge.

III. Microphone Design Parameters

The basic characteristics of a measurement microphone are determined by factors such as size, diaphragm tension, distance between the diaphragm and the

_____
[2] Polytetrafluoroethylene.

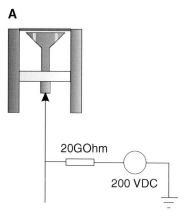

A

20GOhm

200 VDC

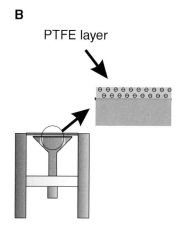

B

PTFE layer

Fig. 2. A externally polarized microphone; **B** pre-polarized microphone

backplate and the acoustic damping within the microphone. These factors will determine the frequency range of the microphone, its sensitivity and dynamic range. The sensitivity is described as the output voltage of the microphone for a given sound pressure excitation, and, in itself, is of little interest for the operation of the microphone, except for calibration purposes. However, the sensitivity of the microphone (together with the electrical impedance of its cartridge) also determines the lowest sound pressure level that can be measured with the microphone. The size of the microphone is the first parameter determining its sensitivity. Generally, the larger the diameter of the diaphragm, the more sensitive the microphone will be. There are, however, limits to how sensitive the microphone can be made by simply making it larger. The polarization voltage causes the diaphragm to be attracted, and deflected, towards the backplate. When the size of the microphone is increased this deflection will also increase and eventually the diaphragm will be deflected so much that it will actually touch the backplate. To avoid this, the distance between the diaphragm and the backplate can be increased or the polarization voltage can be lowered, in both cases accompanied by a decrease in sensitivity. The optimum practical size of a measurement microphone for use up to 20 kHz is very close to $\frac{1}{2}$" (12.7 mm). In some special applications, involving very low-level measurements, 1" (25.4 mm) microphones may be required, but for the majority of applications, $\frac{1}{2}$" or smaller is the best choice.

When the size of the microphone is decreased, the useful frequency range is increased. The frequency range obtainable is determined in part by the size of the microphone. At high frequencies, when the wavelength of sound waves becomes much smaller than the diameter of the diaphragm, the diaphragm will stop behaving like a rigid piston (the diaphragm is said to "break up" – not a destructive phenomenon). Different parts of the diaphragm will start to move with differing magnitude and phase, bringing about a change in the microphone's frequency response. To avoid this, the upper limiting frequency is placed such that the sensitivity of the microphone drops off before the diaphragm starts to break up. For a typical $\frac{1}{2}$" microphone, the upper limiting frequency is placed in the range of 20 kHz to 40 kHz depending on dia-

phragm tension. If the diaphragm is tensioned, i.e. made stiffer, its resonance frequency will be higher – on the other hand the sensitivity of the microphone will be reduced because the diaphragm will deflect less for a given sound pressure level.

The frequency response of a microphone is determined by diaphragm tension, diaphragm mass and the acoustic damping in the airgap between the diaphragm and backplate. This system can be represented by a simple mass-spring-damper system (Fig. 3). In this analogy, the mass represents the mass of the diaphragm and the spring represents the tension in the diaphragm. Thus, if the diaphragm is tensioned to make it stiffer, the corresponding spring becomes stiffer. The damping element in the analogy represents the acoustic damping between the diaphragm and the backplate. This can be adjusted, for example by drilling holes in the backplate. This will make it easier for the air to slip away from the airgap when the diaphragm is deflected, thereby decreasing damping.

This simple mechanical analogy of the microphone is a first order mechanical system with a simple frequency response. At low frequencies (below resonance frequency), the response of the microphone

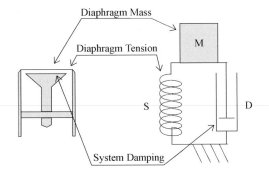

Fig. 3. Spring-Mass-Damper (SMD) analogy of microphone

A

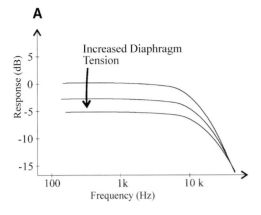

B

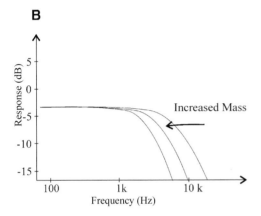

C

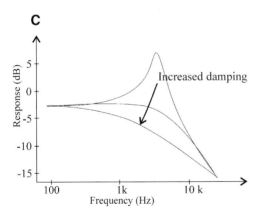

Fig. 4. Influence of microphone parameters on frequency response. **A** Low frequency response determined by diaphragm tension; **B** microphone response around resonance determined by damping; **C** microphone response above resonance determined by mass

is determined by the diaphragm tension (Fig. 4A). When the diaphragm tension is increased, the output of the microphone will decrease, and when it is decreased, the output will increase. The tension and mass of the diaphragm determine the resonance frequency such that an increase in tension increases resonance frequency and an increase in mass decreases resonance frequency. The response around the resonance frequency (Fig. 4B) is determined by the acoustic damping, where an increase in damping will decrease the response. At higher frequencies, above the resonance frequency, the response of the microphone to sound pressure is determined by the mass of the diaphragm (Fig. 4C). The greater the mass of the diaphragm, the smaller will be the output signal from the microphone since the sound pressure has greater difficulty in moving the diaphragm.

IV. Dynamic Range

The dynamic range of a microphone is the range between the highest and the lowest sound pressures that can be handled by the microphone. The highest level is determined by the distance between the backplate and the diaphragm. Usually, this distance is in the order of $20\,\mu m$ $(2 \cdot 10^{-5}\,m)$. If the sound pressure is high enough, it will deflect the diaphragm so much that it will actually touch the backplate. In this situation the microphone will become highly non-linear and it is usual to limit the maximum deflection to about $\frac{1}{10}$ of the distance between the backplate and the diaphragm. For a typical $\frac{1}{2}''$ measurement microphone, this will correspond to around $140\,dB$ (re. $2 \cdot 10^{-5}\,Pa$). As the sound pressure is decreased, diaphragm movement is also decreased

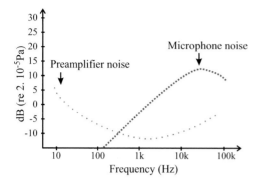

Fig. 5. Inherent noise in microphone and preamplifier

proportionally. Thus, at a sound pressure level of 20 dB (re. $2 \cdot 10^{-5}$ Pa) the movement of the diaphragm is reduced by a factor of 10^{-6} (120 dB) giving a diaphragm movement of around $2 \cdot 10^{-12}$ m.

The lowest level of sound pressure is determined by the inherent noise generated in the microphone itself and the conversion factor from pressure to voltage in connection with the preamplifier (Fig. 5). The inherent noise generated by the microphone is associated with the damping in a simple mechanical model (Wong et al. 1995). In an electrical analogy of this mechanical model, the damping is present in the form of a resistor R_A and the Johnson noise, as power spectral density (PSD) from this, is given by the Nyquist equation:

$$PSD = p_{rms}^2 / \Delta f = 4 \cdot k_B \cdot T \cdot R_A$$

where k_B is the Boltzmann constant and T is the absolute temperature. It can be seen that this causes a frequency-independent thermal noise contribution proportional to the damping.

The Johnson noise contribution will be present on the output terminal of the microphone and is superimposed on the signal generated by the sound pressure on the diaphragm. This signal is given by the

sound pressure p_i on the diaphragm and the sensitivity S_m. The sensitivity, or electromechanical coupling coefficient, is the ratio between the sound pressure and resulting output voltage. The total output V_o from the microphone is therefore given by:

$$V_o = N_s + p_i \cdot S_m$$

where N_s is the thermal noise contribution. It can be seen that the lower limit can be extended downwards by either increasing the sensitivity or by decreasing the damping[3].

V. Directionality

Sound pressure is a scalar quantity and, as such, has no direction or directionality. The propagation of sound or sound field may, however, contain directionality information. The simplest type of wave is the plane one-dimensional wave found far from its sound source free of disturbances or reflections. The plane wave will travel in a certain direction, hence, giving the sound field directionality.

The basic types of microphones are free-field microphones, pressure microphones and random incidence microphones. They are constructed to have different frequency characteristics in order to comply with different requirements. The pressure microphone is constructed to measure the actual sound pressure, as it exists on the diaphragm. A typical application is the measurement of sound pressure in a closed coupler or the measurement of the sound pressure at a boundary such as a wall. In this case the microphone forms part of the wall and measures the sound pressure on the wall itself. The frequency response of

---

[3] The reader is referred to chapter 10 in this volume which deals with similar problems in cricket filiform hairs.

this microphone should be flat over a wide-as-possible frequency range; taking into account that the sensitivity will decrease with increased frequency range. The acoustic damping in the airgap between the diaphragm and backplate is adjusted so that the frequency response is flat up to and a little beyond the resonance frequency.

The introduction of a microphone into a sound field will result in a pressure increase in front of the diaphragm (Fig. 6B) depending on the wavelength. The free-field microphone is designed essentially to measure the sound pressure as it existed before the microphone was introduced into the sound field. At higher frequencies, the presence of the microphone in the sound field will change the sound pressure. Generally, the sound pressure around the microphone cartridge will increase due to reflections and diffraction. The free-field microphone is designed so that its frequency characteristics compensate for this pressure increase. The result-

ing output is proportional to the sound pressure as it existed before the microphone was introduced into the sound field. The free-field microphone should always point towards the sound source (0° incidence) (Fig. 6A). For a typical $\frac{1}{2}$" microphone, the maximum pressure increase will occur at 26.9 kHz, where the wavelength of the sound coincides with the diameter of the microphone ($\lambda = 342 \, \text{ms(i)}/26.9 \, \text{kHz} \approx 12.7 \, \text{mm} \approx \frac{1}{2}$"). The microphone is then designed so that its sensitivity decreases by an amount that compensates for increased acoustic pressure in front of the diaphragm. This is obtained by increasing the internal acoustic damping in the microphone cartridge in order to obtain the desired frequency response (Fig. 7B).

The result is an output from the microphone that is proportional to the sound pressure as it existed before the microphone was introduced into the sound field (Fig. 7C). The curve shown in Figure 7A is also called the free-field

A

B

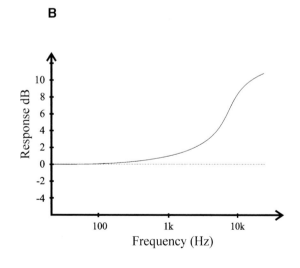

Fig. 6. Free-field microphone

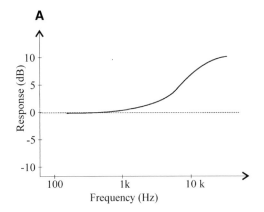

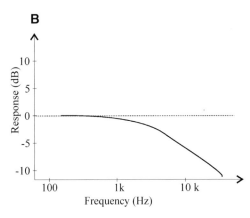

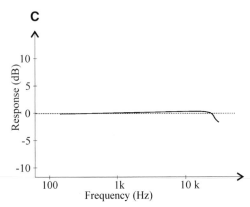

Fig. 7. Frequency response of free-field micro-phone. **A** Increase of sound pressure in front of microphone diaphragm; **B** pressure response of microphone; **C** resulting free field response of microphone

correction curve for the microphone and must be added to the frequency response of the microphone cartridge to obtain the acoustic characteristic of the microphone in a free field.

In principle, the free-field microphone requires to be pointed towards the sound source and that the sound waves are travelling in essentially one direction. In some cases, e.g. when measuring in a reverberation room or other highly reflecting surroundings, the sound waves will not have a well-defined propagation direction, but will arrive at the microphone from all directions simultaneously. The sound waves arriving at the microphone from the front will cause a pressure increase as described above for the free-field microphone, whereas the waves arriving from the back of the microphone will cause a pressure decrease to a certain extent due to the shadowing effects of the microphone cartridge. The combined influence of the waves coming from different directions depends therefore on the distribution of sound waves from the different directions. For measurement microphones, a standard distribution has been defined, based on statistical considerations, resulting in a standardized random incidence microphone.

VI. Conclusion

Modern microphones can measure over wide frequency ranges and can cover dynamic ranges that correspond with pressure fluctuations that range from very small to very large. To give an example of the dimensions involved in microphones consider a standard $\frac{1}{2}''$ microphone. When measuring a sound pressure level of 40 dB (corresponding to the level in a quiet living room), the diaphragm will move approximately 10^{-11} m. This is a very small value and difficult to imagine. If the microphone diameter were as large as the diameter of the Earth (12.700 km), its diaphragm would be about 2 km thick. The distance between the backplate and the diaphragm would be about 20 km and the diaphragm would move only about 10 mm.

References

Beranek L (1949) Acoustic Measurements. John Wiley & Sons, New York

Rasmussen G (1959) A New Condenser Microphone. Brüel & Kjær *Technical Review*, No. 1

Wong G, Embleton et al. (1995) Condenser Microphones, American Institute of Physics

5. The Middle and External Ears of Terrestrial Vertebrates as Mechanical and Acoustic Transducers

John J. Rosowski

vertebrates and discusses the structures that affect the frequency selectivity of these processes.

Abstract

The external and middle ears of terrestrial vertebrates work together to transform sound in free air to sound pressures and volume velocities in the fluid in the vestibule of the auditory inner ear. The primary structural bases for this transformation are the acoustics of the ear canal and the ratio of the areas of the tympanic membrane and the stapes footplate. The efficiency of this transformation and the extreme sensitivity of the auditory inner ear to sound enable vertebrates to hear sounds that produce sub-Angstrom displacements of middle-ear structures. Nonetheless, the absolute size and the form of the tympanic membrane, ossicles and ossicular suspension make the transformation process frequency selective, and it is argued that this selectivity is the primary force shaping what sounds are audible by different animal species. This chapter quantifies sound transformation through the ear of several

I. Introduction

The ears of terrestrial vertebrates show exquisite sensitivity to air-borne sounds. While a large part of this sensitivity is determined by the processes within the inner ear (see chapter 6, this volume), the external and middle ear couple sound power from the air-filled environment to the fluid-filled inner ear. In performing this coupling, these most peripheral parts of the auditory system (Fig. 1) act as the first acoustical-mechanical filter in the transformation of sound energy to sensation. As part of the coupling process the ear transduces ossicular displacements on the order of Ångstroms and forces on the order of 10^{-9} Newtons.

This chapter points out that although the external and middle ears of all terrestrial vertebrates have some features in common, there are also many differences in the size and shape of the external and middle ear. Furthermore those structural differences are related to differences in the sensitivity of the ears of different animals to different sound frequencies.

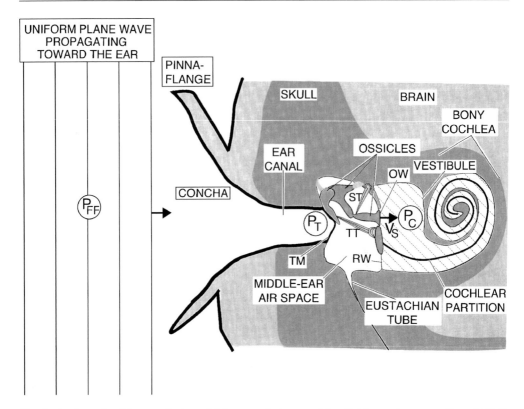

Fig. 1. A schematic of the ear of a terrestrial mammal. A wave of free-field sound pressure $\mathbf{P_{FF}}$ propagating toward the ear first interacts with the head, body and external ear – composed of the pinna-flange or flap, the concha and the tube-like ear canal that is terminated by the tympanic membrane (TM) and middle ear. $\mathbf{P_T}$ is the resultant sound pressure just lateral to the TM. The middle ear is composed of the TM, the middle-ear air space or tympanum, the three ossicles and their supporting ligaments and middle-ear muscles, the tensor tympani (TT) and the stapedius (ST) muscles. The output of the middle ear is the velocity of the stapes $\mathbf{V_S}$, and the sound pressure $\mathbf{P_C}$ at the entrance to the mammalian auditory inner ear or cochlea. OW oval window, RW round window. Bony structures are shaded in dark gray, soft tissues are shaded a lighter gray, the two muscles are striated and the inner ear fluid is noted by stippled lines

Indeed, we argue, the shape of the auditory sensitivity function in different animal species is primarily determined by the frequency selectivity of the different structures of the external and middle ear.

A. Differences in the Frequency Range of Hearing in Different Animals

Auditory threshold functions, determined from behavioral measurements in five dif-ferent vertebrate species, are shown in Fig. 2. Each point represents the minimum sound pressure that is audible at a given frequency. The selected exemplary species illustrate different classes of hearing sensitivity: *Human*, a mammal with good sensitivity to low-frequency sounds and poor sensitivity to high-frequency sounds; *Bat*, a mammal with poor sensitivity to low-frequency sound and good sensitivity to high-frequency sound; *Cat*, a mammal with good sensitivity to both low and high-frequency sounds; *Pigeon*,

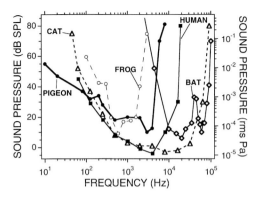

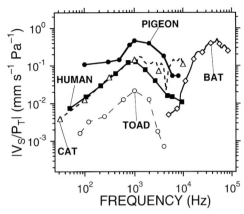

Fig. 2. Behaviorally determined hearing threshold functions for air-borne sound in five terrestrial vertebrates including three mammals: human (Sivian and White 1933), house cat (Heffner and Heffner 1985) and horseshoe bat (Long and Schnitzler 1975), the common pigeon (Hienz et al. 1977), and the bull frog (Megela-Simmons et al. 1985). The left-handed ordinate is scaled in dB SPL (i.e. dB re 2×10^{-5} rms Pascals), the right-handed log ordinate is scaled in rms Pascals. The abscissa scales frequency in Hz

Fig. 3. Differences in the frequency response and absolute sensitivity of the middle ear of 5 terrestrial vertebrates. The abscissa codes sound frequency in Hz. The ordinate is the magnitude of the transfer admittance of the stapes velocity to sound pressure at the tympanic membrane. For loud sounds of 1 Pascal in sound pressure amplitude, near the best frequency, the stapes velocity in the mammals and the bird are on the order of 0.1–0.5 mm/s. Near the lowest auditory thresholds, the linear velocity of the stapes in the mammals and birds is on the order of 10–50 nm/sec. (Human: Kringlebotn and Gundersen 1985; cat: Guinan and Peake 1967; bat: Wilson and Bruns 1983; pigeon: Gummer et al. 1989; toad: Saunders and Johnstone 1972)

a bird with an audiogram similar to that of humans, and a *Frog* that exhibits moderate hearing sensitivity over a narrow band of relatively low sound frequencies. The sensitive areas of the threshold curves cover about 3 decades in cat, 2 decades in humans, 1.5 decades in the pigeon, 1 decade in the bat and less than a decade in the frog. The minimum sound pressure thresholds in the five animals vary from 1 to 6×10^{-5} rms Pa. These pressures interact with tympanic membranes that vary from 80 mm^2 in the human ear to less than 10 mm^2 in the bat, such that the resultant forces acting on the tympanic membranes near threshold are on the order of 10^{-9} to 10^{-10} Newtons.

B. Differences in the Middle-Ear Response in Different Animals

The middle ears of different animals vary widely in their frequency range and absolute sensitivity to sound (Fig. 3). Figure 3 illustrates measurements of the sound-induced velocity of the stapes – the ossicle that delivers sound power to the inner ear – normalized by the stimulus sound pressure (this ratio is a middle-ear transfer admittance) in a similar set of five animals. There are obvious similarities in the frequency ranges of good threshold sensitivity and maximum stapes response between Figs. 2 and 3. However, the ranking of middle-ear sensitivities (estimated from the maximum in the velocity transfer function) are somewhat different from the ranking of threshold sensitivities; pigeon shows superior stapes velocity compared to cat and human, but the pigeon's auditory thresholds are inferior to those two mammals. The similarity of the shapes of the middle-ear transfer admittances and the threshold functions are consistent with the middle-ear acting to filter the sound energy that reaches the inner ear. The differences in sensitivity of the two measurements suggest that the inner ear is important in scaling the auditory response. Since the middle-ear is known to act linearly with stimulus levels between threshold levels (near 0 dB SPL) and about 120 dB SPL (Guinan and Peake 1967, Nedzelnitsky 1980), the data of Fig. 3 suggest that the stapes velocity at threshold is of the order of 10^{-8} m/s. Such velocities correspond to stapes displacements at threshold that are less than Ångstroms (*sinusoidal Displacement magnitude = sinusoidal Velocity magnitude/$(2\pi$ frequency)*).

C. Variations in Ear Structure

The variations among terrestrial vertebrates in external-ear structure are obvious to anyone with an inquiring eye, but the middle-ear structures are hidden from view. Figures 4 and 5 point out that while there are gross similarities in all terrestrial vertebrate middle ears – including the presence of a tympanic membrane TM, an air space and ossicles – the middle-ear structures of different vertebrate species vary greatly in both form and size. The most fundamental differences in form occur between the ears of mammals and other terrestrial vertebrates. The middle ears of mammals (Fig. 4 right) have three ossicles, two middle-ear muscles, a conical concave inward TM, and a closed middle-ear air space (aerated by the

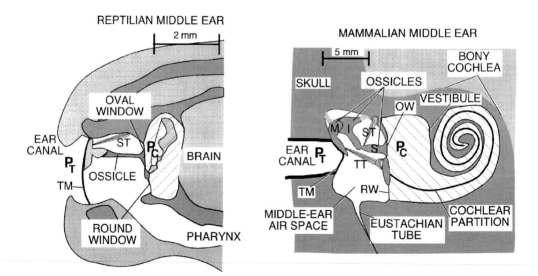

Fig. 4. A comparison of the middle ears of non-mammalian vertebrates and mammals. On the left is a schematic of a reptilian ear (after Rosowski et al. 1985) on the right a mammalian ear (after Rosowski 1994). Note: (1) the difference in the shape of the TM, (2) the difference in the number of ossicles (one in the reptile, three in the mammal including the malleus M, incus I and stapes S), (3) the presence of two middle-ear muscles in the mammal and one in the reptile, (4) the essentially closed middle-ear air space in the mammal and the open space in the reptile

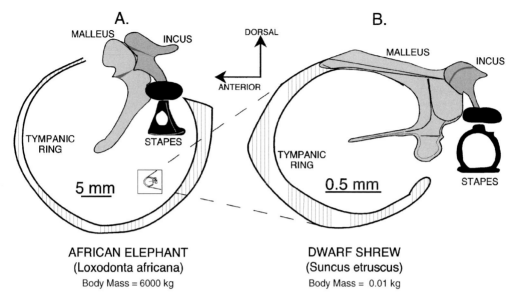

Fig. 5. Examples of variations in the form and size of the TM and ossicles in two mammals of greatly different body size: A. The African elephant; B. the dwarf shrew. A and B are drawn to different scales, however a version of B, scaled as in A, is inset within the tympanic ring of A. In each ear, the malleus, the incus, the oval stapes footplate and the bony tympanic ring that supports the TM are illustrated as one would view them looking out from the inner ear. A second orthogonal view of the stapes is also illustrated to document inter-specific variations in the shape of this ossicle (after Rosowski 1994)

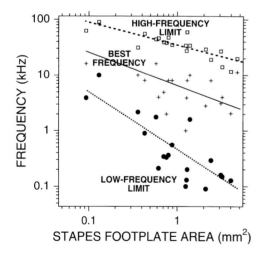

Fig. 6. Plots of the *Best Frequency* (the frequency to which the ear is most sensitive) and the *High- and Low-Frequency Limits* estimated from behavioral measurements of hearing in 20 mammalian species. The lines represent best-fit power functions $(y = a\ x^b)$ and have exponents of -0.40, -0.52, and -1.09 for the high, best and low frequency data respectively (after Rosowski 1992)

Eustachian tube). The middle ears of birds and reptiles (Fig. 4 left) have one ossicle, no more than one middle-ear muscle, a rounded concave outward TM, and a middle-ear air space that communicates widely with other air spaces, e.g. the pharynx in reptiles, and the opposite middle ear in birds (Henson 1974).

Figure 5 illustrates differences in the form and size of mammalian middle ears. The two examples chosen, the African Elephant and the dwarf shrew, are extremes in the body size of terrestrial mammals. Indeed the entire body of the shrew could fit inside the middle-ear air spaces of the elephant. Figure 5 clearly associates those differences in body size with differences in the size of the middle-ear structures, though it should be obvious that the middle ear size does not scale proportionately with body weight. (The ear of the shrew is about $1/1000$ the volume of the elephant ear, while the mass of the shrew is about $1/600,000$ of the elephant's mass.) Differences in area of the TM and stapes footplate in mammals correlate with the limits of hearing. Figure 6 illustrates the relationship between estimates of the highest audible frequency,

the lowest audible frequency, the best frequency (the frequency with lowest threshold), and measurements of area of the stapes footplate in 20 mammalian species. A factor of ten increase in stapes-footplate area is associated with a factor of ten decrease in lowest audible frequency and a factor of 3 decrease in best frequency and highest audible frequency. Similar correlations can be seen between middle-ear areas and the audible limits in non-mammalian vertebrates though the frequency dependence is less severe (Rosowski and Graybeal 1991).

Figure 5 also illustrates large differences in the form of the ossicles in these two species. The most obvious differences are in the relative sizes of the ossicles (in the shrew, the incus is very small compared to the malleus) and in the relationship between the ossicles and the tympanic ring.

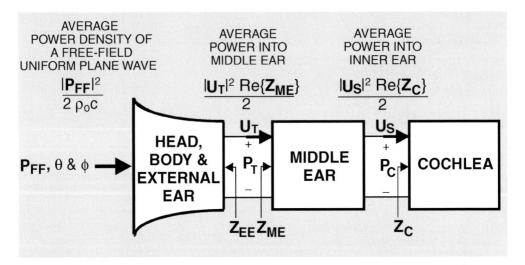

Fig. 7. Acoustic – Electrical analog of external- and middle-ear function (after Rosowski et al. 1986). A free-field sound wave (of power density defined by the peak sound pressure $\mathbf{P_{FF}}$, the density of air ρ_0, the propagation velocity of sound in air c) of direction defined by azimuth θ and elevation ϕ interacts with the head, body and external ear. Some fraction of the average sound power available from the source is coupled to the middle ear at the tympanic membrane, where: $\mathbf{P_T}$ is the sound pressure at the TM, $\mathbf{U_T}$ is the volume velocity at the TM, $\mathbf{Z_{ME}} = \mathbf{P_T}/\mathbf{U_T}$ is the middle-ear input impedance, and $\mathbf{Z_{EE}}$ is the acoustic impedance looking out the external ear from the tympanic membrane. Some fraction of the average power input to the middle ear is then coupled to the inner ear, where $\mathbf{U_S}$ is the stapes volume velocity, $\mathbf{P_C}$ is the sound pressure within the inner ear and $\mathbf{Z_C} = \mathbf{P_C}/\mathbf{U_S}$ is the input impedance of the inner ear or cochlea

In the 'microtype' middle ear of the shrew, the malleus and ring are attached by connective tissue, whereas in the elephant, the malleus is independent of the ring (Fleischer 1973, 1978). This difference in attachment is directly related to a difference in the stiffness and frequency dependence of ossicular motion. While the relationship between hearing limits and the form of the ossicles is complicated by a correlation between the size of the ear and the ossicular type (small mammalian ears tend to look more like those of the shrew in ossicular form), animals with middle ears with an extensive and firm attachment between the malleus and the tympanic ring do not hear low sound frequencies (Rosowski 1992).

II. Function of the Middle and External Ear

Since the time of Helmholtz (1868) the external and middle ear have been described as transformers that gather sound power from the free-field and couple that power to the inner ear. That function is illustrated in Fig. 7, which schematizes the flow of average power from the environment to the inner ear in terrestrial vertebrates.

The gathering and transformation of free-field sound power to power in the fluid-filled inner ear is accomplished via the horn-like properties of the external ear (Shaw 1974, 1988; Rosowski et al. 1988) and the impedance transformation

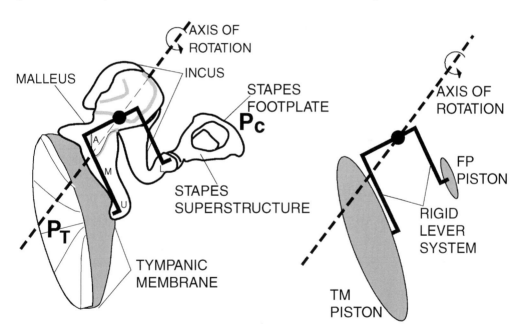

Fig. 8. The *Ideal Transformer Model* of the mammalian middle ear in which the TM and stapes footplate act as rigid pistons that are coupled by a lever-like, rigid, rotating ossicular linkage. In this ideal case the ratio of the sound pressures outside the tympanic membrane PT and inside the inner ear is a simple constant:

$$\frac{P_C}{P_T} = \frac{Area\ of\ TM}{Area\ of\ Footplate}\ \frac{Length\ of\ Mallear\ Lever\ Arm}{Length\ of\ Incus\ Lever\ Arm}$$

While this constant does limit the magnitude of the actual pressure transformation performed by the middle ear, non-idealities in the transformer components make the transformer ratio frequency dependent, see e.g. Aibara et al. 2001

performed by the middle ear. The difference in area between the larger tympanic membrane and the smaller footplate together with the lever function of the rotating ossicles leads to a sound pressure transformation between the TM and inner ear (Fig. 8). Anatomical measurements of the middle-ears of different mammals suggest that the magnitude of the pressure transformation varies between a factor of 10–40 (Fleischer 1978, Hunt and Korth 1980, Rosowski 1994). However, the magnitude of the middle-ear transformation factor is heavily influenced by non-idealities within the middle-ear transformer. These non-idealities include: (1) inherent stiffnesses of the TM and ossicular supports (Rosowski 1992, 1994), (2) the mass of the TM and ossicles, (3) non-rigidities within the ossicular joints (Guinan and Peake 1967) and within the ossicles themselves (Decraemer et al. 1995), (4) complexities in the motion of the tympanic membrane produced by sound (Khanna and Tonndorf 1972). These non-idealities introduce a frequency dependence that greatly restricts middle-ear transformer function.

III. Measurements of Middle-Ear Function

The actual transfer characteristics of the middle ears of several dozen vertebrates have been measured using various velocity measurement techniques and miniature sound pressure transducers. Some of the middle-ear transfer admittances (the ratio of stapes velocity to sound pressure at the TM) are illustrated in Fig. 3. Figure 9 illustrates a second kind of transfer function: the middle-ear sound pressure transfer ratio measured in the cat by Décory (1989). This alternate transfer function is of interest because it is a direct assessment of the *Ideal Transformer Model* of Fig. 8. Contrary to that model, (1) the

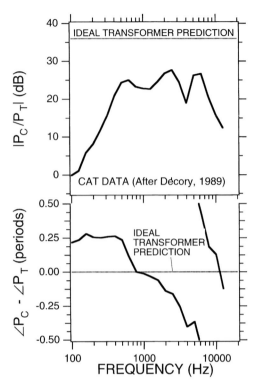

Fig. 9. Measurements of the sound pressure transfer ratio of the cat middle ear. The input stimulus P_T is the sound pressure delivered to the TM. The output response P_C is the resultant sound pressure in the cat cochlea (auditory inner ear). The top panel shows the magnitude of the transfer ratio in dB; the lower panel shows the difference in the phase of the response and stimulus. The dashed lines in the two panels show the theoretical response of an ideal transformer based on cat-like dimensions for the TM and ossicles (data from Décory 1989; after Rosowski 1992)

measured pressure transfer ratio has a magnitude and angle that are frequency dependent, (2) the maximum magnitude is less than that predicted from the anatomy of the cat middle ear, (3) in general, the phase of the sound pressure in the inner ear P_C lags the stimulus sound pressure P_T. These data are a direct confirmation of the idea that the middle ear acts as a mechano-acoustic filter that collects the sound stimulus that it couples to the inner ear.

Another indication of filtering by the middle ear system comes from estimates of middle-ear efficiency, which we define as the ratio of power output from the middle ear over the power input at the tympanic membrane. The results of an analysis of the efficiency of the cat and human middle ears are illustrated in Fig. 10. These results are consistent with a frequency dependent transfer of power through the middle ear, in that the efficiency is highest near 1000 Hz in both cat and human. The middle-ear efficiency in both animals falls as frequency is decreased from 1000 Hz, and the human efficiency drops off rapidly as frequency increases above 1000 Hz. The fine structure in the cat data in figure 10 near 4000 Hz is related to the existence of acoustic resonances within the cat middle-ear cavity (Guinan and Peake 1967, Huang et al. 1997, 2000).

The best evidence for a primary role of the external and middle ear in shaping the auditory response are the results of analyses that match behavioral measurements of auditory sensitivity (like those of Fig. 2) to the sound pressure necessary to maintain a constant level of middle-ear

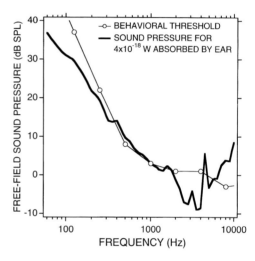

Fig. 11. A comparison of behavioral free-field sound thresholds measured in cats (Heffner and Heffner 1985) and the computer sound level needed to maintain a 'threshold level' of 4×10^{-18} Watts of average power absorbed by the cat inner ear. The ordinate is the free-field sound level in dB SPL (dB re 2×10^{-5} Pa) from a source at $0°$ elevation and azimuth; the abscissa is tone frequency. The behavioral measurements were performed using shock-avoidance conditioning. The sound power threshold was computed from measurements of external and middle ear function in cats together with the power collection scheme of Fig. 7 (Rosowski et al. 1986). The level of the constant inner-ear power constraint was chosen to best fit the behavioral thresholds

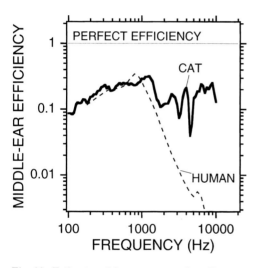

Fig. 10. Estimates of the power transfer efficiency of the middle ears of cats and human. The efficiency is defined by the ratio:

Middle-Ear Efficiency

$$= \frac{Average\ Power\ Delivered\ to\ the\ Inner\ Ear}{Average\ Power\ Absorbed\ at\ the\ Tympanic\ Membrane}$$

This efficiency was calculated from measurements of middle-ear and cochlear input impedance and the transfer admittance of the middle ear. (After Rosowski et al. 1986)

output. Dallos (1973) and Zwislocki (1975) demonstrated that auditory thresholds in cat and guinea pig could be well approximated by measurements of the free-field sound pressure required to produce a constant stapes velocity. Similar matches have been made using either constant cochlear sound pressure (Puria et al. 1997) or constant average sound power into the cochlea (Rosowski et al. 1985, Rosowski 1991a,b). An example of a constant average power comparison is illustrated in Fig. 11. The behavioral thresholds in cat were determined using avoidance-conditioning techniques. The sound pressure required to produce a constant average power absorbed by the cochlea of 4×10^{-18} Watts was computed using measurements of external and middle ear function and the schema of Fig. 7. The relatively good match between the behavioral thresholds and the sound level required to produce a constant sound power absorbed by the inner ear suggests that at threshold, the inner ear acts as a power detector. The extremely low threshold power level is consistent with other estimates that demonstrate that the sensory cells within the inner ear are extremely sensitive to sound (e.g. Khanna and Sherrick 1981).

IV. Conclusions

The external and middle ears of terrestrial vertebrates act to gather and couple sound to the fluid-filled inner ear. This function can be understood in terms of the acoustic and mechanical processes within the ear. At threshold, these processes transduce forces on the order of 10^{-9} Newtons and displacements less than Ångstroms. The mechanics of these processes are limited by the stiffness, inertance, and damping within the middle and inner ear and are inherently frequency selective. This selectivity is influenced by both the size and the form of the ear. Comparisons of the power collection performance and auditory sensitivity of different ears support the idea that the external and middle-ear impart a frequency selectivity to auditory thresholds and suggest that the inner ear, at threshold, functions as a frequency independent power detector at least at sound frequencies lower than 10,000 Hz.

Acknowledgements

I wish to thank all of my co-workers in the Eaton-Peabody Laboratory, including Kelly Brinsko for her editorial help. This work was supported by grants from the U. S. National Institutes of Health.

References

Aibara R, Welsh JT, Puria S, Goode RL (2001) Human middle-ear sound transfer function and cochlear input impedance. Hearing Res 152: 100–109

Dallos P (1973) The Auditory Periphery. Academic Press, New York, 548 p

Décory L (1989) Origine des différences interspecifiques de susceptibilité à bruit. Thèse de Doctorat de l'Université de Bordeaux, France

Decraemer WF, Khanna SM, Funnell WRJ (1995) Bending of the manubrium in cat under normal sound stimulation. Progress in Biomedical Optics 2329: 74–84

Fleischer G (1973) Studien am Skelett des Gehörorgans der Säugetiere, einschliesslich des Menschen. Säugetierkundl Mitt (München) 21: 131–239

Fleischer G (1978) Evolutionary principles of the mammalian middle ear. Adv Anat, Embryol and Cell Biol 55: 3–69

Guinan JJ Jr, Peake WT (1967) Middle-ear characteristics of anesthetized cats. J Acoust Soc Am 41: 1237–1261

Gummer AW, Smolders JWT, Klinke R (1989) Mechanics of a single-ossicle ear: II. The columella footplate of the pigeon. Hearing Res 39: 15–26

Heffner RS, Heffner HE (1985) Hearing range of the domestic cat. Hearing Res 19: 85–88

Helmholtz HL von (1868) Die Mechanik der Gehörknöchelchen und des Trommelfells. Pflüg Arch ges Physiol 1: 1–60

Henson OW (1974) Comparative anatomy of the middle ear. In: Keidel WD, Neff WD (eds) Handbook of Sensory Physiology: Vol V/1 The Auditory System. Springer, Wien New York pp 39–110

Hienz RD, Sinnott JM, Sachs MB (1977) Auditory sensitivity of the redwing blackbird (*Agelaius phoeniceus*) and brown-headed cowbird (*Molothrus ater*). J Comp Physiol Psychol 91: 1365–1376

Huang GT, Rosowski JJ, Flandermeyer DT, Lynch TJ III, Peake WT (1997) The middle ear of a lion: Comparison of structure and function to domestic cat. J Acoust Soc Am 101: 1532–1549

Huang GT, Rosowski JJ, Peake WT (2000) Relating middle-ear acoustic performance to body size in the cat family; measurements and models. J Comp Physiol A 186: 447–465

Hunt RM, Korth WW (1980) The auditory region of Dermoptera: morphology and function relative to other living mammals. J Morphol 164: 167–211

Khanna SM, Sherrick C (1981) The comparative sensitivity of selected receptor systems. In: Gualtierotti T (ed) The Vestibular System: Function and Morphology. Springer-Verlag, New York, pp 337–348

Khanna SM, Tonndorf J (1972) Tympanic membrane vibrations in cats studied by time-averaged holography. J Acoust Soc Am 51: 1904–1920

Kringlebotn M, Gundersen T (1985) Frequency characteristics of the middle ear. J Acoust Soc Am 77: 159–164

Long GR, Schnitzler H-U (1975) Behavioral audiograms from the bat, *Rhinolophus ferrumequinum*. J Comp Physiol A 100: 211–219

Megela-Simmons A, Moss CF, Danile KM (1985) Behavioral audiograms of the bullfrog (*Rana catesbeiana*) and the green treefrog (*Hyla cinerea*). J Acoust Soc Am 78: 1236–1244

Nedzelnitsky V (1980). Sound pressures in the basal turn of the cat cochlea. J Acoust Soc Am 68: 1676–1689

Puria S, Peake WT, Rosowski JJ (1997) Sound-pressure measurements in the cochlear vestibule of human cadavers. J Acoust Soc Am 101: 2745–2770

Rosowski JJ (1991a) The effects of external- and middle-ear filtering on auditory threshold and noise-induced hearing loss. J Acoust Soc Am 90: 124–135

Rosowski JJ (1991b) Erratum: The effects of external- and middle-ear filtering on auditory threshold and noise- induced hearing loss. [J Acoust Soc Am 1991; 90: 124–135] J Acoust Soc Am 90: 3373

Rosowski JJ (1992) Hearing in transitional mammals: Predictions from the middle-ear anatomy and hearing capabilities of extant mammals. In: Popper AN, Fay RR, Webster DB (eds) The Evolutionary Biology of Hearing. Springer-Verlag, New York, pp 625–631

Rosowski JJ (1994) Outer and middle ear. In Popper AN, Fay RR (eds) Springer Handbook of Auditory Research: Comparative Hearing: Mammals. Springer-Verlag, New York, pp 172–247

Rosowski JJ, Graybeal A (1991) What did *Morganucodon* hear? Zool J Linnean Soc 101: 131–168

Rosowski JJ, Carney LH, Lynch TJ III, Peake WT (1986) The effectiveness of external and middle ears in coupling acoustic power into the cochlea. In: Allen JB, Hall JL, Hubbard A, Neely ST, Tubis A (eds) Peripheral Auditory Mechanisms. Springer-Verlag, New York, pp 3–12

Rosowski JJ, Peake WT, Lynch TJ III, Leong R, Weiss TF (1985) A model for signal transmission in an ear having hair cells with free-standing stereocilia, II. Macromechanical stage. Hearing Res 20: 139–155

Rosowski JJ, Carney LH, Peake WT (1988) The radiation impedance of the external ear of cat: Measurements and applications. J Acoust Soc Am 84: 1695–1708

Saunders JC, Johnstone BM (1972) A comparative analysis of middle-ear function in non-mammalian vertebrates. Acta Otolaryngol 73: 353–361

Shaw EAG (1974) The external ear. In: Keidel WD, Neff WD (eds) Handbook of Sensory Physiology: Vol V/1 The Auditory System. Springer-Verlag, New York, pp 455–490

Shaw EAG (1988) Diffuse field response, receiver impedance and the acoustical reciprocity principle. J Acoust Soc Am 84: 2284–2287

Sivian LJ, White SD (1933) On minimum sound audible fields. J Acoust Soc Am 4: 288–321

Wilson JP, Bruns V (1983) Middle-ear mechanics in the CF-bat *Rhinolopus ferrumequinum*. Hearing Res 10: 1–13

Zwislocki J (1975) The role of the external and middle ear in sound transmission. In: Tower DB (ed) The Nervous System, Volume 3: Human Communication and its Disorders. Raven Press, New York, pp 45–55

6. The Outer Hair Cell:
A Mechanoelectrical and Electromechanical Sensor/Actuator

Kenneth V. Snyder, Frederick Sachs, and William E. Brownell

Abstract

The outer hair cell (OHC) is unique to mammals and is required to perceive and discriminate high frequency sounds (> 10 kHz). This sensory cell has the remarkable motor ability to convert sound induced receptor potentials into mechanical energy (electromotility). Although the mechanism of electromotility has remained elusive, a systematic review of its characteristics places many constraints on possible models. A sensor-motor ability resides in the outer hair cell's lateral plasma membrane, the outermost layer of an elegant trilaminate lateral wall. The passive biophysical properties of this wall contribute to the robustness of the electromotile response. Analysis and modeling of OHC electromotility is an important step in understanding the putative role of the outer hair cell as the cochlear amplifier. This role requires that OHCs are both sensors and actuators making the cell a self-assembling micro-electro-mechanical system. Electromotility has been observed in the membranes of other cell types, suggesting a universal mechanism. This review will focus on the characteristics and modeling of OHC electromotility.

I. Introduction

For years, Helmholtz' theory (Helmholtz 1954), based on fundamentals of passive mechanics and hydrodynamics, provided the framework for understanding how we hear. It suggested that the mammalian hearing apparatus, the cochlea, was arranged similar to the strings of a harp, with a number of highly tuned elements arrayed along the frequency scale. These elements split complex sound into its pure tone components (functionally equivalent to a Fourier analysis). Helmholtz' theory was generally correct; however, the viscosity of the fluid-filled cochlea dampens a passive mechanical system's motion degrading its frequency response. Thus, a simple mechanical model cannot explain the cochlea's wide dynamic range, extending to 20 kHz in humans and beyond in other mammals. Gold (Gold 1948), a communications engineer during World War II, realized that for a mechanical system in a fluid environment to have considerable bandwidth and high frequency selectivity, it must have active amplification. He proposed that correctly timed cycle-by-cycle positive feedback of mechanical energy would act as a cochlear amplifier, counteracting viscous damping (Nobili et al. 1998).

Gold's idea was largely abandoned when von Békésy published his classic measurements of cochlear motion in cadavers (von Bekesy 1960) demonstrating (1) broad frequency mapping and (2) linear cochlear responses to physiologically relevant sound intensities. The fact that von Békésy's experiments suppressed the signatures of an active element was not realized until data from live cochleas confirmed Gold's prediction, showing sharp tuning and nonlinear responses (Ruggero and Rich 1991, Ruggero 1992, Ruggero et al. 1997, Robles and Ruggero 2001). If the cochlea's ion gradient is temporarily disturbed, the highly-tuned nonlinearity reversibly degrades until it resembles the broad response demonstrated by von Békésy. This suggests that the active element depends on a biological energy source and not exclusively on the mechanics of the cochlea or the hydrodynamics of the fluid environment.

Thus, the hearing organ can be interpreted as a coupled series of mechanical resonators with tiny sensory/motor "microphones" that actively tune the resonators to specific frequencies. Gold speculated that such a highly tuned system might be triggered to produce feedback oscillations. These were discovered in the form of spontaneous otoacoustic emissions (SOAE) (Wilson 1980) or evoked otoacoustic emissions (EOAEs) (Kemp 1978, 1979). This surprising result is analogous to the eye producing light!

Although otoacoustic emissions (OAEs) occur in all vertebrates, its source is most likely different in the mammalian and non-mammalian hearing organ. Non-mammalian vertebrates use some form of electrical tuning (rather than micromechanical tuning described above) to obtain frequency selectivity. In electrical tuning the ion channel complement of hair cell membranes creates electrical resonances at preferred frequencies (Fettiplace and Fuchs 1999). The selectivity of an individual cell is achieved by varying the number and/or kinetics of its ion channels. Comparative studies suggest that the unspecialized hearing organs of the earliest amniote vertebrates (i.e. turtles), which consist of a single hair cell type, have the ability to produce cochlear amplification (Manley and Koppl 1998). Since branching from stem reptiles ∼ 320 million years ago, the hearing organ has developed along three parallel lines: mammals, birds/crocodiles, and lizards/snakes. Within each group, specific hair cell specializations have enhanced tuning and led to unique morphologies (Manley 2000). Regardless of their evolutionary differences, sensory hair cells from all vertebrates detect sound via mechanically sensitive hair bundles. Although the molecular identity of the hair bundle mechanosensitive ion channel (MSC) is unknown, extensive research over the past twenty years has clarified many elements of the mechanoelectrical transduction apparatus (Fettiplace et al. 2001). Within the hair bundle, adjacent stereocilia are connected by fine extracellular links so deflection of a single tip moves the whole bundle (Hackney 2000). This tip-link has been suggested to transmit force directly to the MSCs (Pickles et al. 1984). When the hair bundle is deflected by a shear force toward the tallest stereocilia, the channels open allowing an influx of K^+ and Ca^{2+} ions; deflection towards the shortest results in closure. Interestingly, bundle compliance is non-linear due to the opening and closing of the MSCs, known as gating compliance (Howard and Hudspeth 1988). Thus, forces incurred from channel conformational changes can influence hair bundle motion.

As in other sensory receptors, an adaptation mechanism is required to maintain the mechanotransduction channels within a narrow operating range to optimize sensitivity to incoming signals (Eatock 2000). This adaptation occurs in two stages: a fast, sub-millisecond closure that is correlated with the amount of Ca^{2+} entering the MSCs (Ricci et al. 2000), and a slower process which becomes apparent for large stimuli. One hypothetical adaptation mechanism is that tension in the tip link is adjusted by moving its attachment point, suggested to contain myosin Iß, along the side of the sterocilium (Gillespie and Corey 1997; Manley and Gallo 1997; Garcia-Anoveros 1998; Gillespie et al. 1999). In this scheme, Ca^{2+} influx via MSCs detaches the myosin molecules from the actin core of the stereocilium, allowing translation of the transduction channel. This reduces tip-link tension favoring channel closure. Although this myosin-based mechanism might regulate bundle position on a slow time scale, it does not act quickly enough to account for fast adaptation. In support of this, fast adaptation is unaffected by myosin motor inhibitors (Wu 1999). The speed of fast adaptation most likely requires a direct interaction between Ca^{2+} and the MSC (Crawford 1991).

Active bundle oscillations have been observed in several non-mammalian species (Crawford and Fettiplace 1985; Martin and Hudspeth 1999; Martin et al. 2000). Similar movement has also been demonstrated in reaction to bundle displacement via elastic probes (Crawford and Fettiplace 1985; Howard and Hudspeth 1987; Ricci et al. 2000; Ricci et al. 2002). The evoked movements occur with time constants similar to channel adaptation. Hence, the active bundle movements are intimately related to adaptation of the mechanotransducer channels (Ricci et al. 2000, 2002). The fast response is sufficient to work on a cycle-by-cycle basis at auditory frequencies, a requirement of the cochlear amplifier. The proposed oscillatory mechanism is that MSCs open during bundle deflection, Ca^{2+} enters and binds to the channel intracellularly, closing them in a manner that produces sufficient force to return the bundle back to its original position. Ca^{2+} entry is reduced permitting channels to re-open and begin the cycle again (Hudspeth 1997). This negative feedback mechanism is common in biological systems. Using this model, depending on stiffness parameters and Ca^{2+} binding and flux rates, oscillations at a specific frequency might be produced. With certain model parameters, the feedback system becomes unstable and can create spontaneous limit cycle oscillations (Choe et al. 1998; Camalet 2000). Such conditions not only account for cochlear amplification but also otoacoustic emissions (Koppl 1995; Martin and Hudspeth 1999). Although negative feedback is responsible for bundle oscillations, it is positive feedback of the bundle motion relative to the sound induced shear forces that produces cochlear amplification.

The mammalian cochlea encodes a much wider frequency range than the auditory organs of lower vertebrates. Critical to high frequency hearing was the emergence of a new type of hair cell, the outer hair cell (OHC). Although both mammalian hair cells convert bundle deflections into receptor potentials, each has a distinct role. The inner hair cells (IHC) IHCs relay signals to the brain

(a)   (c)

(b)   (d)

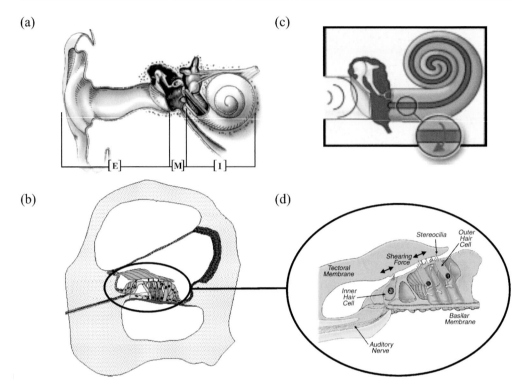

Fig. 1. Fundamentals of mammalian hearing apparatus. **(a)** Sound waves captured by the external ear [E] pass through the ear canal and impinge on the tympanic membrane, the beginning of the middle ear [M]. The bones of the middle ear match the impedance of air to the water-filled cochlea, located in the inner ear [I]. **(b)** A cross section through the cochlear spiral shows the three cochlear ducts. The middle compartment, scala media, contains the hearing organ and is bounded below by the basilar membrane and above by the thin Reissner's membrane. **(c)** The pressure difference between cochlear compartments produces a traveling wave along the elastic basilar membrane. Crude frequency mapping along the length of the cochlea occurs from systematic differences in geometry, elastic properties, and mass of the basilar membrane. **(d)** Resting on the basilar membrane is the hearing organ, the organ of Corti. It consists of sensory hair cells supported by an elegant matrix of cellular and acellular structures. There are three rows of outer hair cells and single row of inner hair cells. 'Inner/outer' refers to their location relative to the central axis of the cochlear spiral. The basilar membrane traveling wave produces a shear force on the hair bundles. When this occurs, hair cell stereocilia are bent, modulating the flow of ions through mechanically sensitive ion channels (MSCs) located near tip links connecting adjacent stereo-cilia (Eatock 2000, Jaramillo and Hudspeth 1991). Thus, both inner and outer hair cells are primarily mechanoreceptors, but only IHC receptor potentials trigger afferent nerve activity sending signals to the brain. OHCs have the specialized role of enhancing local micromechanical tuning. (Fig. 1a is Netter Plate 87, 1b is modified from *http://depts.washington.edu/hearing/*, 1c is a copyright of Stephan Blatrix, Montpellier (France), taken from the web site "Promenade round the cochlea" *http://www.iurc.montp.inserm.fr/cric/audition/english/*, and 1d adapted from (Brownell 1999))

and OHCs are responsible for micro-mechanical tuning (Fig. 1). The OHCs specialization became clear when it was shown that OHCs display electromotility (EM); the cells change length with membrane potential (Fig. 2) (Brownell 1983,

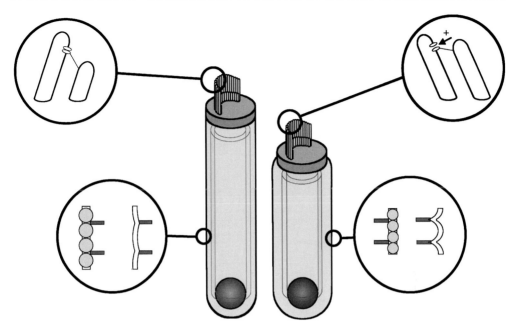

Fig. 2. The outer hair cell is both a sensor and actuator. Classic mechanoelectric transduction (MET) takes place in the hair bundle on the apical surface of the cell. Upper inserts depict the effect of bending the bundle on postulated MET channels attached to tip-links. In addition to the hair bundle mechano-electric transducer common to all hair cells, there is a somatic sensor/motor located in its lateral wall. This motor is believed to be the mammalian cochlear amplifier that provides force feedback to the basilar membrane to overcome viscous damping. The two principle models for the motor's mechanism are an (a) area motor, or a (b) bending motor (depicted in the lower inserts, the left for the elongated/hyperpolarized state and right for the shortened/depolarized state). Although the figure shows the electromotile length changes that occur in an isolated OHC, *in vivo*, the OHC may affect local mechanics by either changing shape or changing stiffness

1984; Brownell et al. 1985; Ashmore 1987; Santos-Sacchi and Dilger 1988). A variety of arguments support the notion that this somatic motility is responsible for mammalian cochlear amplification and OAEs. Abolishing EM with salicylate (Dieler et al. 1991) eliminates SOAEs (McFadden and Plattsmier 1984) and EOAEs (Quaranta et al. 1999). Stimulating the efferent nerves that innervate the OHCs modulates the strength of OAEs (Mountain 1980; Siegel and Kim 1982) and also stimulates neurotransmitter release from inner hair cells (Brown et al. 1983). When OHCs are absent due to disease, the dynamic range and frequency discrimi-nation are reduced (Risberg et al. 1990). To satisfy the role of the cochlear amplifier, OHC electromotility must not only occur throughout the dynamic range, a formidable task at high frequencies, but also generate enough force to affect local mechanics. It has been established that OHCs are able to generate phase locked forces at frequencies up to and exceeding 80 kHz (Frank et al. 1999). Electrical stimulation can produce basilar membrane motion in the physiological range and this effect is abolished when OHCs are removed (Nuttall and Ren 1995; Reyes et al. 2001). In addition, *in vivo* electromotility of a single OHC produces receptor potentials in adjacent OHC (Zhao and Santos-Sacchi 1999). Although the role of somatic EM in cochlear amplification and OAEs is clear, the discovery of electromotility has meant less attention paid to the OHC's stereociliar bundle. Therefore, even though no evidence for active bundle motion in mammals exists, involvement of hair bundle mechanoelectrical transduction channels should not be ruled out (Yates and Kirk 1998; van Netten and Kros 2000).

It has been established that electromotility can be elicited *in vivo* by moving the OHC's stereocilia (Evans and Dallos 1993), however the exact mechanism by which OHCs carry out cochlear amplification is not yet clear. The proposed mechanism is that receptor potentials generated by bundle deflection drives electromotility which feeds back mechanically on the basilar membrane helping overcome viscous forces. Along the cochlea, the positive feedback forces must occur with varying phase and amplitude to produce micromechanical tuning, a formidable task. The resulting local mechanics and hydrodynamics stimulate inner hair cell stereociliar motion and produce receptor potentials which lead to afferent nerve activity. This view of cochlear mechanics requires that OHCs serve as both sensors and actuators. This review will focus on the characteristics and modeling of OHC electromotility- the basolateral voltage sensor/actuator.

II. Structure of the Outer Hair Cell

The cylindrically shaped OHC is functionally divided into apical, lateral, and basal regions. These regional divisions are identifiable by the distribution of membrane particles, organelles, and other membrane specializations (Gulley

and Reese 1977; Brownell 2001). The 100 nm thick lateral wall has a unique trilaminate organization, with the plasma membrane (PM), cortical lattice (CL), and the subsurface cisternae (SSC) forming three axially concentric cylinders at the perimeter of the cell's axial core (Fig. 3).

A. Plasma Membrane

The lateral wall plasma membrane (PM) appears rippled when viewed with transmission electron microcopy (TEM) (Smith 1968; Ulfendahl and Slepecky 1988; Dieler et al. 1991). Support for *in vivo* folding comes from micropipette aspiration, where the apparent (optically resolved) OHC surface area before and after removal of membrane vesicles appears the same (Morimoto et al. 2002). Similar results are obtained using laser tweezers to pull tethers (Li et al. 2002). The plasma membrane contains a variety of proteins. Freeze fracture studies reveal a closely packed array of 2500–6000/μm^2 large (9–12 nm) particles in the plasma membrane cytoplasmic face (Gulley and Reese 1977; Forge 1991; Kalinec et al. 1992; Tolomeo et al. 1996). It has been suggested that these particles represent an integral membrane protein important to electromotility, but they may also represent lipid phase separations that form when membranes are frozen (Brownell et al. 2001).

B. Subsurface Cisternae

Separated from the plasma membrane by a ~30 nm wide extracisternal space (ECS) (Saito 1983), the subsurface cisternae (SSC) are an additional membrane bound organelle that surrounds the axial core. This unique organelle may be single or multilayered with the outermost layer continuous, extending from the cuticular

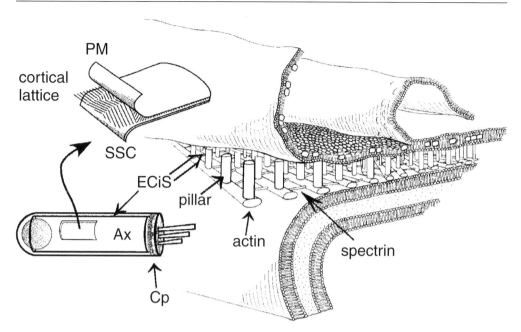

Fig. 3. The lateral wall contains the cytoskeletal cortical lattice (CL) sandwiched between the plasma membrane (PM) and the subsurface cisternae (SSC). Circumferential filaments of F-actin (4–7 nm diameter) (Holley and Ashmore 1988a, Holley et al. 1992) are spaced ∼ 40 nm apart and cross-linked with molecules of spectrin (2–3 nm diameter and up to 80 μm long) (Drenckhahn et al. 1985, Holley and Ashmore 1990, Holley et al. 1992). The pillars are 25–35 nm long, 7–10 nm diameter, and spaced 30–35 nm apart (Gulley and Reese 1977, Flock et al. 1986, Arima et al. 1991, Holley 1991, Holley et al. 1992). Figure taken from Brownell and Popel (1998)

plate to the nucleus. Communication between the ECS and the axial core occurs only at the basal end of the cell (Brownell et al. 1997). The function, composition, and contents of the SSC are unknown. It stains positive for endoplasmic reticulum (ER), Golgi, and mitochondrial markers (Zajic et al. 1993) and sequesters ceramide and $DiOC_6$ (Pollice and Brownell 1993; Oghalai et al. 1998), but shows no evidence of belonging to the PM-ER-Golgi membrane pool (Siegel and Brownell 1986). The mitochondrial staining character is interesting since most of the cell's mitochondria are located beneath the SSC. Unlike the plasma membrane, the SSC appears smooth under TEM (Saito 1983, Dieler et al. 1991).

C. Cortical Lattice

The cortical lattice (CL) is a matrix of crosslinked actin and spectrin located in the ECS between the plasma membrane and the SSC. The CL extends the length of the lateral wall and is organized in microdomains defined by regions of parallel actin filaments (Holley et al. 1992). The orientation of the actin differs between the microdomains but on average is circumferential with a mean pitch of 10 degrees (Holley et al. 1992). Radially oriented pillars of unknown composition tether the PM to circumferential bands of actin (Holley and Ashmore 1988a). There is no apparent correlation between the pillars and particles in the PM. The axial core of the cell is typically devoid of filamentous cytoskeletal proteins and other subcellular organelles (Raphael et al. 1994). This may reduce energy dissipation associated with electromotility (Brownell 1990; Brownell and Popel 1998; Brownell et al. 2001).

III. Mechanical Properties

A detailed understanding of the cell's passive mechanics provides a framework for testing various models of electromotility. It is therefore important to quantify the biophysical properties of the cell.

A. Turgor Pressure

The OHC is the only known *vertebrate* hydrostat; an elastic outer shell enclosing a pressurized core. OHC turgor pressure (∼1 kPa) (Ratnanather et al. 1993; Chertoff and Brownell 1994) often causes cells to burst when impaled with microelectrodes (Brownell 1983; Brownell et al. 1985; Holley 1991). Counter intuitively, if internal pressure is reduced the cell elongates and thins, as expected from Laplace's law. If internal pressure is rapidly increased, the cell can shorten by as much as 70% before bursting. After bursting, the cell returns to its original shape, showing that the lateral wall structure contains the information on cell shape. In addition, although shape is maintained, the cell is elongated by 6% (Ashmore 1987). The difference in length is the result of turgor pressure induced lateral wall stress. Data on non-linear capacitance (see below) suggest that wall tension is shared between the SSC, CL and plasma membrane. Part of the wall stress appears to be in the bilayer since the OHC can elongate up to 12% when demembranated using TritonX-100 (Holley and Ashmore 1988a; Tolomeo et al. 1996; Oghalai et al. 1998). Thus, in the resting OHC, the CL and plasma membrane are prestressed (Tolomeo et al. 1996).

A variety of interventions can alter the OHC's pressure and volume and result in what has been termed "slow motility," based on its time constant relative to electromotility (Table 1). In addition, isosmotic high external K and isosmotic low ionic strength each increase turgor

Table 1. Correlation of Cell Properties with Various Perturbations

Manipulation	↑ Direct pressure	↑ Osmotic pressure	Sustained hyperpol	SAL	CPZ	Gd^{3+}	Diamide	Chol	Ach
Pressure	↑	↑[c,l,m,n,o,p,q]	↑[r,s,t]	↓[s,t,u]		↓[v]			
Length	↓	↓	↓	↑		↑[k]			
Axial stiffness	↑[c]	↑[d]		↓[j] or NE[c]		NE[f]↓[g]	↓[a,e]		↓[b]
Aspiration stiffness				↓[f] or ↑[g]				↑[h]	
PM fluidity	↓[i]	↑[i]		↓[i]	↓[i]			↑[h]	
EM Force				↓[j,l]		↓[a]			

SAL = salicylate, CPZ = chlorpromazine, Ach = acetylcholine, Chol = cholesterol, PM = plasma membrane, EM = electromotility, NE = no effect. Length changes correspond to changes in turgor pressure. [a](Adachi and Iwasa 1997), [b](Dallos et al. 1997), [c](Hallworth 1997), [d](He and Dallos 1999), [e](Kalinec and Kachar 1993), [f](Lue and Brownell 1999), [g](Morimoto et al. 2002), [h](Nguyen and Brownell 1998), [i](Oghalai et al. 2000), [j](Russell and Schauz 1995), [k](Santos-Sacchi 1991), [l](Belyantseva et al. 2000b), [m](Chertoff and Brownell 1994), [n](Crist et al. 1993), [o](Dulon et al. 1988), [p](Gale and Ashmore 1994), [q](Harada et al. 1994), [r](Brownell et al. 1985), [s](Brownell et al. 1990), [t](Shehata et al. 1991), [u](Dieler et al. 1991), [v](Kakehata and Santos-Sacchi 1995)

Table 2. Axial Stiffness Measurements

Method	Axial stiffness k [mN/m] (range)	Unit compliance [stiffness^{-1}l^{-1}] or [m/N/μm]	dF [pN/mV]
Glass fiber compression (Holley and Ashmore 1988a)	0.544 (0.15–1.1)	70	
Glass fiber compression (Ulfendahl and Chan 1998)	1.1 (0.13–3.3)	13	
Glass fiber compression held by wide mouth pipette (Russell and Schauz 1995)	0.755 (0.27–1.8)	20	4.1
Glass fiber compression, cell held by narrow mouth pipette (Hallworth 1995, 1997)	5–6 (0.8–25)	4	
Glass fiber 3 point bending (Tolomeo et al. 1996)	*1.2	16	
Glass fiber extension, cell held by WCR pipette (Iwasa and Adachi 1997, Adachi and Iwasa 1997)	*10	2	100
Driven glass fiber compression in microchamber (He and Dallos 1999)	(0.4–8)	2–33	
Pressure induced distortion in microchamber (ext and comp) (Zenner et al. 1992)	2.8	7	
Suction induced extension from microchamber (Gitter et al. 1993)	2	10	
Direction WCR inflation (extension) (Iwasa and Chadwick 1992)	*13	1.5	**500
Osmotic swelling (extension) (Ratnanather et al. 1996)	*20	1	
Micropipette aspiration (extension) (Sit et al. 1997, Nguyen and Brownell 1998, Lue and Brownell 1999)	KS = 0.5–1 *30–40	0.5–0.6	
Atomic force lever compression and microchamber (Frank et al. 1999)	*5	2.4	50
Prestin			
Atomic force lever compression and microchamber (Ludwig et al. 2001)			0.3

The moduli (E or Ka, see text) were converted to axial stiffness; note: Aspiration stiffness (K$_s$) is not equal to axial stiffness; **theoretical estimate

pressure (Zenner et al. 1985; Kachar et al. 1986). Turgor pressure plays a critical role in the robustness of the electromotile response and may be responsible for maintaining sensitivity of the electromo- tor (Brownell 1990; Santos-Sacchi 1991; Shehata et al. 1991; Kakehata and Santos-Sacchi 1995; Kakehata and Santos-Sacchi 1996; Adachi et al. 2000). The origin of turgor pressure is presumably osmotic. The mechanisms controlling it are unknown but they may involve regulating the cytoplasmic levels of glucose (Brownell et al. 1990; Brownell 1990; Crist et al. 1993; Chertoff and Brownell 1994; Geleoc et al. 1999).

B. Stiffness Measurements

OHC stiffness (or mechanical imped- ance) is a measure of the cell's resis- tance to an externally applied force, equivalent to the spring constant of a Hookean spring ($k = F/x$). There is con- siderable variability in the reported val- ues of axial stiffness of intact cells (Brownell et al. 2001) (Table 2), some of which arises from not normalizing to cell length. Consequently, for comparison, the data is expressed as displacement/ unit length or strain. Some author's have reported "stiffness" as elastic moduli, however these are *model dependent quantities*. For example, in analyzing ax- ial stiffness data, the wall Young's modu- lus ($E = $ stress/strain) or area expansion modulus ($Ka = $ the integration of E over the thickness of the wall), will depend on the mechanical model used to define the OHC (i.e., as a solid cylinder, a spring, or a thin walled hollow cylinder) and there may be multiple moduli de- pending on whether the model is isotro- pic, anisotropic, or orthotropic.

For low strain ($< 5\%$) measurement techniques, glass fiber or AFM cantilever bending, the stiffness is $\sim 1\,mN/m$. When high strain techniques are used, osmotic swelling or direct pressure, the axial stiff- ness rises by a factor of ten. This suggests that the system is nonlinear, but the origin of this nonlinearity has yet to be determined. Nonlinearity also appears when comparing stiffness estimated from compression or extension. It is difficult to draw conclusions from these various stud- ies because in most studies turgor pres- sure is not maintained and, as a general

rule, changes in axial stiffness are corre- lated with changes in pressure (Table 1).

IV. Basic Properties of Outer Hair Cell Electromotility

OHC electromotility has attracted particu- lar attention because of its suggested role as the cochlear amplifier and its correla- tion with mammalian OAEs. OHC somatic motility was unexpected. It was observed while studying the population of voltage gated ion channels in the cell's lateral wall. The OHC visibly changed length when the membrane potential was altered (Brownell et al. 1985; Ashmore and Meech 1986; Kachar et al. 1986a,b; Dallos and Evans 1995a,b). The electromotile mechanism differs from known me- chanisms since it does not depend on calcium (Ashmore and Meech 1986; Kachar et al. 1986; Holley and Ashmore 1988b; Santos-Sacchi and Dilger 1988; Santos-Sacchi 1989) or intracellular stores of ATP (Ashmore and Meech 1986; Kachar et al. 1986; Holley and Ashmore 1988b). A distinctive feature of EM is that it follows membrane potential and not ionic currents (Ashmore 1987; Santos- Sacchi and Dilger 1988; Iwasa and Kachar 1989; Santos-Sacchi 1992). The voltage dependence and short time constant sug- gest that the transduction process is similar to piezoelectricity (Mountain and Hubbard 1994; Iwasa 2001).

A. Motor Characterization

Depolarization from a resting potential produces both a decrease in cell length (4–5%) and increase in cell diameter (Brownell et al. 1985; Zenner et al. 1985; Zenner 1986; Kachar et al. 1986; Ashmore 1987; Santos-Sacchi and Dilger 1988; Iwasa and Kachar 1989; Dallos et al. 1991). Hyperpolarization produces the opposite response. When a micropipette, used to control the membrane potential of the cell interior, is inserted at different locations along the cell length, the apical and basal ends move in opposite phase with graded amplitude. The more distance between the end and the pipette, the larger the movement. Thus, the motor element appears distributed roughly linearly over the cell length. A plot of the cell length vs. voltage (Fig. 4A) shows Boltzmann characteristics, Eq. 1,

$$\Delta L = \Delta L_{max}/1 + e^{b(V-V_0)} \qquad (1)$$

where ΔL_{max} is the maximum length change and V is the membrane potential. V_0, the voltage at half maximal length change, and the sensitivity, b, are obtained by curve fitting. Table 3 summarizes literature values from Boltzmann fits to OHC displacement-voltage data. Note that V_0 is influenced by turgor pressure (Adachi et al. 2000). Therefore, different holding potentials result in changes in the set point, thereby resetting the motor to its most sensitive region. The sensitivity from single-point measurements, i.e. the slope between two points located somewhere along the Boltzmann curve, have been reported in the literature. Not knowing of the location of the data points relative to the complete curve (Eq. 1) makes interpretation difficult. The intrinsic response time and force output of the electromotile motors is best measured isometrically. Under these conditions, the frequency response is nearly flat to 80 kHz and the maximum force is ~ 50 pN/mV (Frank et al. 1999). A response time $< 10\,\mu s$ suggests that the intrinsic electromotor is capable of cycle-by-cycle feedback at the highest acoustic frequencies. Taking the system as Hookean, $F = -k(x - x_0)$, the voltage dependence is given by $dF/dV = -dk/dV(x - x_0) - k\,dx/dV$. Thus, voltage dependent changes in stiffness or in dimension can account for the observed data.

Similar to the excess compliance associated with hair bundle gating, conformational changes of the motor will contribute to the axial stiffness, and may account for the voltage dependent stiffness reported by He and Dallos (1999). In those experiments, however, the voltage-induced changes in turgor pressure were not controlled and

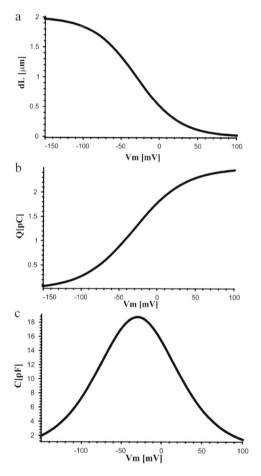

Fig. 4. (a) Displacement-voltage function (Eq. 1) using parameters from Table 3. **(b)** Charge-voltage function (Eq. 2) using parameters from Table 4. Note the negative dependence of the sensitivity term (b). **(c)** Nonlinear capacitance function (Eq. 3) using parameters from Table 4

Table 3. Parameters from Boltzmann Fits to Displacement-Voltage Data

Reference	n	V_0 [mV]	b [mV]$^{-1}$	ΔL_{max} [µm]
Ashmore (1987)	7		20*	
Santos-Sacchi (1991)	1	− 38	0.033	1.5
			30*	
Santos-Sacchi (1992)	1	− 25	0.037	2.25
			21*	
Kakehata and	1 at + 0.54	− 28	0.035	2
Santos-Sacchi (1995)	kPa		20*	
Kakehata and	1	− 36	0.037	2
Santos-Sacchi (1996)	3		19*	
Wu and Santos-Sacchi (1998)	7	− 15	0.035	2
Prestin				
Zheng et al. (2000)			0.5**	
Native Membranes				
Mosbacher et al. (1998)			0.1**	
Zhang et al. (2001)			0.01**	

n = the number of cells used to obtain data points; *Max gain (slope) [nm/mV]; **Single point sensitivity measurement, [nm/mV]

that could also account for the results.

B. Properties of the Voltage Sensor

Cell membrane capacitance is $\sim 1\,\mu F/cm^2$ and is independent of voltage. However, in the presence of mobile charge bound to the membrane, a voltage dependent capacitance is added. This nonlinearity is commonly called a "gating current" in the context of voltage dependent ion channels (Bezanilla 2000). The OHC exhibits a large nonlinear capacitance (NLC) not associated with any ionic current (Santos-Sacchi 1991, 1992). The NLC can be modeled as a two state system, where the amount of charge in each state is a Boltzmann function of voltage (Fig. 4B), Eq. 2,

$$\Delta Q = \Delta Q_{max}/(1 + e^{-b(V - V_0)}) \quad (2)$$

where ΔQ_{max} is the total amount of available charge. The sensitivity term can be expressed as $b = ze/k_B T$ where z is the mobile charge, e is the unitary electric charge, and $k_B T$ is Boltzmann's constant times temperature ($k_B T \sim 25$ meV $= 4.1$ pNnm at room temperature). Eq. 2 has the same form as Eq. 1 suggesting a similarity between the motor and the sensor. Taking the derivative of Eq. 2 with respect to voltage, we obtain an expression for the nonlinear capacitance (Fig. 4C), Eq. 3,

$$C_{nlc} = \frac{Q_{max} \cdot b}{e^{b(V - V_0)}\left(1 + e^{-b(V - V_0)}\right)^2} \quad (3)$$

Table 4 summarizes parameters obtained from the literature. Calculation of the membrane-charge density required to account for the maximal measured displacement currents yields numbers that are similar to the PM particle density.

Gale and Ashmore (1997) measured the frequency response of the NLC and predicted a frequency limit of 26 kHz, corresponding to a step response of 6 µs. The time constant of the sensor and the motor are within an order of magnitude, but it is important to repeat this experiment at higher time resolution.

Table 4. Parameters from Boltzmann Fits to Nonlinear Capacitance

Reference	n	V_0 [mV]	Q_{max} [pC]	b [mV^{-1}]	z	C_{nlc}max [pF]	Charge density [C/μm^2]
Santos-Sacchi (1991)	4	− 25 (DC)	2.5	0.034	0.87	17	4000
	4	− 40 (AC)		0.030	0.76	23.2	
Huang and Santos-Sacchi (1993)	12	− 34	2.4	0.036	0.92	21.55	5542
Gale and Ashmore (1994)	7	− 33	1.6–2.4*	0.037	0.95	20*	
Huang and Santos-Sacchi (1994)	9	− 15	1.7	0.038	0.98		
Tunstall et al. (1995)	11	− 27	2.5–2.7	0.015– 0.035	0.78	18	
Kakehata and Santos-Sacchi (1995)	4	− 17	2.4	0.029	0.73		
Kakehata and Santos-Sacchi (1996)	1	− 15	4.7	0.025	0.65		
Wu and Santos-Sacchi (1998)	7	− 21.9	3.69	0.026	0.67	44.8	
Adachi and Iwasa (1999)	1	− 46		0.029	0.75	20	
Geleoc et al. (1999)	1	− 10**	2.56	0.030	0.77		
Prestin							
Zheng et al. (2000)	7	− 60	1.76	0.036	0.91		5360
Ludwig et al. (2001)	4	− 66	0.060	0.030	0.77		
Santos-Sacchi et al. (2001)	5	− 74	1	0.034	0.88		
Oliver et al. (2001)	16	− 75		0.028	0.73		

n = number of cells used to obtain data points. The charge density is calculated assuming 10 fF/μm^2. DC refers to a voltage step subtractive protocol (P/ − 4), and AC refers to analysis in the frequency domain (using phase tracking); *measurements on single cells; **dependent on dominant sugar used in bath solution (fructose produced positive shift)

C. Are the Sensor and Motor the Same Entity?

The similarity of the parameters from the Boltzmann equation and the frequency response of the sensor and motor suggest that the mechanism responsible is a single unit. Consistent with this hypothesis is their similar distribution pattern. NLC and mechanical displacement both vary proportionally with the length of the exposed lateral wall in a microchamber (Huang and Santos-Sacchi 1993). The sensory and motor functions are maintained in excised patches of plasma membrane (Kalinec et al. 1992a,b). However, in these experiments it is not possible to exclude the fact that the excised patches contained CL and SSC (Ruknudin et al. 1991). Thus, the sensor and actuator elements are confined to the lateral wall

plasma membrane that is adjacent to the SSC (Dallos et al. 1991; Kalinec et al. 1992a,b; Huang and Santos-Sacchi 1993). Exploring the local strain, Kalinec placed microbeads along the cell as fiducial marks and tracked their displacements (Kalinec and Kachar 1995). Although the beads generally moved parallel to the long axis of the cell, the displacement vectors differed from point to point on the lateral wall. Thus, the motors do not appear to be in a continuous crystalline array.

A variety of factors have been tested for affects on EM and NLC (Table 5). Interventions that affect the sensor are likely to affect the motor. Membrane tension shifts the midpoint of both the EM and NLC functions without affecting the form of the voltage dependence. The fact that pulling on the motor generates a voltage

Table 5. Factors that Affect EM and NLC

Manipulation	Motor			Sensor		
	V_0	b	ΔL_{max}	V_0	b	Q/C_{max}
↑Membrane tension	↑[n]	NE[n]	↑[n,x,y]	↑[n,k,h,o,a,r]	NE[n]	↓[n,k,h,o,a]
Negative prepulse				↓[aa]		
Pharmacology						
Lanthanides*			↓[x,n,o]	↓[n,o] ↑[x]	↓[o,n]	↓[x,o]
Trypsin			NE***[p]	↑[n]		↓[a] NE***[j]
SAL$^-$**	↑	↓[o]	↓[y,t,o]	↑[o]	↓[o,z]	↓[z,o]
CPZ$^+$	↑		NE	↑[s]	NE	
TPP$^+$				↓[aa]		NE[aa]
TPB$^-$	↓[aa]		NE[aa]	↓[aa]		↑[aa]
↓Ionic strength		↑[m]				
Phosphorylating agents				↑/↓[j]		NE[j]
Ach		↑[d]	↑[d]			
↓$[Cl^-]_{int}$						↓[u]
Digitonin			↓[l]			
NEM, Diamide			NE[q]			
EA, mersalyl, pHMPS, pCMPS			↓[q]			
Negative results						
Dinitrophenol, iodo acetic acid			NE[m]			
Metabolic			NE[i,c,m,b]			
↓$[Mg^{2+}]_i$			NE[i]			
SITS, DIDS			NE[v]			
Furosemide			NE[q]			
Cytoskeletal			NE[i,e]			
↓ $[Ca^{2+}]_o$			NE[v,c,m,w,b]			
Cs, Ba, or Cd Ca substitution			NE[v]			
Co			NE[f]			
K channel blockers			NE[g,v,f]			
Ca-act K channel blockers			NE[g]			
L type Ca channel blockers			NE[g]			
Non-L-type Ca channel blockers			NE[g]			

*Cells able to respond to changes in wall stress during lanthanide treatment; **SAL effect unaffected by simultaneous trypsin treatment; ***on isolated membrane patch

Membrane tension = induced osmotically, by aspiration pressure, or by direct stretch. Lanthanides = Gd^{3+}, Lu^{+3}, Ce^{3+}; SAL = salicylate, benzoate, and butyrate; CPZ = chloropromazine; TPP = tetraphenyl phosphonium, TPB = tetraphenyl boron; Metabolic = CCCP, FCCP, iodoacetamide, NaN_3, AMP, PNP, no ATP; SITS = 4-acetamido-4-isothiocyano-2,2'-disulfonic acid stilbene, DIDS = 4,4'-diisothiocyano-2,2'-disulfonic acid; Cytoskeletal = *Actin agents*: Phalloidin, cytochalasin B and D, *Tubulin agents* (cochicine, nocodazone, colcemide); K channel blockers = agiotoxin, α-,β-,γ-,δ-dendrotoxin, dendrotoxin-I, margatoxin, MCD peptide, stichodactila toxin, tityustoxin Kα, tetraeltyl ammonium; L type Ca channel blockers = calcicludine, calciseptine, FS-2, TaiCai toxin, waglerine-I; Ca-act K channel blockers = apamin, charibdotoxin, iberiotoxin, kaliotoxin, noxiustoxin, paxilline, penitrem; Non-L-type Ca channel blockers = ω-agatoxin TK, ω-conotoxin, FTX-3.3, sFTX-3.3, PLTX-II

[a](Adachi and Iwasa 1999), [b](Ashmore 1987), [c](Ashmore and Meech 1986), [d](Dallos et al. 1997), [e](Dieler et al. 1991) [f](Evans et al. 1991), [g](Frolenkov et al. 1998), [h](Gale and Ashmore 1994), [i](Holley and Ashmore 1988b), [j](Huang and Santos-Sacchi 1993), [k](Iwasa 1993), [l](Iwasa and Kachar 1989), [m](Kachar et al. 1986), [n](Kakehata and Santos-Sacchi 1995), [o](Kakehata and Santos-Sacchi 1996), [p](Kalenic et al. 1992), [q](Kalinec and Kachar 1993; Kalinec and Kachar 1993), [r](Ludwig et al. 2001), [s](Lue et al. 2001), [t](Oghalai et al. 2000), [u](Oliver et al. 2001), [v](Santos-Sacchi and Dilger 1988), [w](Santos-Sacchi 1989), [x](Santos-Sacchi 1991), [y](Shehata et al. 1991), [z](Tunstall et al. 1995), [aa](Wu and Santos-Sacchi 1998)

is consistent with the interpretation of biological piezoelectricity. Voltage and tension work additively characterize independent energy sources. The reduction of the membrane capacitance with an increase in tension is further reduced by hyperpolarization (Iwasa 1993). However, contrary to a maximally efficient coupling between the sensor and the motor, the sensor shows hysteresis, i.e. the sensor sensitivity can be reset by a negative prepulse. The mechanism for this hysteresis is unknown, but may be related to changes in turgor pressure.

Amphipaths, lipid soluble compounds that can alter membrane curvature, surface charge, and order parameters, affect the midpoint and shape of the Boltzmann function similarly for both the sensor and the motor. The lipophilic ions TPP^+ (tetraphenyl phosphonium) and TPB^- (tetraphenyl borate) alter the surface potential, but only TPB^- also increases the NLC. The difference between the two ions suggests that they are not serving as the intended set of symmetric mobile carriers, but are interacting with intrinsic fields within the membrane, most likely the dipole potentials of the head groups.

To test the importance of the cortical cytoskeleton in electromotility, Kalinec et al. (1992) dialysed cells with trypsin. The cells become spherical and the mean cell shape didn't change with voltage, although EM persisted as a rippling of the membrane. The lack of changes in shape suggests that, at least after trypsin treatment, there is no other asymmetric ordering. Thus, the cortical lattice is necessary for constraining cell motion to produce length changes (Holley and Ashmore 1988b; Huang and Santos-Sacchi 1994; Adachi and Iwasa 1999).

To summarize, a model of electromotility must contain a voltage dependent sensor and motor in the lateral wall PM. The model must predict (1) the total force generation, (2) the effects of in-plane tension on the sensor and motor parameters, (3) surface charge effects on the sensor and motor, and (4) a voltage dependent axial stiffness.

V. Molecular Basis of Electromotility and Nonlinear Capacitance

Although there may be multiple components of the sensor-motor unit, it was appropriate to search for the molecule(s) involved. A membrane protein capable of endowing generic cells with both EM and NLC was recently discovered (Zheng et al. 2000; Ludwig et al. 2001). The protein, *prestin*, was found using a differential screen between outer and inner hair cells. *Prestin* shows homology to sulfate transporters such as *pendrin*. When generic cells were transfected with prestin, the cells displayed a number of properties similar to outer hair cells: EM and NLC became apparent and both were suppressed by salicylate. Quantitative comparison of EM and NLC in transfected cells and EM and NLC in OHCs is difficult because values will be functions of expression efficiency and differences in the passive mechanical constants and constraints. Q_{max} associated with transfection was smaller than seen in OHCs, but the sensitivity term, b, and the midpoint, V_0, were similar. To produce measurable displacements, cells were constrained within a cup shaped pipette and a photodetector was used to track the exposed edge of the cell. The average sensitivity was $\sim 0.5\,nm/mV$. The significance of the sensitivity and phase is unclear, because within a microchamber the membrane voltage on one side of the pipette was hyperpolarized while the other side is depolarized. Further, in the absence of a proper mechanical model, the intrinsic values of the motor cannot be calculated. For example, holding the cell in a rigid pipette will tend to sum the displacements of individual motors, relative to an unconstrained cell. In general, it is not possible to characterize the motor without specifying the constitutive properties of the environment and the boundary conditions. However, the qualitative results with prestin support its central role in a piezoelectric-like motor mechanism Using prestin antibodies, Belyantseva et al. (2000a) were able to show that during development of the rat cochlea, prestin is expressed with a time course that matches the emergence of EM.

Ludwig et al. (2001) measured forces generated by prestin in transfected CHO cells. With an atomic force microscope cantilever pressed against cells held in a microchamber pipette, they found an isometric force of $\sim 0.3\,pN/mV$. The amplitude and phase of force generation was largely constant over the frequency range 0.2 to 20 kHz. The force produced by control cells was smaller and of opposite phase from transfected cells. Salicylate reduced the force by a factor ~ 2.5 and shifted the phase by 180 degrees as though the motor had been blocked. It is curious that this background motion with opposite phase has not been reported for OHCs treated with salicylate.

Despite the elegance of the experiments, the significance of the isometric force in terms of the intrinsic properties of prestin is not clear. In a microchamber, the membrane potential changes on the cis and trans side of the cell are opposite. However, the two sides of the cell are coupled by membrane tension and by the constraint of constant volume. Adding to the confusion, the areas and volumes of each side are not equal, nor discrete. To assume that half of the cell can be measured independently of the other is like having a cylindrical water balloon and changing the area of one end and not

being able to detect that at the other end. Alternatively, if the two sides of the cell are truly independent in a microchamber, one should get identical results by disrupting half the cell or, by expressing prestin in only half the cell. Finally, the movement of the free face of a cell constrained in a rigid microchamber will be exaggerated relative to an unconstrained cell. This amplification of motion of a free edge is the method used to measure membrane elasticity in aspiration experiments (Evans et al. 1979; Evans 1989; Fung 1993).

OHCs show a coupling of turgor pressure to the voltage dependent capacitance (Table 5). This is expected because both turgor and voltage influence cell shape. With outside-out patches from prestin transfected cells, positive pipette pressure shifted the NLC along the voltage axis toward more depolarized potentials without affecting the slope (Ludwig et al. 2001). These experiments show that the reciprocal electromechanical property induced by prestin does not require the particular environment provided by OHCs. Santos-Sacchi et al. (2001) extended Ludwig et al.'s result of mechanical sensitivity of NLC to include "voltage memory" seen in OHCs (Wu and Santos-Sacchi 1998). This is an effect where pre-pulses can shift V_0 with a time constant of $\sim 0.2\,s$. The origin of this effect is not known, but the parallel between the OHC and prestin transfected cells is striking.

Oliver et al. (2001) examined the mechanism underlying the voltage sensitivity of prestin. To determine the location of prestin's voltage sensor, each of the non-conserved charged residues in the putative membrane domain was mutated. In no case was NLC abolished, but in some cases V_0 was shifted (without changing b). This led to the idea that the voltage sensor is extrinsic to the protein. Replacing intracellular Cl^- with sulfate or pentane-sulfate reversibly decreased the electromotility and NLC. This finding was confirmed in excised patches. Replacement of Cl^- on the extracellular side had no effect. NLC was not affected by substituting cations (K, Na, NMDG, TEA) on either side of the membrane. After removal of cytoplasmic Cl^-, the cells remained in a contracted state. The selectivity of the Cl^- binding site(s) was explored with various anions. All monovalent anions could support NLC in transfected cells. The ionic selectivity of NLC in transfected cells and electromotility in OHCs followed the Hofmeister series $(I \sim Br > NO_3 > Cl > HCO_3 > F)$, that is characteristic of chaotropic effects on water. This selectivity is similar to that observed for anion-binding to pendrin and some chloride channels. Of all the monovalents, only Cl^- and bicarbonate are thought to be present in the cytoplasm at $> mM$ concentrations suggesting they may both work as the voltage sensors of prestin. Application of salicylate (an anion) to inside-out patches changed the voltage sensitivity suggesting competition with Cl^- for the anion binding site, with salicylate having a 300 fold higher affinity than Cl^-. This important study presents an elegant sensor mechanism for prestin; however, the molecular mechanism of how prestin produces force remains unclear.

The ability of control cells to display electromotility (Ludwig et al. 2001) is consistent with an early report that showed membrane movements correlated with action potentials (Iwasa et al. 1980). Using atomic force microscopy (AFM) and electrophysiology on "generic" HEK293 cells, Mosbacher et al. (1998) reported a sensitivity of $\sim 0.1 nm/mV$. The movement was outward with depolarization. Our lab has recently extended these results (Zhang et al. 2001). At negative potentials the motion was proportional to voltage $(\sim 0.01 nm/mV)$, and the direction of movement was reversed by lowering external ionic strength. Salicylate reversibly blocked motility in normal and low ionic strength conditions. These results demonstrate that membrane based motor activity can occur in the absence of prestin. Although the magnitude of the response is lower than reported in prestin transfected cells, these cells are not mechanically constrained as occurs when held in a microchamber. Even though salicylate has been shown to specifically interact with prestin (Oliver et al. 2001), its ability to reduce the magnitude of EM may occur by a more general mechanism. Further, these findings suggest that prestin alters the inherent motile ability of cells rather than confers it. Recently, we found that transfection of HEK cells with prestin reversed the sign of the voltage dependence and the motility depended upon the internal Cl^- concentration (Zelenskaya et al. 2002).

VI. Motor Models

The motor mechanism has many of the characteristics of biological piezoelectricity: a voltage generates a force, and conversely, pulling on the motor generates a voltage. Electromotility can be described phenomenologically using the mean constitutive relations of the cell and piezoelectric components (Iwasa 2001; Brownell et al. 2001). By its nature, the fit of a Boltzmann function to data suggests an ensemble of two-state units whose transition probability changing with a given stimulus. Currently, there are two hypotheses of what constitutes the piezoelectric unit: the area motor and the folding motor (Fig. 2).

A. The Area Motor

The model assumes that uniformly distributed membrane proteins, currently assumed to be prestin, change their in-plane area with voltage (Dallos et al. 1991; Kalinec et al. 1992; Iwasa and Chadwick 1993; Evans and Dallos 1993; Iwasa 1993; Huang and Santos-Sacchi 1994; Iwasa 1994; Iwasa and Adachi 1997). The structural basis for the model is the high density of membrane particles in the lateral PM. It is assumed that a membrane charge is transferred across the membrane during the conformational change. In essence, this model began as a two state model where the Helmholtz free energy of the Boltzmann equations (the qV term in Eqs. 1 and 2) was extended to contain the mechanical energy of the motor conformational change (T*dA) (Iwasa 1993). A quantitative issue with this model is that all of the elastic energy and translational work is assumed to be contained only in the motor, and the elastic energy stored in lipid membrane was ignored. This produced an error is estimating area change of the motor. This bias was corrected in the next iteration of the model (Iwasa 1994) that included the anisotropy of membrane tension of a cylinder and possibly of the motor unit itself. This model successfully predicts the changes in NLC due to changes in membrane tension and voltage, and predicts a force generation of $100 \, pN/mV$. This model was further extended from mechanical isotropy to mechanical orthotropy to accommodate axial stiffness data (Iwasa and Adachi 1997), but assumed that the elasticity of the membrane and motor are the same.

The area motor model qualitatively predicts the voltage dependent decrease in axial stiffness consistent with He and Dallos (1999), but it predicts a minimum that is not apparent in the data. In fact, the model shows a region of negative stiffness (similar to hair bundle stiffness data), corresponding to the NLC peak where the motor open probability equals $\frac{1}{2}$. Recently, the area motor model has been

extended to account for the phenomenological piezoelectric parameters of the cell (Iwasa 2001; Dong et al. 2002).

Incorporating molecular data, the area model can be interpreted as prestin's 12 transmembrane α-helixes forming a compact structure within the membrane. When Cl^- enters the transporter, it changes the orientation of the α-helical segments and thus produces an in-plane area change. The Cl^- gating charge moves deep into a cup shaped pore within the electric field. Thus, the area motor can be thought of as a transporter with an incomplete transport cycle.

In summary, the area motor model is successful in predicting the NLC changes with membrane tension and voltage, the force generation of the cell, and the voltage dependent axial stiffness.

B. Folding Motor

In this model, the effective motor is a small, curved, region of membrane that includes proteins and lipids (radius of curvature of $\sim 30 \, nm$) (Raphael et al. 2000; Morimoto et al. 2002). Force is generated by changes in curvature induced by changes in voltage. The plasma membrane of the OHC is hypothesized to be rippled with accordion-like folding creating excess membrane (Smith 1968, Ulfendahl and Slepecky 1988, Dieler et al. 1991, Li et al. 2002, Morimoto et al. 2002). The model is based upon the general, piezoelectric-like property of polarized interfaces to change their curvature with applied voltage known as flexoelectricity (Petrov 1999). The force balance involves either a "naturally flat" bilayer that is bent by tension in the cytoskeleton, or a naturally curved, asymmetric, bilayer that compresses the cytoskeleton. Flexoelectricity is linear in potential, yet the OHC has a distinctive saturable relation between potential and movement and potential and capacitance. By adding a saturable dipole (prestin?) to the membrane, the flexoelectric and capacitive response will become saturable as well. Raphael et al. (2000) showed that this kind of flexoelectric model can explain many of the properties of OHCs. For example, the effect of salicylate on reducing motility can be explained if it alters the mobile charge of the dipole.

One can think of flexoelectricity as the result of differential tension on the two surfaces of the bilayer producing a bending moment. The voltage dependence of tension at a polarizable interface was described more than a hundred years ago in the classic work of Lippmann. Intuitively, the tension arises from the lateral repulsion of charges at the interface (Zhang et al. 2001). The applied voltage changes the local charges, and these change the tension. There is maximal tension when the interface is uncharged. This phenomenon can account for the electromotility of generic cells and the inhibition with salicylate. In this case, salicylate inhibits electromotility by making the membrane have a symmetric surface charge. Recent experiments have shown that prestin transfection reverses the sign of electromotility. This result is predicted if prestin adds relative negative charge to the outside of the membrane. The Lippmann model predicts that electromotility has a parabolic dependence on voltage, in disagreement with data from OHCs. However, the midpoint of this parabola may not be physiologically accessible because of the presence of large surface charges. An interesting feature of the Lippmann approach is that it predicts the selectivity of anions on electromotility (Cacace et al. 1997). In this case the effect is driven by complicated interactions of hydration and local electric fields. One other appealing aspect of the interfacial approach is that it

provides an evolutionary path for developing electromotility. Prestin need not have appeared *de novo* as a key protein to make cells electromotile, but the evolution could follow a conservative path of endogenous motility leading to gradually improved efficiency.

C. Testing Models

It is difficult to distinguish these two classes of models since both can explain most of the observed features of OHC electromotility. The area motor theorists suggest their model is validated by the fact that constraining the cells surface area inhibits NLC (Adachi and Iwasa 1999). Although true, both models predict this result. The nonlinear aspects of both models can be explained with a quantized charge carrier. At first it appears possible to distinguish the two models by altering the surface charge with amphipaths or ionic strength and looking for a sign reversal of motility. A positive result would be striking, but a negative result would not be conclusive without a separate measurement of the true surface charge. Furthermore, one would have to correct for the way in which surface charges affect the local anion concentration since mobile anions are required for function in either model. It is possible that in the folding motor models there are local domains, such as where the cytoskeleton is linked to the membrane, where the forces produced by surface charges are critical. Perhaps the most ideal method would ask directly if prestin changes area with voltage. We could imagine an AFM imaging experiment in which prestin is placed in an immobilized bilayer and changes its in-plane shape with the applied voltage (Engel et al. 1999, Muller et al. 2000). However, if prestin changes its dimensions normal to the plane of the membrane, it might cause bending and leave the matter un-resolved.

VII. Conclusion

The OHC evolved with mammals some 220 million years ago. Its appearance was a response to the selection pressure for detecting and analyzing ever-higher frequencies. The fluid environment of vertebrate sensory organs required their sensory cells be both sensors and actuators. The speed and voltage dependence of OHC electromotility demanded a motor mechanism different from all known cellular force generating mechanisms. Mechanical energy in most cells comes from the conversion of cellular stores of ATP by proteins known as motor molecules. The relatively slow time constants of these protein based motor molecules precluded their use for processing signals with time constants on the order of $10\,\mu s$. The mammalian OHC adopted a membrane based motor mechanism. Whether the integral membrane protein prestin converts the transmembrane potential energy directly into a mechanical force or modifies the piezoelectric-like force transduction machinery inherent to all biological membranes remains to be determined. The answers will ultimately involve determination of prestin's tertiary structure and experiments on membranes reconstituted with prestin. Appreciating the function of OHC as an actuator provides a fresh perspective on sensors and sensing in biology and suggests provocative engineering approaches for future microphone design in acoustics and speech processing.

Acknowledgements

We benefited from useful discussions with Drs Farrell, Raphael, Petrov, Popel, Spector and Zhao. Funded by research grants DC00354 & DC02775 (WEB) and NIH (FS).

References

Adachi M, Iwasa KH (1997) Effect of diamide on force generation and axial stiffness of the cochlear outer hair cell. Biophys J 73: 2809–2818

Adachi M, Iwasa KH (1999) Electrically driven motor in the outer hair cell: Effect of mechanical constraint. Proc Natl Acad Sci USA 96: 7244–7249

Adachi M, Sugawara M, Iwasa KH (2000) Effect of turgor pressure on outer hair cell motility. J Acoust Soc Am 108: 2299–2306

Arima T, Kuraoka A, Toriya R, Shibata Y, Uemura T (1991) Quick-freeze, deep-etch visualization of the 'cytoskeletal spring' of cochlear outer hair cells. Cell Tissue Res 263: 91–97

Ashmore JF (1987) A fast motile response in guinea-pig outer hair cells: the cellular basis of the cochlear amplifier. J Physiol 388: 323–347

Ashmore JF, Meech RW (1986) Ionic basis of membrane potential in outer hair cells of guinea pig cochlea. Nature 322: 368–371

Belyantseva I, Adler H, Curi R, Frolenkov G (2000a) Expression and localization of prestin and the sugar transporter GLUT-5 during development of electromotility in cochlear outer hair cells. J Neurosci 20: RC116

Belyantseva I, Frolenkov G, Wade J, Mammano F, Kachar B (2000b) Water Permeability of Cochlear Outer Hair Cells: Characterization and Relationship to Electromotility. J Neurosci 20: 8996–9003

Bezanilla F (2000) The voltage sensor in voltage-dependent ion channels. Physiol Rev 80: 555–592

Brown MC, Nuttall AL, Masta RI (1983) Intracellular recordings from cochlear inner hair cells: effects of stimulation of the crossed olivocochlear efferents. Science 222: 69–72

Brownell WE (1983) Observations on a motile response in isolated outer hair cells. In: Webter WR, Aitken LM (eds) Mechanisms of Hearing. Monash University Press, Monash, pp 5–10

Brownell WE (1984) Microscopic observation of cochlear hair cell motility. Scanning Microscopy 3: 1401–1406

Brownell WE (1990) Outer hair cell electromotility and otoacoustic emissions. Ear Hearing 11: 82–92

Brownell WE (1999) How the ear works – nature's solution for listening. Volta Rev 99: 9–28

Brownell WE (2002) On the origins of outer hair cell electromotility. In: Berlin CI, Hood LJ, Ricci AJ (eds) Hair Cell Micromechanics and Otoacoustic Emissions. Delmar Learning, San Diego, pp 25–46

Brownell WE, Popel AS (1998) Electrical and mechanical anatomy of the outer hair cell. In: Palmer AR, Rees A, Summerfield AQ, Meddis R (eds) Psychophysical and Physiological Advances in Hearing. Whurr, London, pp 89–96

Brownell WE, Bader CR, Bertrand D, de Ribaupierre Y (1985) Evoked mechanical responses of isolated cochlear outer hair cells. Science 227: 194–196

Brownell WE, Shehata WE, Imredy JP (1990) Slow electrically and chemically evoked volume changes in guinea pig outer hair cells. In: Akkas N (ed) Biomechanics of Active Movement and Deformation of Cells. Springer, Heidelberg, pp 493–498

Brownell WE, Spector AA, Raphael RM, Popel AS (2001) Micro- and nanomechanics of the cochlear outer hair cell. Annu Rev Biomed Eng 3: 169–194

Brownell WE, Zhi M, Halter JA (1997) Outer hair cell electro-anatomy. In: Lewis ER et al. (eds) Diversity in Auditory Mechanics. World Scientific Publishing Co., Singapore, pp 573–579

Cacace MG, Landau EM, Ramsden JJ (1997) The Hofmeister series: salt and solvent effects on interfacial phenomena. Quart Rev Biophys 30: 241–277

Camalet S (2000) Auditory sensitivity provided by self-tuned critical oscillations of hair cells. Proc Natl Acad Sci USA 97: 3183–3188

Chertoff ME, Brownell WE (1994) Characterization of cochlear outer hair cell turgor. Am J Physiol 266: C467–79

Choe Y, Magnasco MO, Hudspeth AJ (1998) A model for amplification of hair-bundle motion by cyclical binding of calcium to mechano-electrical transducer channels. Proc Natl Acad Sci USA 95: 15321–15326

Crawford AC (1991) The actions of calcium on the mechano-electrical transducer current of turtle hair cells. J Physiol 434: 369–398

Crawford AC, Fettiplace R (1985) The mechanical properties of ciliary bundles of turtle cochlear hair cells. J Physiol 364: 359–379

Crist JR, Fallon M, Bobbin RP (1993) Volume regulation in cochlear outer hair cells. Hearing Res 69: 194–198

Dallos P, Evans BN (1995a) High-freqeuency motility of outer hair cells and the cochlear amplifier. Science 267: 2006–2009

Dallos P, Evans BN (1995b) High-frequency outer hair cell motility: Corrections and addendum. Science 1420–1421

Dallos P, Evans BN, Hallworth R (1991) Nature of the motor element in electrokinetic shape changes of cochlear outer hair cells. Nature 350: 155–157

Dallos P, He DZZ, Lin X, Sziklai I, Mehta S, Evans BN (1997) Acetylcholine, outer hair cell electromotility, and the cochlear amplifier. J Neurosci 17: 2212–2226

Dieler R, Shehata-Dieler WE, Brownell WE (1991) Concomitant salicylate-induced alterations of outer hair cell subsurface cisternae and electromotility. J Neurocytol 20: 637–653

Dong XX, Ospeck M, Iwasa KH (2002) Piezoelectric reciprocal relationship of the membrane motor in the cochlear outer hair cell. Biophys J 82: 1254–1259

Drenckhahn D, Schafer T, Prinz M (1985) Actin, myosin, and associated proteins in the vertebrate auditory and vestibular organs: Immunocytochemical and biochemical studies. In: Drescher DG (ed) Auditory Biochemistry. Charles C. Thomas, Springfield, IL, pp 317–335

Dulon D, Aran JM, Schacht J (1988) Potassium-depolarization induces motility in isolated outer hair cells by an osmotic mechanism. Hearing Res 32: 123–130

Eatock RA (2000) Adaptation in hair cells. Ann Rev Neurosci 23: 285–314

Engel A, Gaub HE, Muller DJ (1999) Atomic force microscopy: a forceful way with single molecules. Curr Biol 9: R133–R136

Evans BN, Dallos P (1993) Stereocilia displacement induced somatic motility of cochlear outer hair cells. Proc Natl Acad Sci USA 90: 8347–8351

Evans BN, Hallworth R, Dallos P (1991) Outer hair cell electromotility: The sensitivity and vulnerability of the DC component. Hearing Res 52: 288–304

Evans EA (1989) Structure and deformation properties of red blood cells: concepts and quantitative methods. Method Enzymo 173: 3–35

Evans EA, Skalak R, Hochmuth RM (1979) Mechanics and thermodynamics of biomembranes: Part 1. Critical Reviews in Bioengineering 3: 180–330

Fettiplace R, Fuchs PA (1999) Mechanisms of hair cell tuning. Ann Rev Physiol 61: 809–834

Fettiplace R, Ricci AJ, Hackney CM (2001) Clues to the cochlear amplifier from the turtle ear. Trends Neurosci 24: 169–175

Flock A, Flock B, Ulfendahl M (1986) Mechanisms of movement in outer hair cells and a possible structural basis. Arch Otorhinolaryngol 243: 83–90

Forge A (1991) Structural features of the lateral walls in mammalian cochlear outer hair cells. Cell Tissue Res 265: 473–483

Frank G, Hemmert W, Gummer AW (1999) Limiting dynamics of high-frequency electromechanical transduction of outer hair cells. Proc Natl Acad Sci USA 96: 4420–4425

Frolenkov GI, Atzori M, Kalinec F, Mammano F, Kachar B (1998) The membrane-based mechanism of cell motility in cochlear outer hair cells. Molecular Biology of the Cell 9: 1961–1968

Fung YC (1993) Biomechanics: Mechanical Properties of Living Tissues, 2nd ed., Springer, Berlin Heidelberg New York

Gale JE, Ashmore JF (1994) Charge displacement induced by rapid stretch in the basolateral membrane of the guinea pig outer hair cell. Proc R Soc Lond B 255: 243–249

Gale JE, Ashmore JF (1997) An intrinsic frequency limit to the cochlear amplifier. Nature 389: 63–66

Garcia-Anoveros J (1998) Localization of myosin-1ß near both ends of tip links in frog saccular hair cells. J Neurosci 18: 8637–8647

Geleoc GSG, Casalotti SO, Forge A, Ashmore JF (1999) A sugar transporter as a candidate for the outer hair cell motor. Nat Neurosci 2: 713–719

Gillespie PG, Corey DP (1997) Myosin and adaptation by hair cells. Neuron 19: 955–958

Gillespie PG, Gillespie SK, Mercer JA, Shah K, Shokat KJ (1999) Engineering of the myosin-1ß nucleotide-binding pocket to create selective sensitivity to N(6)-modified ADP analogs. J Biol Chem 274: 31373–31381

Gitter AH, Rudert M, Zenner HP (1993) Forces involved in length changes of cochlear outer hair cells. Pflügers Archiv – Europ J Physiol 424: 9–14

Gold T (1948) Hearing II. The physical basis of the action of the cochlea. Proc R Soc Lond B 135: 492–498

Gulley RL, Reese TS (1977) Regional specialization of the hair cell plasmalemma in the organ of corti. Anat Rec 189: 109–124

Hackney CM (2000) The composition of linkages between stereocilia. In: Recent Developments in Auditory Mechanics. World Scientific, pp 302–306

Hallworth R (1995) Passive compliance and active force generation in the guinea pig outer hair cell. J Neurophysiol 74: 2319–2328

Hallworth R (1997) Modulation of outer hair cell compliance and force by agents that affect hearing. Hearing Res 114: 204–212

Harada N, Ernst A, Zenner HP (1994) Intracellular calcium changes by hyposmotic activation of cochlear outer hair cells in the guinea pig. Acta Otolaryngol 114: 510–515

He DZZ, Dallos P (1999) Somatic stiffness of cochlear outer hair cells is voltage-dependent. Proc Natl Acad Sci USA 96: 8223–8228

Helmholtz HLF (1954) On the sensations of tone as a physiological basis for the theory of music. Dover Publications, Inc., New York

Holley M (1991) High frequency force generation in outer hair cells from the mammalian ear. Bioessays 13: 115–120

Holley MC, Ashmore JF (1988a) A cytoskeletal spring in cochlear outer hair cells. Nature 335: 635–637

Holley MC, Ashmore JF (1988b) On the mechanism of a high-frequency force generator in outer hair cells isolated from the guinea pig cochlea. Proc R Soc Lond B 232: 413–429

Holley MC, Ashmore JF (1990) Spectrin, actin and the structure of the cortical lattice in mammalian cochlear outer hair cells. J Cell Sci 96: 283–291

Holley MC, Kalinec F, Kachar B (1992) Structure of the cortical cytoskeleton in mammalian outer hair cells. J Cell Sci 102: 569–580

Howard J, Hudspeth AJ (1987) Mechanical relaxation of the hair bundle mediates adaptation in mechanotransduction by the bullfrog's saccular hair cell. Proc Natl Acad Sci USA 84: 3064–3068

Howard J, Hudspeth AJ (1988) Compliance of the hair bundle associated with gating of mechanoelectrical transduction channels in the bullfrog's saccular hair cell. Neuron 1: 189–199

Huang G, Santos-Sacchi J (1993) Mapping the distribution of the outer hair cell motility voltage sensor by electrical amputation. Biophys J 65: 2228–2236

Huang G, Santos-Sacchi J (1994) Motility voltage sensor of the outer hair cell resides within the lateral plasma membrane. Proc Natl Acad Sci USA 91: 12268–12272

Hudspeth AJ (1997) Mechanical amplification of stimuli by hair cells. Curr Opin Neurobiol 7: 480–486

Iwasa K, Tasaki I, Gibbons RC (1980) Swelling of nerve fibers associated with action potentials. Science 210: 338–339

Iwasa KH (1993) Effect of stress on the membrane capacitance of the auditory outer hair cell. Biophys J 65: 492–498

Iwasa KH (1994) A membrane motor model for the fast motility of the outer hair cell. J Acoust Soc Am 96: 2216–2224

Iwasa KH (2001) A two-state piezoelectric model for outer hair cell motility. Biophys J 81: 2495–2506

Iwasa KH, Adachi M (1997) Force generation in the outer hair cell of the cochlea. Biophys J 73: 546–555

Iwasa KH, Chadwick RS (1992) Elasticity and active force generation of cochlear outer hair cells. J Acoust Soc Am 92: 3169–3173

Iwasa KH, Chadwick RS (1993) Factors influencing the length change of an auditory outer hair cell in a tight-fitting capillary [letter]. J Acoust Soc Am 94: 1156–1159

Iwasa KH, Kachar B (1989) Fast in vitro movement of outer hair cells in an external electric field: Effect of digitonin, a membrane permeabilizing agent. Hearing Research 40: 247–254

Jaramillo F, Hudspeth AJ (1991) Localization of the hair cell's transduction channels at the ahir bundle's top by iontophoretic application of a channel blocker. Neuron 7: 409–420

Kachar B, Brownell WE, Altschuler R, Fex J (1986) Electrokinetic shape changes of cochlear outer hair cells. Nature 322: 365–368

Kakehata S, Santos-Sacchi J (1995) Membrane tension directly shifts voltage dependence of outer hair cell motility and associated gating charge. Biophys J 68: 2190–2197

Kakehata S, Santos-Sacchi J (1996) Effects of salicylate and lanthanides on outer hair cell motility and associated gating charge. J Neurosci 16: 4881–4889

Kalinec et al. (1992) A membrane-based force generation mechanism in auditory sensory cells. Proc Natl Acad Sci USA 89: 8671–8675

Kalinec F, Kachar B (1993) Inhibition of outer hair cell electromotility by sulfhydryl specific reagents. Neurosci Lett 157: 231–234

Kalinec F, Kachar B (1995) Structure of the electromechanical transduction mechanism in mammalian outer hair cells. In: Flock A, Ottoson D, Ulfendahl M (eds) Active Hearing. Elsevier Science Ltd, pp 181–193

Kemp DT (1978) Stimulated acoustic emissions from within the human auditory system. J Acoust Soc Am 64: 1386–1391

Kemp DT (1979) Evidence of mechanical nonlinearity and frequency selective wave amplification in the cochlea. Arch Otorhinolaryngol 224: 37–45

Koppl C (1995) Otoacoustic emissions as an indicator for active cochlear mechanics: a primitive property of vertebrate auditory organs. In: Manley GA (ed) Advances in Hearing Research. World Science Publishers, pp 207–216

Li Z, Anvari B, Takashima M, Brecht P, Torres JH, Brownell WE (2002) Membrane tether formation from outer hair cells with optical tweezers. Biophys J 82: 1386–1395

Ludwig J, Oliver D, Frank G, Klocker N, Gummer AW, Fakler B (2001) Reciprocal electromechanical properties of rat prestin: The motor molecule from rat outer hair cells. Proc Natl Acad Sci USA 98: 4178–4183

Lue AJ, Brownell WE (1999) Salicylate induced changes in outer hair cell lateral wall stiffness. Hearing Res 135: 163–168

Lue AJ, Zhao HB, Brownell WE (2001) Chlorpromazine alters outer hair cell electromotility. Otolaryngology-Head and Neck Surgery 125: 71–76

Manley GA (2000) Cochlear mechanisms from a phylogenetic viewpoint. Proc Natl Acad Sci USA 97: 11736–11743

Manley GA, Gallo L (1997) Otoacoustic emissions, hair cells, and myosin motors. J Acoust Soc Am 102: 1049–1055

Manley GA, Koppl C (1998) Phylogenetic development of the cochlea and its innervation. Curr Opin Neurobiol 8: 468–474

Martin P, Hudspeth AJ (1999) Active hair bundle movement can amplify a hair cell's response to oscillatory mechanical stimuli. Proc Natl Acad Sci USA 96: 14306–14311

Martin P, Mehta AD, Hudspeth AJ (2000) Negative hair-bundle stiffness betrays a mechanism for mechanical amplification by the hair cell. Proc Natl Acad Sci USA 97: 12026–12031

McFadden D, Plattsmier HS (1984) Aspirin abolishes spontaneous oto-acoustic emissions. J Acoust Soc Am 76: 443–448

Morimoto N, Raphael RM, Nygren A, Brownell WE (2002) Excess plasma membrane and effects of ionic amphipaths on the mechanics of the outer hair cell lateral wall. Am J Physiol – Cell Physiol 282: C1076–C1086

Mosbacher J, Langer M, Horber JK, Sachs F (1998) Voltage-dependent membrane displacements measured by atomic force microscopy. J Gen Physiol 111: 65–74

Mountain DC (1980) Changes in endolymphatic potential and crossed olivocochlear bundle stimulation alter cochlear mechanics. Science 210: 71–72

Mountain DC, Hubbard AE (1994) A piezoelectric model of outer hair cell function. J Acoust Soc Am 95: 350–354

Muller DJ, Heymann JB, Oesterhelt F, Moller C, Gaub H, Buldt G, Engel A (2000) Atomic force microscopy of native purple membrane. Biochimica et Biophysica Acta 1460: 27–38

Nguyen TVN, Brownell WE (1998) Contribution of membrane cholesterol to outer hair cell lateral wall stiffness. Otolaryngology-Head and Neck Surgery 119: 14–20

Nobili R, Mammano F, Ashmore J (1998) How well do we understand the cochlea?. Trends Neurosci 21: 159–167

Nuttall AL, Ren T (1995) Electromotile hearing: evidence from basilar membrane motion and otoacoustic emissions. Hearing Res 92: 170–177

Oghalai JS, Patel AA, Nakagawa T, Brownell WE (1998) Fluorescence-imaged microdeformation of the outer hair cell lateral wall. J Neurosci 18: 48–58

Oghalai JS, Zhao HB, Kutz JW, Brownell WE (2000) Voltage- and tension-dependent lipid mobility in the outer hair cell plasma membrane. Science 287: 658–661

Oliver D, He DZZ, Klocker N, Ludwig J, Schulte U, Waldegger S, Ruppersberg JP, Dallos P, Fakler B (2001) Intracellular anions as the voltage sensor of prestin, the outer hair cell motor protein. Science 292: 2340–2343

Petrov AG (1999) Lyotropic State of Matter: Molecular Physics and Living Matter Physics. Gordon & Breach Publishing Group

Pickles JO, Comis SD, Osborne MP (1984) Cross-links between stereocilia in the guinea pig organ of Corti, and their possible relation to sensory transduction. Hearing Res 15: 103–112

Pollice PA, Brownell WE (1993) Characterization of the outer hair cell's lateral wall membranes. Hearing Res 70: 187–196

Quaranta A, Poralatini P, Camporeale M, Sallustio V (1999) Effects of salicylates on evoked otoacoustic emissions and remote masking in humans. Audiology 38: 174–179

Raphael RM, Popel AS, Brownell WE (2000) A membrane bending model of outer hair cell electromotility. Biophys J 78: 2844–2862

Raphael Y, Athey BD, Wang Y, Lee MK, Altschuler RA (1994) F-actin, tubulin, and sprectrin in the organ of Corti: comparative distribution in different cell types and mammalian species. Hearing Res 76: 173–187

Ratnanather JT, Brownell WE, Popel AS (1993) Mechanical properties of the outer hair cell. In: Duifhuis H, Horst JW, van Dijk P, van Netten SM (eds) Biophysics of Hair Cells Sonsory Systems. World Sci, Singapore, pp 199–206

Ratnanather JT, Zhi M, Brownell WE (1996) The ratio of elastic moduli of cochlear and outer hair cells derived from osmotic experiments. J Acoust Soc Am 99: 1025–1028

Reyes S, Ding D, Sun W, Salvi R (2001) Effect of inner and outer hair cell lesions on electrically evoked otoacoustic emissions. Hearing Res 158: 139–150

Ricci AJ, Crawford AC, Fettiplace R (2000) Active hair bundle motion linked to fast transducer adaptation in auditory hair cells. J Neurosci 20: 7131–7142

Ricci AJ, Crawford AC, Fettiplace R (2002) Mechanisms of active hair bundle motion in auditory hair cells. J Neurosci 22: 44–52

Risberg A, Agelfors E, Lindstrom B, Bredberg G (1990) Electrophonic hearing and cochlear implants. Acta Otolaryngol S469: 156–163

Robles L, Ruggero MA (2001) Mechanics of the mammalian cochlea. Physiol Rev 81: 1305–1352

Ruggero MA (1992) Responses to sound of the basilar membrane of the mammalian cochlea. Curr Opin Neurobiol 2: 449–456

Ruggero MA, Rich NC (1991) Application of commercially-manufactured Doppler-shift laser velocimeter to the measurement of basilar-membrane. Hearing Res 51: 215–230

Ruggero MA, Rich NC, Recio A, Narayan SS, Robles L (1997) Basilar membrane responses to tones at the base of the chinchilla cochlea. J Acoust Soc Am 101: 2151–2163

Ruknudin A, Song MJ, Sachs F (1991) The ultrastructure of patch-clamped membranes: a study using high voltage electron microscopy. J Cell Biol 112: 125–134

Russell I, Schauz C (1995) Salicylate ototoxicity: effects on stiffness and electromotility of outer hair cells isolated from the guinea pig cochlea. Audit Neurosci 1: 309–320

Saito K (1983) Fine structure of the sensory epithelium of guinea-pig organ of Corti: Subsurface cisternae and lamellar bodies in the outer hair cells. Cell Tissue Res 229: 467–481

Santos-Sacchi J (1989) Asymmetry in voltage-dependent movements of isolated outer hair cells from the organ of Corti. J Neurosci 9: 2954–2962

Santos-Sacchi J (1991) Reversible inhibition of voltage-dependent outer hair cell motility and capacitance. J Neurosci 11: 3096–3110

Santos-Sacchi J (1992) On the frequency limit and phase of outer hair cell motility: effects of a membrane filter. J Neurosci 12: 1906–1916

Santos-Sacchi J, Dilger JP (1988) Whole cell currents and mechanical responses of isolated outer hair cells. Hearing Res 35: 143–150

Santos-Sacchi J, Shen W, Zheng J, Dallos P (2001) Effects of membrane potential and tension on prestin, the outer hair cell lateral membrane motor protein. [see comments]. J Physiol 531: 661–666

Shehata WE, Brownell WE, Dieler R (1991) Effects of salicylate on shape, electromotility and membrane characteristics of isolated outer hair cells from guinea pig cochlea. Acta Otolaryngol 111: 707–718

Siegel JH, Brownell WE (1986) Synaptic and golgi membrane recycling in cochlear hair cells. J Neurocytol 15: 311–328

Siegel JH, Kim DO (1982) Efferent neural control of cochlear mechanics? Olivocochlear bundle stimulation affects cochlear biomechanical nonlinearity. Hearing Res 6: 171–182

Sit PS, Spector AA, Lue AJ, Popel AS, Brownell WE (1997) Micropipette aspiration on the outer hair cell lateral wall. Biophys J 72: 2812–2819

Smith CA (1968) Ultrastructure of the organ of Corti. Adv Sci 122: 419–433

Tolomeo JA, Steele CR, Holley MC (1996) Mechanical properties of the lateral cortex of mammalian auditory outer hair cells. Biophys J 71: 421–429

Tunstall MJ, Gale JE, Ashmore JF (1995) Action of salicylate on membrane capacitance of outer hair cells from the guinea-pig cochlea. J Physiol 485: 739–752

Ulfendahl M, Chan E (1998) Axial and transverse stiffness measures of cochlear outer hair cells suggest a common mechanical basis. Pflügers Archiv – Europ J Physiol 436: 9–15

Ulfendahl M, Slepecky N (1988) Ultrastructural correlates of inner ear sensory cell shortening. J Submicrosc Cytol Pathol 20: 47–51

van Netten SM, Kros CJ (2000) Gating energies and forces of the mammalian hair cell transducer channel and related hair bundle mechanics. Proc R Soc Lond B 267: 1915–1923

von Békésy G (1960) Experiments in Hearing. McGraw Hill, New York

Wilson JP (1980) Evidence for a cochlear origin for acoustic re-emissions, threshold fine structure and tonal tinnitus. Hearing Res 2: 233–252

Wu M, Santos-Sacchi J (1998) Effects of lipophilic ions on outer hair cell membrane capacitance and motility. J Membrane Biol 166: 111–118

Wu YC (1999) Two components of transducer adaptation in auditory hair cells. J Neurophysiol 82: 2171–2181

Yates GK, Kirk D (1998) Cochlear electrically evoked emissions modulated by mechanical transduction channels. J Neurosci 18: 1996–2003

Zajic G, Forge A, Schacht J (1993) Membrane stains as an objective means to distinguish isolated inner and outer hair cells. Hearing Res 66: 53–57

Zelenskaya A, Snyder KV, Zhang PC, Sachs F (2002) Nanometer scale movement of prestin-transfected hek-293 cells. Biophys J 82: 275

Zenner HP (1986) Motile responses in outer hair cells. Hearing Res 22: 83–90

Zenner HP, Gitter AH, Rudert M, Ernst A (1992) Stiffness, compliance, elasticity and force generation of outer hair cells. Acta Otolaryngol 112: 248–253

Zenner HP, Zimmermann U, Schmitt U, Fromter E (1985) Reversible contraction of isolated mammalian cochlear hair cells. Hearing Res 18: 127–133

Zhang PC, Keleshian AM, Sachs F (2001) Voltage induced membrane movements. Nature 413: 428–432

Zhao HB, Santos-Sacchi J (1999) Auditory collusion and a coupled couple of outer hair cells. Nature 399: 359–362

Zheng J, Shen W, He DZZ, Long KB, Madison LD, Dallos P (2000) Prestin is the motor protein of cochlear outer hair cells. Nature 405: 149–154

7. The Silicon Cochlea

Rahul Sarpeshkar

Abstract

The debate about whether we hear through a mechanism involving a bank of bandpass filters or through a traveling-wave mechanism was settled by von Bekesy in favor of traveling waves. Experience gained in the engineering design and analysis of a silicon cochlea suggests why we may have evolved a traveling-wave mechanism for hearing: Distributed traveling-wave amplification is a vastly more efficient way of constructing a wide-dynamic-range frequency analyzer than is a bank of bandpass filters. Traveling-wave mechanisms are, however, more susceptible to parameter variations and noise. Collective gain control and an exponentially tapering set of filters can solve both of these problems; the biological cochlea implements both of these solutions. These engineering studies suggest that the biological cochlea, which is capable of sensing sounds over 12 orders of magnitude in intensity while dissipating only a few microwatts, is an extremely well-designed sensing instrument. We illustrate the engineering principles in the cochlea by demonstrating a 117-stage adaptive silicon cochlea that operates over six orders of magnitude in intensity over a 100 Hz–10 kHz frequency range while only consuming 0.5 mW of power. This
artificial cochlea with automatic gain control and a low-noise traveling-wave amplifier architecture has the widest dynamic range of any artificial cochlea built to date.

I. Introduction

The biological cochlea or inner ear performs two important functions: Firstly, it decomposes sounds via a frequency-to-place transformation much like a constant-Q wavelet-like filter bank does; however, the transformation is accomplished via a traveling-wave mechanism and not via a bank of bandpass filters. Secondly, it compresses an extremely large input dynamic range of 120 dB[1] into a narrow output dynamic range of 30–40 dB in auditory nerve firing rates. It accomplishes both these tasks with an impressively low estimated power dissipation of 14 μW (Johnstone 1967; Sarpeshkar et al. 1998). Our ears are capable of performing at least 100 MFLOPs of computation[2] and operating on a pair of AA batteries for about 15 years. In addition, at their most sensitive frequency, our ears are capable

---

[1] This dynamic range is attained at the cochlea's most sensitive frequency of approximately 3 kHz.
[2] Sarpeshkar et al. (1998) provide an estimate of the computational capacity of a silicon cochlea which is far inferior to that of the biological cochlea; thus 100 MFLOPs may be viewed as a lower bound on the computational capacity of the cochlea.

of sensing sub Ångström displacements of the ear drum; they report information with enough fidelity such that the auditory system can make a sound-location discrimination that corresponds to an interaural time difference of a few microseconds.

A silicon cochlea models a significant fraction of the biophysics of the biological cochlea in an analog Very Large Scale Integration (VLSI) chip. A silicon cochlea has several applications. For example, it may be used as a front end for speech recognition in noisy environments, as a low-power front end for bionic ears (cochlear implants) in the deaf, and as a wide-dynamic-range frequency analyzer in applications like sonar target recognition or in diagnostic fault analysis of vibrating machinery. In this chapter, we shall not focus much on the applications of the silicon cochlea. Rather, we shall emphasize how biology has inspired a good engineering architecture for building a low power wide-dynamic-range frequency analyzer. In turn, an analysis of the engineering architecture yields insight into why the biological cochlea may have evolved to being a distributed traveling wave amplifier rather than a bank of bandpass filters.

II. Low-Power Wide-Dynamic-Range Systems

The dynamic range of operation of a system is measured by the ratio of the intensities of the largest and smallest inputs to the system. Typically, the dynamic range is quoted in the logarithmic units of decibel (dB), with 10 dB corresponding to 1 order of magnitude. The largest input that a system can handle is limited by nonlinearities that cause appreciable distortion or failure at the output(s). The smallest input that a system can handle is limited by the system's input-referred noise floor.

At the *same* given bandwidth of operation, a low-current system typically has a higher noise floor than does a high-current system: The low-current system averages over fewer electrons per unit time than does the high-current system, and, consequently, has higher levels of shot or thermal noise (Sarpeshkar et al. 1993). Thus, it is harder to attain a wide dynamic range in low-current systems than in high-current systems. A low-voltage system does not have as wide a dynamic range as a high-voltage system because of a reduction in the maximum voltage of operation.[3] Low-power systems have low-current or low-voltage levels; consequently, it is harder to attain a wide dynamic range in low-power systems than in high-power systems. The biological cochlea is impressive in its design because it attains an extremely wide dynamic range of 120 dB (at 3 kHz), although its power dissipation is only about 14 μW; *in engineering, such low-power wide-dynamic-range systems are extremely hard to build.* The biological cochlea has a wide dynamic range because it has an adaptive traveling-wave amplifier architecture, and also because it uses a low-noise electromechanical technology.

We have built a 117-stage adaptive silicon cochlea that operates over six orders of magnitude in intensity (61 dB dynamic range) and over a 100 Hz–10 kHz frequency range while only consuming 0.5 mW of power. This artificial cochlea with automatic gain control and a low-noise traveling-wave amplifier architecture has the widest dynamic range of any artificial cochlea built to date. Prior

---

[3] We are assuming that the supply voltage limits the range of operation of the system. If there is some other voltage that limits the range of operation of the system, then power is wasted through an unnecessarily high supply voltage. We choose not to operate the system in this nonoptimal situation.

designs do not usually cite their dynamic range, but from our own experience with designs that do not pay any attention to dynamic range, we know that the dynamic range for such designs is typically 35 dB. Our cochlea is described in great detail in Sarpeshkar et al. (1998). In this chapter, we will only present an overview of this work.

III. Theoretical Ideas

Figure 1 and its associated caption explain why a traveling-wave medium may be approximated by a cascade of filters (Lyon 1998). The electronic cochlea models the traveling-wave amplifier architecture of the biological cochlea as a cascade of nonlinear-and-adaptive second-order filters with corner frequencies that decrease exponentially from approximately 10 kHz to 100 Hz (Mead 1989). The exponential taper is important in creating a

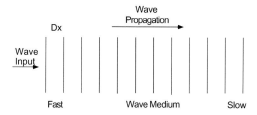

Fig. 1. The Wentzel-Kramer-Brillouin (WKB) method of modeling traveling wave solutions. Each block of the traveling-wave medium causes a frequency-dependent phase shift and an amplification or attenuation of the wave. Thus each block may be modeled by a filter with the requisite gain and phase characteristics. Since the gain or phase changes may depend on the input amplitude of the wave, the filter is in general a nonlinear adaptive filter. The effect of the entire traveling-wave medium on the wave may be computed by multiplying the gains from each block, and by adding the phase shifts from each block. The entire traveling-wave medium is thus modeled by a cascade of nonlinear adaptive filters. For simplicity, we do not consider backward-propagating waves

cochlea that is roughly scale invariant at any time scale; it is easily implemented in subthreshold Complementary Metal Oxide Semiconductor (CMOS), or in bipolar technology. Figure 2 and its caption illustrate these ideas and outline the basic theoretical facts behind the silicon cochlea.

One of the key and underappreciated ideas that emerges from this theory is that a traveling-wave mechanism is well suited for doing distributed gain control and dynamic-range compression: Due to the successive compounding of gains, a change in the individual filter gains of a few percent can alter the gain of the composite transfer function by many orders of magnitude. For example, $(1.1)^{45} = 73$ while $(0.9)^{45} = 0.009$. It is almost impossible to accomplish such wide-dynamic-range gain control with one localized amplifier without changing the amplifier's bandwidth, temporal resolution, and power dissipation drastically as well. However, in distributed gain-control systems like the cochlea, each individual filter does not change its gain appreciably although the collective system does change its gain appreciably. Thus, the system can maintain its bandwidth, temporal resolution, and power dissipation to be relatively invariant with amplitude.

IV. The Single Cochlear Stage

Figure 3 shows a schematic for a single cochlear stage. The arrows indicate the direction of information flow (input to output). The second-order filter (SOS) is composed of two Wide-Linear-Range amplifiers (WLRs), two capacitors, and offset-compensation circuitry (LPF and OCR). The topology of the SOS enables it to operate with lower noise than previously described topologies (Mead 1989). The corner frequency $1/\tau$ and quality factor Q of the filter are proportional to sqrt($I_1 I_2$) and

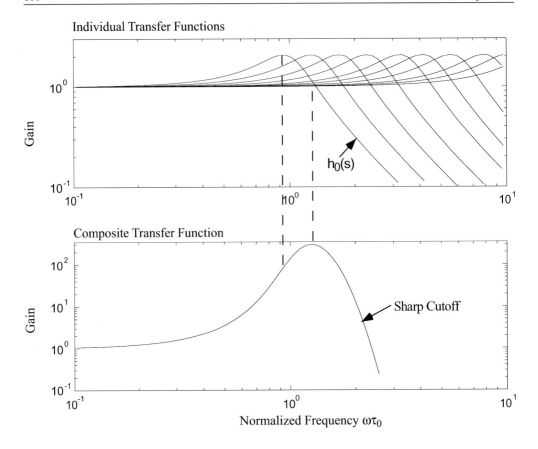

Individual Transfer Functions

$$h_n(s) = 1/(1 + \tau_n s/Q + \tau_n^2 s^2)$$

$$\tau_n = \tau_0 \exp(-n/N_{nat})$$

Composite Transfer Function

$$H_0(s) = \prod_{n=0}^{n=\infty} h_n(s)$$

Fig. 2. Basic theory behind the silicon cochlea. The exponentially tapering time constants of the individual cochlear filters give rise to a convergent cochlear filter shape after 3–4 octaves of the cochlea have been traversed. Thus, successive cochlear outputs have similar composite transfer functions but at different peak frequencies. The composite gain is high because of the successive compounding of small gains from the individual filters, and, similarly, the composite rolloff is very steep because of the successive compounding of attenuations from the individual filters. At any point in the cochlea, the peak frequency of the composite transfer function is half an octave greater than the peak frequency of the last individual filter in the chain

sqrt(I_1/I_2), respectively, where I_1 and I_2 are the bias currents of the WLR amplifiers. The tau-and-Q control circuit controls the values of the currents I_1 and I_2 such that the value of $1/\tau$ depends only on the bias voltage V_T, and the small-signal value of Q depends only on the bias voltage V_Q. An Automatic Gain Control

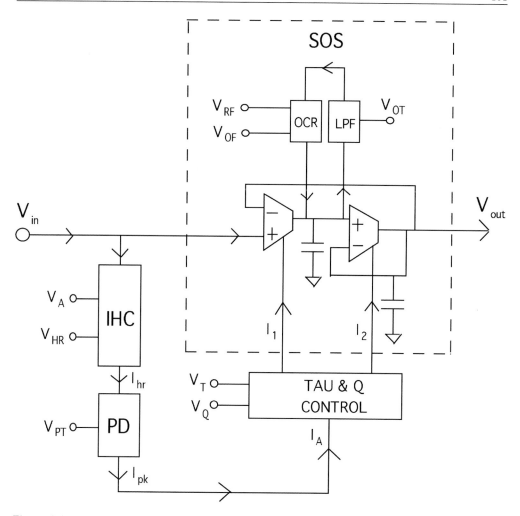

Fig. 3. Schematic for a cochlear stage. A single cochlear stage is composed of a filter (second order system) with offset-adaptation circuitry (Lowpass filter and offset correction circuit), an inner-hair-cell and peak-detector circuit (IHC and PD), and a tau-and-Q control circuit

(AGC) correction current I_A attenuates the small-signal value of Q at large-signal levels in a graded fashion.

The inner-hair-cell circuit (IHC) rectifies, differentiates, and transduces the input voltage to a current I_{hr}. The voltage V_A controls the value of an internal amplifier bias current in the IHC. The voltage V_{hr} controls the transduction gain of the IHC. The peak detector (PD) extracts the peak value of I_{hr} as a DC current I_{pk}. The current I_{pk} becomes the AGC correction-current input (I_A) to the tau-and-Q control circuit. The bias voltage V_{PT} determines the time constant of the peak detector, and thus the response time of the automatic gain control. The peak detector is designed such that it can respond to increases in input intensity within one cycle of a sinusoidal input at V_{in}; its response to decreases in input intensity is much slower and is determined by V_{PT}.

The offset-compensation circuit is composed of a lowpass filter (LPF) whose time constant is determined by V_{OT}. The lowpass filter extracts the DC voltage of the filter's intermediate node, and compares this voltage with a global reference voltage V_{RF} in the offset-correction block (OCR). The offset-correction block applies a correction current to the intermediate node to restore that node's voltage to a value near V_{RF}. The DC voltage of the output node is then also near V_{RF}, because the systematic offset voltage of a wide-linear-range amplifier is a small negative voltage. The maximal correction current of the offset-correction block scales with the bias current I_1; the bias voltage V_{OF} controls the scaling ratio. Since the restoration is performed at every cochlear stage, the output voltage of each stage is near V_{RF}, and offset does not accumulate across the cochlea. If there were no offset adaptation, a systematic offset voltage in any one stage would accumulate across the whole cochlea.

Since the gain-control topology is feedforward, rather than feedback, we avoid instabilities or oscillations in the Q. However, since raising Qs lowers the DC voltage, and the DC voltage does have a mild influence on the Q, the DC and AC output-voltage dynamics are weakly dependent on each other.

The details of each of the circuits of Fig. 3, and their theoretical and experimental characterization have been described in Sarpeshkar et al. (1998). The wide-linear-range (WLR) circuit has been described in detail in Sarpeshkar et al. (1997). Sarpeshkar et al. (1998) discuss the details behind the construction of a cochlea with a dynamic range greater than 60 dB with 0.5 mW of power consumption with a total of 117 stages. For technical reasons, the cochlea of the latter reference was constructed out of three 39-

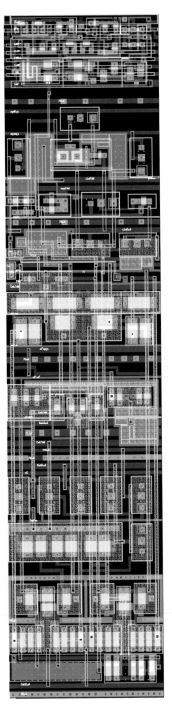

Fig. 4. VLSI layout of a single cochlear stage

stage cochleas in an architecture referred to as *overlapping cochlear cascades*. The detailed description of this architecture is beyond the scope of this chapter.

Figure 4 shows the VLSI (very large scale integration) layout of this cochlear stage in a 1.2 μm MOSIS process; it measures approximately 400 μm by 100 μm.

V. Experimental Results

Figure 5(a) shows the frequency response at different input amplitudes at tap 30 in a cochlea with 11.2 filters per octave. The data were taken from a 2.2 mm × 2.2 mm analog VLSI chip built in a 1.2 μm MOSIS process. The adaptation in Q with increasing input amplitude is evident. Figure 5(b) shows data from a biological chinchilla cochlea provided through the courtesy of Mario Ruggero (Ruggero 1992). We see that both in electronics and biology, the output is approximately linear at frequencies lower than the best frequency (BF), is compressive at the BF, and is even more compressive at frequencies higher than the BF. Such compression effects arise because of the accumulated effects of gain adaptation over several cochlear stages. Figure 6 illustrates how, in both silicon and biology, active amplification and gain control mechanisms extend the dynamic range of the cochlea.

VI. Traveling Waves versus Bandpass Filters: What Engineering May Reveal about Biology

The debate about whether we hear through a mechanism involving a bank of bandpass filters or through a traveling-wave mechanism was settled by von Békésy in favor of traveling waves. Experience gained in the engineering design and analysis of a silicon cochlea suggests why we may have evolved a traveling-wave mechanism for hearing: Distributed traveling-wave amplification is a vastly more efficient way of constructing a wide-dynamic-range frequency analyzer than is a bank of bandpass filters. Traveling-wave mechanisms are, however, more susceptible to parameter variations and noise. Collective gain control and an exponentially tapering set of filters can solve both of these problems; the biological cochlea implements both of these solutions. Our engineering studies suggest that the biological cochlea, which is capable of sensing sounds over 12 orders of magnitude in intensity while dissipating only a few microwatts, is an extremely well-designed sensing instrument.

Why did nature choose a traveling-wave architecture that is well modeled by a filter cascade instead of a bank of bandpass filters? We suggest that nature chose wisely, for the following three reasons:

1. To adapt to input intensities over a 120 dB dynamic range, a filter bank would require a tremendous change in the Q of each filter. To compress 120 dB in input intensity to about 40 dB in output intensity the filter Q's must change by 80 dB; a dynamic-range problem in the input is merely transformed into a dynamic-range problem in a parameter. In contrast, in a filter cascade, due to the exponential nature of gain accumulation, enormous changes in the overall gain for an input can be accomplished by small distributed changes in the Q of several filters.

2. Large changes in the Q of a filter are accompanied by large changes in the filter's window of temporal integration. Thus, in filter banks, faint inputs would be sensed with poor temporal resolution, and loud inputs would be sensed with good temporal resolution. In contrast, in a filter cascade, the

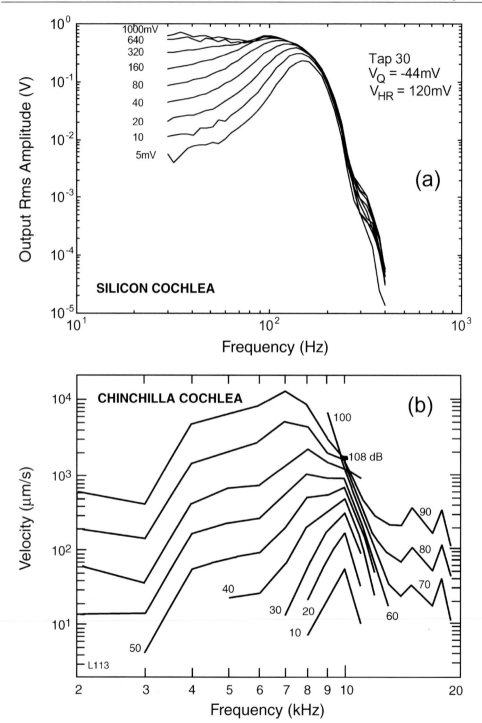

Fig. 5. Frequency-response curves of the cochlea: (a) the silicon cochlea; (b) the biological cochlea

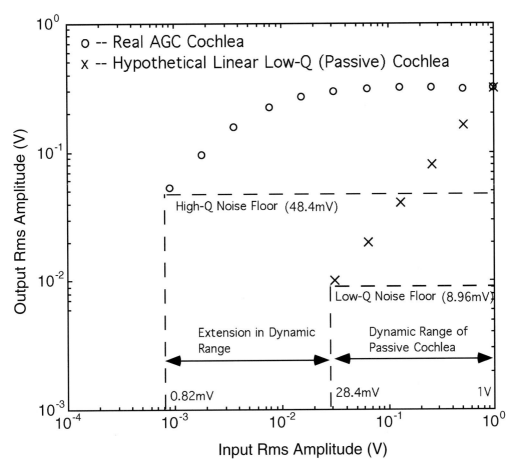

Fig. 6. Extension of dynamic range at the best frequency of a cochlear tap. A hypothetical low-Q cochlea that is completely linear would have a dynamic range of only 30 dB due to the uniformly low gain of 0.315 at all mplitudes; such a cochlea is analogous to the passive biological cochlea with no outer hair cells. Our automatic gain control cochlea has a dynamic range of 60 dB because faint signals at 0.82 mV are amplified by a factor of 59 to be just above the noise floor of 48.4 mV, whereas loud signals at 1V rms amplitude are attenuated by a factor of 0.315 to prevent distortion; such a cochlea is analogous to the active cochlea with outer hair cells.

shifts in temporal resolution with intensity change only in a logarithmic fashion with intensity, as opposed to in a linear fashion in the filter bank.

3. A sharp rolloff slope in a filter is extremely useful in limiting distortion, and in enhancing spectral contrasts. A sharp rolloff slope arises naturally in the cochlear filter cascade. To

accomplish such a rolloff slope in a filter bank requires very high-order filters, and consequently an enormous amount of circuitry at each tap. In contrast, in the filter cascade, the burden of creating a high-order rolloff is shared collectively in an exponentially-tapered cascade such that only one new filter needs to be added for each desired corner frequency. For example, to create 150 15th-order filters with corner frequencies spanning 3 decades of frequency (10 octaves) requires circuitry on the order of 15 (order of filter) $\times$ 150 filters = 2250 first-order filters in the bank-of-bandpass filters approach; in contrast, the traveling-wave approach needs approximately 15 filters/ octave $\times$ 14 octaves (4 octaves to account for edge effects at the beginning of the cochlea) = 210 first-order filters, a savings of more than one order of magnitude! In general, if we want to create an array of filters with corner frequencies spanning N octaves, with m filters/octave with each filter having an order of m, the bank-of-bandpass filters approach needs Nm^2 first-order filters; in contrast, the traveling-wave approach needs approximately $(N + 4)m$ first-order filters.[4] The improvement from an $O(m^2)$ to $O(m)$ algorithm arises because the traveling-wave architecture exploits increasing filter density (number of filters per octave) to achieve high-order filtering in an exponentially-tapered cascade for free; the bank-of-bandpass simply repeats the high-order filter computations at each corner frequency.

There are two problems that need to be addressed in a filter cascade:

1. A filter cascade is prone to noise accumulation and amplification. The solution to this problem is either to have an exponential taper in the filter time constants such that the output noise converges (the solution found at high corner frequencies in the biological cochlea and documented in the silicon cochlea as well; Sarpeshkar et al. 1998), or to limit the length of the cascade (the solution at low corner frequencies in the biological cochlea and in the overlapping-cascades architecture; Sarpeshkar et al. 1998). The exponential taper also results in elegant scale-invariant properties.

2. The overall gain is quite sensitive to the value of each filter's Q. The solution to this problem is to have gain control regulate the value of the Qs in the cascade. If the gain control is sufficiently strong, then the collective adaptation in Q across many filters will compress a wide input dynamic range into a narrow output dynamic range.

VII. Summary

We described a 117-stage 100 Hz-to-10 kHz silicon cochlea that attained a dynamic range of 61 dB while dissipating 0.5 mW of power. This cochlea has the widest dynamic range of any artificial cochlea built to date. The analysis of our electronic cochlea suggests that nature preferred a traveling-wave mechanism over a bank of bandpass filters to decompose sounds because a traveling-wave mechanism is an efficient collective method of implementing a large number of high-order filters with corner frequencies over a wide frequency range. A traveling-wave mechanism is also a natural method for performing distributed gain control to create a wide-dynamic-range frequency analyzer.

References

Johnstone BM (1967) Genesis of the cochlear endolymphatic potential. Curr Top Bioenerg 2: 335–352

Lyon RF (1998) Filter cascades as analogs of the cochlea. In: Lande TS (ed) Neuromorphic Systems Engineering. Kluwer Academic Publishers, The Netherlands, pp 3–18

---

[4] For simplicity, we have assumed that an octave (factor of 2) and an e-fold (factor of 2.7) are approximately the same; such an assumption merely changes a few constants; an exact discussion would obscure our main point while adding little in the way of understanding.

Mead CA (1989) Analog VLSI and Neural Systems, pp 179–192, 279–302, Addison-Wesley Publishing Co., Reading, MA

Ruggero MA (1992) Response to sound of the basilar membrane of the mammalian cochlea. Curr Opin Neurobiol 2: 449–456

Sarpeshkar R, Delbruck T, Mead CA (1993) White noise in MOS transistors and resistors. IEEE Circuits Device 9(6): 23–29

Sarpeshkar R, Lyon RF, Mead CA (1997) A low-power wide-linear-range transconductance amplifier. Analog Integr Circ S 13(1/2): 123–151

Sarpeshkar R, Lyon RF, Mead CA (1998) A low-power wide-dynamic-range analog VLSI cochlea. Analog Integ Circ S 16(3): 245–274

8. Biologically-Inspired Microfabricated Force and Position Mechano-Sensors

Paolo Dario, Cecilia Laschi, Silvestro Micera, Fabrizio Vecchi,
Massimiliano Zecca, Arianna Menciassi, Barbara Mazzolai,
and Maria C. Carrozza

Abstract

The aim of this paper is to discuss an ideal design procedure for biologically-inspired mechano-sensors. The main steps of this procedure are the following: (1) analysis of force and position sensors in humans; (2) analysis of technologies available for MEMS (Micro Electro Mechanical Systems) and (3) design and implementation of biologically-inspired sensors in innovative mechatronic and biomechatronic systems (e.g., anthropomorphic robots, prostheses, and neuroprostheses).

According to this sequence, the first part of the paper is dedicated to the presentation of some features of force and motion sensors in humans. Then, technologies for fabricating miniaturized force and motion sensors (and some examples of such sensors) are briefly presented. Finally, some applications of biologically-inspired systems developed by the authors to sense force and position in anthropomorphic robots and in prosthetics are illustrated and discussed.

I. Introduction

The main reason which determined the failure of early approaches to machine intelligence probably was the assumption that intelligence is essentially abstract thinking, and that therefore human intelligence could be reproduced by developing powerful computers capable of high speed calculations and logic (sequential) reasoning (Brooks and Stein 1994). As the animal world vividly demonstrates, the reality is that intelligence evolves and is built primarily on the availability and processing of sensory information. Recent research on intelligent machines and systems assumes that sensors are key components for obtaining adaptive and intelligent behavior. This assumption holds not only for complex, sometimes human-like machines, such as the robot, but virtually for every modern ("mechatronic") product (e.g., camcorders, washing machines, car subsystems, etc.) (Fujimasa 1996, Dario and Fukuda 1998), to which embedded computers and many different sensors provide the capability to react to variable working conditions (Dario et al. 1996). In fact, whereas in the past the need for sensors was widely recognized, but their practical use in machines was difficult because of the lack of processing capabilities, today embedded computing makes sensors not only useful but usable, and better sensors strongly needed.

An interesting and important question is: which properties a "better" sensor should possess? Obviously a general answer is not possible, but it is certainly true that, along with many engineering properties very important for practical application, such as measuring range, sensitivity, linearity, accuracy, repeatability, robustness, and so forth, some properties inspired by the observation of biological sensors and their use by plants and animals might also be very useful. Recent developments of microfabrication technologies for MEMS (Micro Electro Mechanical Systems) (Fukuda and Menz 1998) and for "BioMEMS" (hybrid artificial-biological microstructures) (Ferrari 1999) make it possible to design and fabricate a range of new miniaturized sensors made out of different (inorganic and organic) materials and with sensing and processing capabilities previously impossible to obtain. This opens up new perspectives to research on bionic sensors, that is on sensors which aim either at imitating or at substituting biological sensors for application in robotics, prosthetics and, in general, in current and future

mechatronic products and biomechatronic systems (Dario et al. 1989).

In this paper we start from the analysis of some representative biological sensors, focusing on force and motion sensors in humans. Then, we describe microfabrication technologies for MEMS and Bio-MEMS, and examples of microfabricated force and motion sensors inspired by animal sensors. Finally, we discuss some applications of these sensors in anthropomorphic robotics, and in prosthetics, with reference to research projects carried on in the authors' laboratory.

II. Sensors of Force and Motion in Humans

In this section we outline the main features of typical sensors of touch and position in humans (Kandel et al. 1991), which represent a very sophisticated and thus inspiring model for mechano-reception in artificial machines. Human skin, in fact, is an active sensory organ that is both highly sensitive and resistant. The glabrous skin has about 17,000 tactile units composed of five major types of receptors: *Meissner corpuscles, Merkel's disks, Pacinian corpuscles, Ruffini endings*, and free receptors. These receptors are found at all depths below the skin surface and can be described as follows (Webster 1988):

– *Meissner's corpuscles* comprise 43% of the tactile units in the hand. Their average size for an adult is about 80 µm × 30 µm, with the long axis perpendicular to the surface of the skin. They are velocity detectors or touch receptors, since they move with the ridged skin of the fingers and the palm and provide the best reception of movement across the skin;
– *Pacinian corpuscles* are from 1 to 4 mm in length and from 0.5 to 1 mm in

Table 1. Classification of Human Cutaneous Mechanoreceptors According to Rate of Adaptation and Adequate Stimulus

	Adaptation to constant pressure stimuli		
	Slow	Moderately rapid	Very rapid
Hairless skin	Merkel's disk	Meissner corpuscle	Pacinian corpuscle
Hairy skin	Tactile disk, Ruffini ending	Hair-follicle receptor	Pacinian corpuscle
		Classification by adequate stimuli	
	Intensity detector	Velocity detector	Acceleration detector

diameter. There are about 2000 Pacinian corpuscles distributed all around the body, with one third of them localized in the digits. They are acceleration detectors and provide vibration reception. They cannot detect steady pressure, but they are responsible for the threshold detection of light touch;

– *Ruffini's endings* are composed of a fusiform structure with a definite capsule in the subcutaneous tissue of the pulp of the human finger. They represent 19% of the tactile units in the hand, and they are detectors of intensity and pressure. They are also responsible of detecting shear on the skin;

– *Merkel's disks* represent 25% of the tactile units in the hand. They are composed of a disk-like nerve ending and a specialized receptor cell, and they are about 70–90 nm in diameter. They provide excellent detection of intensity and tactile and vibration information;

– *Free receptor endings* are the most important cutaneous receptors. They permeate the entire thickness of the dermis. They have diameters ranging from 0.5 to 2.5 µm, and generally they form *thermoreceptors* and *nociceptors* (pain receptors).

Mechanoreceptors in the human skin detect pressure, touch, vibration, and tactile stimuli. They are divided into three main classes: *slowly adapting* (SA), *rapidly* or *fast adapting* (RA or FA) *and very rapidly adapting* (VRA). Each adaptation class can be further divided into two types, namely, *type I* and *type II*, according to their receptive field: small with sharp borders for the *type I*, large with obscure borders for the *type II*. Table 1 shows the classification of mechanoreceptors according to their rate of adaptation.

Other important mechanoreceptors of the human body are the *proprioreceptors*. They are sensory receptors that respond to stimuli arising within the body. Proprioreception provides information on the orientation of our limbs with respect to one another. More generally, proprioreception is the perception of the body's movement and its position in space, whether still or in motion. Proprioreceptors generate the sense of position, the sense of movement, and the sense of force. The first let us know the position of our limbs without visual feedback; the second enables us to perceive the speed of retraction as well as controlled extension of our limbs; the latter is the ability to know how much force to use to push, pull, or lift.

The cutaneous receptors previously discussed could also be used as proprioreceptors, since position and movement can be perceived from the deformation of the skin.

Joint receptors detect position, velocity, and acceleration occurring at the joint capsule. This is possible because whenever a joint is moved, the joint capsules are either compressed or stretched. Physiologically, the rate of change of impulse frequency yields the angular speed, and the magnitude of the impulse frequency yields the position of a joint.

Musculotendinous receptors are divided into tendon receptors and muscle receptors. Golgi tendon organs are located between the muscle and its tendon. Their function is to inhibit muscle contraction when the associated muscle is stretched. Golgi tendon organs are sensitive detectors of tension in distinct, localized regions of their host muscle. Probably they are important contributors to fine motor control. On the other hand, muscle spindles are located throughout the muscle between parallel individual muscle fibers. They detect the stretch of their adjacent muscle fibers. Their functional substructure provides constant monitoring and regulation of sensory-motor functions that enable appropriate body movements.

III. Technologies for the Microfabrication of Artificial Force and Motion Sensors

Technologies for biologically-inspired sensors aim essentially to merge sensing and processing capabilities in order to obtain a smart, adapting, intelligent device. Microsensors that measure physical parameters are common on the market, as they exploit mature, state of the art technologies. Technologies popular for microsensor fabrication are: (1) silicon bulk micromachining and (2) silicon surface micromachining (Madou 1997).

In *bulk micromachining*, mechanical devices are fabricated by selectively removing material from a silicon wafer which is used as substrate. Membranes, cavities, masses and bridges are the basic

mechanical structures etched (isotropically or anisotropically) by this technology. Most pressure and acceleration sensors rely on this technology that is usually compatible with microelectronic circuit fabrication.

Surface micromachining allows the fabrication of thin structures by the deposition and selective etching of thin layers of appropriate materials on a silicon wafer which serves mainly as support. Selectively etched layers are called sacrificial layers; they permit the fabrication of freestanding membranes, beams or mobile parts separated from the substrate by very thin air gaps. Surface micromachining is compatible, in some cases, with IC (integrated circuit) and even with bulk micromachining. Therefore, the main advantage of silicon technologies is the possibility to combine sensing and processing capabilities in a reliable way and often at low cost. In fact, many industrial IC lines based on CMOS (Complementary Metal Oxide Semiconductor) technology can be used for, or can be converted into, lines for sensor fabrication.

Other techniques, derived from precision machining or from hybrid fabrication technologies, are also being developed for sensor applications, although they are not yet suitable for large scale production. For example LIGA (LIthography, Galvo und Abformung), micro-electro discharge machining and micromolding techniques are used to fabricate high aspect-ratio microstructures made out of polymers, ceramics, metals, etc. Machining materials different from silicon is an attractive opportunity and opens new scenarios for microsensors. However, the main obstacle to this is that hybrid and precision machining technologies are not compatible with IC fabrication, although some significant steps have been made towards this goal, e.g. the introduction of the so called SLIGA (Surface Micromachining and LIGA) technique (Muller et al. 1991).

Many types of sensors can be fabricated using *electrically conducting or semiconducting polymers* as sensing layers. Some conductive polymer composites have mechanical properties close to those of biological tissues. There are two types of conducting polymers: polymers that are intrinsically conducting or can be made so by doping; composites that contain an electrically insulating polymer matrix loaded with a conductive filler (Harsanyi 1995).

In general, by applying force or pressure to a conductive polymer composite, the filler concentration locally changes and the resistivity changes too. By exploiting this effect (and related side effects), force and pressure sensors can be fabricated, with the advantage of being easily integrated in biomimetic devices. In fact, conductive polymers and electroactive polymers can be used also as actuators for biological applications and substitutions (e.g. artificial muscles).

In particular, conducting polymers can be used as actuating systems by exploiting electro-mechano-chemical phenomena which occur when ionic species are exchanged with the surrounding medium. These systems are promising candidates for pseudo-muscular actuators because they exhibit large active stresses and low drive voltages.

Polymeric gels are biphasic systems composed of a solid phase or elastic matrix permeated by a fluid, generally water, and a number of different types have been developed and studied. These systems exhibit plastic contraction with changes in temperature, pH, magnetic or electrical field, and have a large number of applications. For example, soft actuators in the biomedical field or for controlled drug release. Gel actuators are characterised by large strains (50% or more) and lower forces than conducting polymers, and several prototype actuating systems have been realised for biomedical use.

Biological tissues are composed of a cellular component in a natural gel matrix, both of which are permeated with water. Bi-phasic materials which possess both rheological characteristics of both solids and fluids are common in nature. In this respect, biological tissues and hydrogels are very similar, and the two can be represented by analogous mathematical models (De Rossi and Ahluwalia 2000).

A very promising technology, called *Shape Deposition Manufacturing* (SDM), has been recently proposed. SDM is not a technique to fabricate specifically sensors and precision structures, but it is a "manufacturing philosophy" which allows to integrate sensors and actuators in many complex structures with a strong biomimetic approach (Bailey et al. 2000). The basic SDM cycle consists of alternate deposition and shaping (often, machining) of layers of part material and sacrificial support material (Stanford University). This cycle of material deposition and removal results in three key features: To build parts in incremental layers allows complete access to the internal geometry of any mechanisms; this access allows to embed actuators, sensors and other prefabricated functional components inside the structure; by changing the materials used in the deposition process, it is possible to spatially vary the material properties of the mechanism itself.

A completely different approach is attempting to engineer biological, cell-based tissues *in vitro* in order to restore, maintain, or improve tissue functions (including sensing function). Instead of fabricating artificial devices for sensing and actuating substitution, "*tissue engineering*" (Langer and Vacanti 1993) aims at growing tissues in a physiological environment. First of all, the development of functional tissues requires to fabricate a bioresorbable 3D scaffold with seeding cells in static culture. Then cells proliferate and differentiate in a dynamic environment (e.g., spinner). Growth of mature tissue happens in a physiological environment (bioreactor), until surgical transplantation. All these steps are quite critical but, from an engineering point of view, the fabrication of an effective scaffold is probably the most interesting problem. Photolithography is often used to pattern biomolecules on glass substrates which mediate cell adhesion. However, other scaffold materials are being investigated for tissue engineering, including ceramics, polyimides, polyphosphazenes and biological polymers, such as collagen.

Although necessarily condensed, this paragraph illustrates a trend in biomimetic sensor technology, that is the evolution from technologies suitable for the development of discrete sensors to technologies suitable for obtaining "composite" structures (including tissues) where different functions (structural sensing, actuation, signal processing) are integrated.

IV. Examples of Force and Motion Sensors

A. Force/Pressure/Tactile Sensors

Measuring force essentially means measuring the displacement or strain induced by force in an instrumented deformable structure (e.g., membrane, cantilever) (Fatikow and Rembold 1997).

Due to their simple construction and wide applicability, mechanical sensors play the most important part in the field of MEMS (Micro Electro Mechanical Systems) and MST (Micro System Technology). Pressure microsensors were the first to be developed and used by industry. Miniaturized pressure sensors must be inexpensive and have a high resolution, accuracy, linearity and stability. Pressure sensors are largely used also as force sensors: By considering the area where load is applied it is possible to shift between force and pressure measurements. A few examples which represent

well the state of the art in the field of microfabricated contact sensors are presented in the following Sections.

1. Piezoresistive Pressure Sensor

Pressure is most often measured via a thin membrane which deflects when pressure is applied. Either the deflection of the membrane or its change in resonance frequency is measured. Both these values are proportional to the pressure applied. These mechanical changes are transformed into electric signals. Pressure sensors usually employ capacitive or piezoresistive measuring principles.

Heuberger (1991) shows the design of a typical piezoresistive pressure sensor. The piezoresistors are integrated in the membrane, and change their resistance proportionally to the applied pressure. The resistance change indicates how far the membrane is deflected and the deflection is proportional to the pressure.

2. Resonance Sensor for Measuring Pressure

In both the sensing principles introduced above, the sensor signal is generated by a deflecting membrane or a displaced mass. It is also possible to get a signal from a change of resonance frequency of the membrane caused by pressure. The main advantage of this measurement principle is that the transmission of the measured value in form of a frequency is practically noiseless and the signals can be digitally processed. An example of pressure resonance sensor is shown in Tilmans and Bouwstra (1993).

The device consists of a silicon substrate, a diaphragm and three transducers equally spaced on the annular diaphragm. Each transducer consists of two resonators which oppose each other.

When a pressure is applied to the diaphragm, the deformation causes the resonant frequencies of the resonators 1 and 2 to increase or decrease, respectively. The frequency difference between the two resonators serves as the output signal of the sensor. The sensor has the following dimensions and performances: diaphragm diameter, 1.2 mm and thickness, 3 μm; resonator length, 100 μm and thickness, 0.5 μm; maximum diaphragm lift, 0.7 μm; pressure range, up to 1000 Pa; accuracy, 0.01 Pa and non-linearity, 0.1%.

3. Capacitive Pressure Sensors

Capacitive sensors make use of the change of capacitance between two metal plates. The membrane deflects when pressure is applied, which causes the distance between the two electrodes to be changed. The capacitance change is measured and the pressure value can be calculated from the amount of membrane deflection. Figure 1a shows a silicon-based capacitive pressure sensor with integrated CMOS components including sensor, transformer, amplifier and temperature compensator (Mehlhorn et al. 1992). The sensor chip size is 8.4 mm × 6.2 mm.

Another example is shown in figure 1b (Despont et al. 1992). The electrodes are made of a planar comb structure. Here, the applied force is exerted parallel to the sensor surface. The sensor element mainly consists of two parts: first, a movable elastic structure which transforms a force into a displacement, and second, a transformation unit consisting of the electrodes which transform the displacement into a measurable change of capacitance. Thanks to the separate measurement of the capacitance changes on both sides, high linearity and sensitivity are obtained. Compared to piezoresistive sensors, capacitive sensors have no hysteresis, better

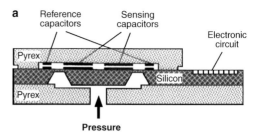

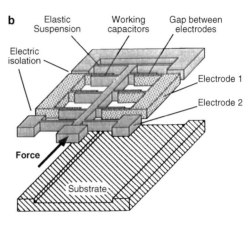

Fig. 1. Silicon-based capacitive pressure sensors, **a** Mehlhorn 1992; **b** Despont 1992

long-term stability and higher sensitivity. However, the advantages of capacitive pressure sensors go along with more complex signal processing and higher production costs.

4. Mach-Zehnder Interferometer

Many physical quantities can be measured by optical sensors, making use of the change of light which is sent through fiber optical cables. A Mach-Zehnder interferometer can be used as pressure sensor (Fischer 1991).

Laser light is brought into the device by a fiber optical cable. The light is split and channeled via two waveguides to a photodiode. One of the light branches crosses a microstructured membrane which can be exposed and serves as a reference signal.

When the sensor membrane is actuated by pressure, the waveguide deforms and changes the properties of the light beam. The modulated light beam has a different propagation speed than the reference light beam, resulting in a phase shift which is detected by the integrated photo-diodes. A sensor prototype with four membranes produced an output signal of $14\,\mu V/mbar$. The entire chip size was $0.3\,mm \times 5\,mm$ and the size of the individual membranes was $200\,\mu m \times 200\,\mu m$.

5. Array of Tactile Sensors

The key aspects of the neurophysiology of touch consist in the representation and coding of spatial and temporal patterns of mechanical stimuli, as perceived by various mechanoreceptor populations.

One example of a sensor which tries to imitate the human skin was presented in Dario et al. (1984) (Fig. 2). The sensor consists of three main layers. The lower or "dermal" layer is a relatively thick layer ($110\,\mu m$) of PVDF (PolyVinyliDene Fluoride). This corresponds to the dermis of the skin, and it measures pressure applied at the surface of the sensor. This layer is bonded to a printed circuit board patterned with an array of electrodes, which allows the detection of location as well as magnitude of the stimulus. The second layer is made of conductive rubber, which is used to enhance the recording of applied force by adding an additional output as a result of applied force. The top or "epidermal" layer is a thin film of PVDF (about $60\,\mu m$), and is used to measure small variation in pressure. This layer can be used to sense slippage, texture, or lighter tactile contacts. The final element of this tactile sensor is a layer of resistive paint between the rubber and the epidermal layer. This layer can be electrically heated so that when the sensor contacts an object, heat is drained off and this can be measured by the pyro-electric effect of the upper PVDF layer.

This sensor alone, however, is a poor replication of the human tactile system, as it can sample data only on one point at a time, thus making impossible the accurate sensing and reconstruction of local indentation profiles (*fine-forms*) that is one of the key aspects in which the sense of touch is superior to vision. In order to overcome this problem, thus making a better replication of the human tactile sense, in recent years a variety of technological solutions have been proposed to obtain high resolution bidimensional maps of contact displacements or forces. Reviews on this subject are found in Nicholls and Lee (1989, 1999).

An example of a sensor array which attempts to emulate the piezoelectric texture of the biological skin by using an array of synthetic polymer (PVDF) elements is presented in Caiti et al. (1995). In this array 42 sensing elements are assembled in seven small hexagonal zones, each containing 6 polymeric strips. These strips have been cut along an appropriate axis, in such a way that each strip presents a piezoelectric response to one particular direction of the stress field.

Another example of tactile sensors is the "KIST Tactile Sensor" developed in the authors' laboratory by using Force

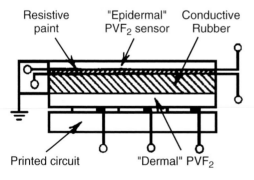

Resistive paint   "Epidermal" PVF$_2$ sensor   Conductive Rubber

Printed circuit   "Dermal" PVF$_2$

Fig. 2. Example of sensor which imitates the human skin (Dario et al. 1984)

Sensing Resistor technology (Section B.4) (Lazzarini et al. 1995).

A further example of an array sensor is a capacitive tactile sensor. In this case, polymide is used, which is highly flexible and has excellent electrical, mechanical and chemical properties. The sensor consists of a polymide base to which a $4\,mm^2$ inner electrode is deposited. Over this there is an outer electrode separated from the inner electrode by a $25\,\mu m$ thick air gap. When a force is applied, the capacitance changes allow the force and location of the force to be determined. The entire signal processing circuit could also be integrated on the substrate.

B. Position and Speed Microsensors

Position and speed microsensors are essential for many practical applications, especially for use in automobiles, robots and medical instruments. In particular, position and speed control is also a major concern in artificial hands, where it is important to measure finger joint position, and in biomedical microdevices, such as microendoscopes, where it is important to find the exact position of the tip and of the instrumented end-effector at any time of examination.

1. Magnetic Sensor to Measure Angular Displacement

A classical sensor used in robotics in order to accurately control the movements of arms, hands and legs or other components having rotating joints, consists of a rotor which has a row of teeth on its bottom (Fatikow and Rembold 1997). The rotor faces a stator which contains several Hall sensors and electronic circuits. A permanent magnet is located under the Hall sensors, thus producing a magnetic field. When the rotor moves, the teeth passing by the Hall sensor change the magnetic field. This change is picked up by the Hall sensors which produce voltage signals.

The developed prototype of the sensor matrix was produced on a GaAs substrate having a $1\,\mu m$ thick silicon dioxide layer. The prototype is about $4\,mm$ long and can measure rotational angles with an accuracy of 0.028 degrees at temperatures ranging between $-10°C$ and $+80°C$.

2. Inductive Position Sensors

Inductive position sensors combine the advantages of silicon integration and inductive sensing principle. Using coil-on-chip technology, the bulky coil windings of the related resolvers and LVDT's (Linear Variable Differential Transformer) are reduced to the size of a silicon chip. In the same package, an application specific integrated circuit (ASIC) provides signal conditioning and a robust interface, suitable for motion control and industrial control applications (Mannion 1999).

3. Vibratory Microgyroscope

Gyroscopes are widely used to detect orientation in space. Standard technology has a number of drawbacks, since it only allows to fabricate gyroscopes that are expensive and too bulky. Furthermore they have high power consumption and limited lifetime. MEMS technologies could improve such limitations and provide for a low cost solution with the capability to merge sensing and processing functions in a synergetic way.

An interesting example of microfabricated gyroscope is the one developed at the University of California at Los Angeles (Fig. 3). This silicon micromachined

Fig. 3. The gyroscope developed at UCLA

vibratory microgyroscope depends on the Coriolis force to induce energy transfer between oscillating modes in order to detect rotation. The advantages of this gyroscope are significant, since it is compact and inexpensive. Furthermore it has low power consumption, long lifetime, negligible turn-on time and large dynamic range.

4. Implantable Joint Angle Transducer (IJAT)

Joint angle transducers for biological applications are particularly important for implantation and prosthetics. A perma-

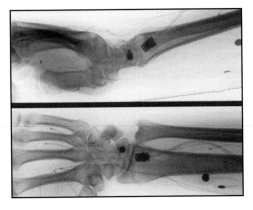

Fig. 4. Xerographic picture of an IJAT

nent magnet can be implanted in one bone of an articulating joint and an array of Hall-effect sensors can be implanted in the opposing bone (Johnson et al. 1999). A typical sensor array consists of three Hall-effect sensors arranged in an equilateral triangle (Fig. 4). As the joint is moved, the relative position and orientation between the sensors and the magnet changes, thus producing sensor voltage changes related to the joint movement. The implantable transducer allows measurement of wrist position in two degrees of freedom, flexion/extension and radial/ulnar deviation. The IJAT is sufficiently small in size to allow implantation in the wrist joint, does not restrict the joint movement, and measures up to 135° of wrist motion.

V. Applications in Humanoid Robotics, Prosthetics, and Neuroprosthetics

A. "Animaloid" and "Humanoid" Robotics

Robotics has always wandered between the need of developing useful machines with high mechanical dexterity and "smart" behavior deriving from their mechatronics design, and the dream of mimicking shape and functions of animals ("Animaloid Robotics") and even humans ("Humanoid Robots"). Humanoid robotics aims at developing biologically-inspired components, such as sensors, actuators, and behavioral schemes, as well as biologically-inspired robots.

Artificial creatures have been developed at different levels of the evolutionary scale (Menzel et al. 2000): from insects (Brooks 2000) to fishes, reptiles (Hirose 1993) and mammals (Fujita 1999, Shibata and Tanie 1999), up to humans (Brooks 1997, Hashimoto et al. 2000, Inoue 2000).

Focussing on sensory systems, some attempts have been made to apply the working principles of animal sensors to detect contact, position and strain (Howe and Cutkosky 1992, Dario et al. 1998b). Interesting examples of biologically-inspired sensors are whiskers, inspired by the vibrissae of some mammals and/or or by insect antennae. These sensors are realized through thin conductive sticks or wires, passing through a hole in a conductive plate, on which they close a circuit when bent by the contact with an external body (Russel 1990).

Often animal sensors are replicated in their functionality, not necessarily in their working principle. For example, the proprioreceptive sensors needed to control robotic limbs (just like animal limbs) can be realized by various means, such as potentiometers or Hall-effect sensors or optical encoders, based on principles very different from biological proprioreceptors. However, these artificial sensors provide essentially the same information for the same purpose of motion control in biological systems. Examples of artificial sensors used in animaloid and in humanoid robots are described in Table 2.

A research goal for developing animaloids and humanoids is to obtain a physical platform to test theoretical models on low level sensory-motor behavior and even on "high level" brain functions. For example, insect-like robots have been developed for two main purposes: 1) to study and replicate six-legged locomotion and 2) to replicate reactive behavior. Especially for the latter purpose, simple and fast sensors play a very important role. For example, artificial whiskers replicate the functionality of antennae in detecting contact; simple load cells detect contact forces (for examples at the feet) while other types of contact sensors can be used to detect contact with the robot body. Potentiometers are often used to measure joint position, while little more sophisticated mercury sensors have been used for tilt detection (Johnson et al. 1999).

The class of pet robots considered here includes a quite wide (and fast growing) range, such as artificial dogs (Fig. 5) (Hirose 1993), cats (Shibata and Tanie 1999), rats, seals (Shibata and Tanie 1999), and many more (Brooks 2000). Robots in this category are characterized by relatively high capabilities of interaction with the environment, very much related to the perception of external stimuli and to their interpretation. Sensory systems in this case include, together with whiskers and touch sensors, more sophisticated tactile arrays and acceleration sensors for joint motion detection.

Table 2. Different force and position sensors used in "animaloid" robots

Robot class	Force sensors	Motion sensors
Insect-like robots	Whiskers Load cells (force on feet) Bump sensors	Potentiometers (joint position) Mercury tilt sensors
Pet robots	Whiskers Microswitches Pressure sensors with balloons Touch sensors Piezo-electric contact sensors Tactile arrays	Acceleration sensors
Humanoids	Contact sensors Tactile arrays Force/torque sensors	Encoders Slip sensors Vibration sensors

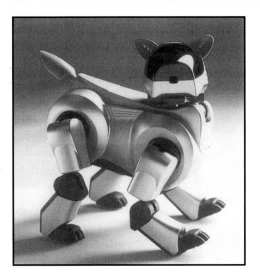

Fig. 5. The AIBO Robot. AIBO means "companion" or "partner" in Japanese. It is also an acronym for Artificial Intelligence Robot

The most ambitious goal of biologically-inspired robotics is certainly the development of humanoid robots. This goal is attracting increasing research efforts. Force sensors for humanoids include contact sensors, tactile arrays, and force/torque sensors, mostly used in manipulation. Motion information relevant in humanoids includes joint angle detection, obtained mostly through encoders, the measurement of acceleration and direction, usually obtained by means of gyroscopes, and slip and vibration detection.

In the authors' laboratory, a multifunctional sensorized fingertip has been developed for integration in a hand designed as a "cybernetic" prosthesis as well as for use in humanoids (Dario et al. 1998b). The integrated fingertip consists of a phalanx equipped with a tactile array, a thermal sensor and a dynamic sensor, with processing electronic circuitry located inside the supporting case. Fingertip design and implementation is presented in more detail

in Section B.4, in relation to its application to human prostheses and neuroprostheses.

The next Section discusses how microsensors can be used for applications which can be seen as an interface (either physical or conceptual) between the biological nervous system and its artificial counterpart in prosthetics and humanoids.

B. Prosthetics and Neuroprosthetics

1. Sensorized Glove

An important application of microsensors, bioengineering and robotics methods and technologies is the development of solutions for persons with spinal cord lesions. Functional Electrical Stimulation (FES) may represent a valuable tool in order to restore some simple limb movements in these patients. An implantable neuroprosthesis (i.e., a system which uses FES to restore motor or sensory functions) has been developed in the authors' laboratory for hand motor function restoration (Fig. 6). The FES system comprises a sensorized glove, used to implement a closed-loop control of the stimulation parameters during grasp.

The glove is sensorized with 20 Hall-effect position sensors and 7 piezoresistive force sensors, which provide information on the rotation of 20 joints of the patient's hand, as well as on contact forces generated during grasp tasks (Freschi et al. 2000, Vecchi et al. 2000).

2. Nerve Regeneration and Interfacing

Peripheral nerve regeneration and interfacing with microsystems is a very promising field of application of tissue engineering. In fact, tactile and position information extracted from the electroneurographic (ENG) signals recorded by

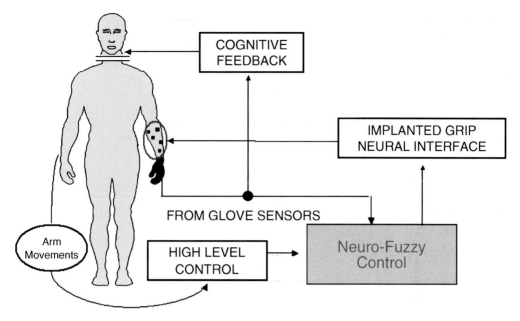

Fig. 6. Scheme of the GRIP final demonstrator (GRIP PROJECT)

means of implantable microelectrodes can be processed and used in order to control neuroprostheses (Sinkjaer et al. 1999).

The first example of microelectrodes for nerve recording and stimulation is the silicon probe developed by Najafi et al. (1985), for high-amplitude multichannel monitoring of neural activity in the cortex (Fig. 7).

Silicon micromachined structures, in which cultured neurons can be implanted and grown, have been described in Bove et al. (1997). The silicon structure is composed of an array of microchannels (24 and 48 µm wide, and 7 µm deep), in which the neurons are guided and grown.

Smart microchips for culture, stimulation, and recording of neural cells arrays have also been developed (Heuschkel et al. 1998). A typical microchip includes a 4 × 4 array of indium-tin oxide electrodes, passivated outside of measurement areas by polymeric layers.

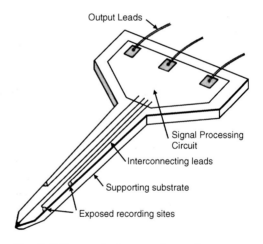

Fig. 7. Silicon probe for neural recording

A class of implantable, regeneration type Neural Interfaces (NIs) for mammalian peripheral nerve recording and stimulation has been developed by the authors in collaboration with European partners (INTER Project). The interface

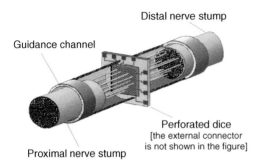

Distal nerve stump

Guidance channel

Perforated dice
[the external connector
is not shown in the figure]

Proximal nerve stump

Fig. 8. Schematic view of the regeneration-type Neural Interfaces

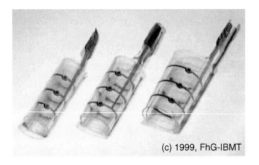

(c) 1999, FhG-IBMT

Fig. 9. Self-sizing cuff electrodes (Schuettler et al. 1999)

comprises 3 components (Fig. 8): 1. a microfabricated silicon die with microelectrode array on multiple through holes; 2. a polymer guidance channel housing the die; and 3. a flexible flat cable connecting the die to an external electronic circuitry (Dario et al. 1998).

A different approach for interfacing peripheral nerves involves cuff-type connectors, as those fabricated by Schuettler et al. (1999). The connectors are based on micromachined polyimide substrate and insulation layers with embedded thin-film metallization. A rolled cuff-type electrode of this kind is shown in Fig. 9.

3. Prosthetic Hand

The implantable neural connector has an additional and very important future ap-

plication: controlling in a natural way a future sensorized prosthetic hand. The authors are developing such a prosthetic hand according to a "bio-mechatronic" approach, that is by designing all the components of the hand (mechanisms, sensors, actuators, and controller) in a strongly integrated way, and by assuming that the hand will be controlled by the amputee through an implantable neural interface. At the moment we have designed and fabricated a prosthetic finger with two degrees of freedom, whose motion resembles closely the motion of the human finger (Carrozza et al. 2000, 2002). The MP (metacarpal-phalangean) joint and PIP (proximal inter-phalangean) joint are actuated by two separate micromotors, while the DIP (distal inter-phalangean) joint is moved passively by the motion of the PIP joint. Although this may not be true for all the activities of the human finger, such reduction of the number of degrees of freedom simplifies the design and the control of the finger and, consequently, of the whole hand.

The finger incorporates a complex and rich sensory system in order to be properly controlled. Joint sensors based on Hall-effect and skin-like fingertip sensors are located in each finger. The main characteristics of the skin-like tactile sensor are illustrated in the following paragraph.

4. Skin-Like Tactile Sensor

The fingertip skin-like tactile sensor developed in the authors' lab (Dario et al. 1984, 1996; Lazzarini et al. 1995) integrates three tactile modalities: pressure sensing (like Ruffini endings or Merkel's disks), thermal flux sensing (like thermoreceptors) and slip sensing (like Pacinian corpuscle).

The pressure is sensed by a fovea-like array of 64 tactile piezoresistive units based on FSR (Force Sensing Resistors) technology, designed by the authors and manufactured by Interlink Inc., Camarillo, CA. The maximum resolution of the fovea-like structure (at the center of the sensor) is set by the current limit for FSR technology, that is about 1 mm. The minimum spatial resolution in the periphery is about 5 mm. This disposition increases the size of the tactile sensing area, for the same number of sensing sites.

The thermal flux sensor behaves like the thermoreceptors in the human hand. However, the artificial sensors do not discriminate between cold and warmth, but they just detect the amount of heat flowing from the fingertip towards the touched object.

Lastly, the dynamic sensor is a small piezoelectric ceramic bimorph element, mounted on the fingertip frame and connected electrically to the preprocessing electronics. The sensor detects the vibrations caused by the movement of the fingertip along a surface.

VI. Conclusions

In this paper we have presented the main characteristics of force and motion sensors in humans. The analysis of these characteristics is an important source of inspiration for the design of artificial sensors of force and motion applicable to robotics and prosthetics. Recent advances in microfabrication technologies make it possible to convert biological inspiration into guidelines for the design of real sensors and sensory systems. These biologically-inspired mechano-sensors could become not only a mere demonstration of technology capabilities or a scientific "virtuosity", but tools usable in engineering practice. In areas like robotics, intelligent machines, and prosthetics, the increasing computing performance of embedded microcontrollers and the continuous miniaturization of the system ("micromechatronics") are key factors for the potential use of more and more microsensors in general, and of biologically-inspired sensors in particular.

In conclusion, the authors believe that there are concrete motivations for promoting interactions and creating links between biologists, engineers and physical scientists, with the aim of cross-fertilizing research on new biologically-inspired mechano-sensors, which can be fabricated using MEMS and Bio-MEMS technologies, exploited using current high performance, miniaturized embedded controllers, and applied in many current and future systems, including robots, prostheses and micromachines.

Acknowledgements

Some of the work described in this paper has been supported by the Commission of the European Community: European Project BIOMED2 "MEDEA – Microscanning Endoscope with Diagnostic and Enhanced Resolution Attributes", Contract no. BMH4-CT97-2399; European Project ESPRIT "MINIMAN – A Miniaturised Robot for Micro Manipulation", Contract no. 33915, European ESPRIT LTR Project "GRIP – An Integrated System for the Neuroelectric Control of Grasp in Disabled Persons", Contract no. 26322; European Project ESPRIT Basic Research "INTER – Intelligent Neural Interface", Contract no. 8897.

Research on prosthetic hands has been supported by INAIL (Istituto Nazionale Assicurazione Infortuni sul Lavoro).

References

Bailey SA, Cham JG, Cutkosky MR, Full RJ (2000) Biomimetic robotic mechanisms via shape deposition manufacturing. In: Hollerbach J, Koditschek D (eds) Robotics Research: the Ninth International Symposium, Springer-Verlag, London

Bove M, Grattarola M, Verreschi G (1997) In vitro 2-D networks of neurons characterized by processing the signals recorded with a planar microtransducer array. IEEE Trans Biomedical Eng 44: 964–977

Brooks RA (1997) The cog project. Advanced Robotics 15(7): 968–970

Brooks RA (2000) Cambrian Intelligence. MIT Press, Cambridge

Brooks RA, Stein LA (1994) Building brains for bodies. Autonomous Robots 1: 7–25

Caiti A, Canepa G, De Rossi D, Germagnoli F, Maganes G, Parisini T (1995) Towards the realization of an artificial tactile system: fine-form discrimination by a tensorial tactile sensor array and neural inversion algorithms. IEEE Transaction on System, Man and Cybernetics 25: 933–946

Carrozza MC, Dario P, Lazzarini R, Massa B, Zecca M, Roccella S, Sacchetti R (2000) An actuating system for a novel biomechatronic prosthetic hand. Actuator 2000: 19–21 June 2000, Bremen

Carrozza MC, Massa B, Dario P, Lazzarini R, Zecca M, Micera S, Pastacaldi P (2002) A Two-DOF finger for a biomechatronic artificial hand. Technology and Healthcare, 10: 77–89

Dario P, Carrozza MC, Allotta B, Guglielmelli E (1996) Micromechatronics in Medicine. IEEE/ASME Trans on Mechatronics 1: 137–148

Dario P, De Rossi D, Domenici C, Francesconi R (1984) Ferroelectric polymer tactile sensors with anthropomorphic features. In: IEEE Int Conf Rob, pp 332–340

Dario P, Fukuda T (1998) Guest editorial, IEEE/ASME Transaction on Mechatronics 3: 1–2

Dario P, Garzella P, Toro M, Micera S, Alavi M, Meyer J-U, Valderrama E, Sebastiani L, Ghelarducci B, Mazzoni C, Pastacaldi P (1998a) Neural interfaces for regenerated nerve stimulation and recording. IEEE Transactions on Rehabilitation Engineering 6: 353–363

Dario P, Laschi C, Guglielmelli E (1998b) Sensors and actuators for 'humanoid' robots. Advanced Robotics, Special Issue on Humanoid 11(6): 567–584

Dario P, Lazzarini R, Magni R (1996) An integrated miniature fingertip sensor. Machine Human Science, Nagoya, pp 91–97

Dario P, Sandini G, Abisher P (eds) (1989) Robots and Biological Systems: Towards a New Bionics?, NATO ASI Series

De Rossi D, Ahluwalia A (2000) Biomimetics: new tools for an old myth. In: 1st Annual International IEEE-EMBS Special Topic Conference on Microtechnologies in Medicine & Biology, October 12–14, 2000, Lyon, France, pp 15–17

Despont M (1992) A comparative study of bearing designs and operational environments for harmonic side-drive micromotors. In: proceedings of MEMS 92: 171–176

Fatikow S, Rembold U (1997) Microsystem Technology. Springer-Verlag, Berlin Heidelberg New York

Ferrari M (1999) Editorial, Journal of Biomedical Microdevices 1: 97–98

Fischer K (1991) Mikromechanik und Mikroelektronik vereint mit Optik. Technische Rundschau 106–108

Freschi C, Vecchi F, Micera S, Sabatini AM, Dario P (2000) Force control during grasp using FES techniques: preliminary results. 5th Annual Conference of the International Functional Electrical Stimulation Society (IFESS 2000), 17–24 June 2000, Aalborg

Fujimasa I (1996) Micromachines, a New Era in Mechanical Engineering. Oxford University Press, Oxford

Fujita M (1999) AIBO: Towards the era of digital creatures. In: International Symposium on Robotics Research, Snowbird UH, October 9–12, pp 257–262

Fukuda T, Menz W (1998) Handbook of Sensors and Actuators. Micro Mechanical Systems. Elsevier Science, Amsterdam

GRIP Esprit LTR Project #26322 An integrated system for the neuroelectric control of grasp in disabled persons

Harsanyi G (1995) Polymer Films in Sensor Applications. Technomic Publishing Co., Basel

Hashimoto S (2000) Humanoid robots in Waseda University – Hadaly-2 and Wabian. In: First IEEE-RAS International Conference on Humanoid Robots – Humanoids 2000, Cambridge, MA, September 7–8

Heuberger A (1991) Mikromechanik: Mikrofertigung mit Methoden der Halbleitertechnologie. Springer-Verlag, Berlin Heidelberg New York

Heuschkel MO, Guerin L, Buisson B, Bertrand D, Renaud P (1998) Buried microchannels in photopolymer for delivering of solutions to neurons in a network. Sensors and Actuators B48: 356–361

Hirose H (1993) Biologically Inspired Robots (Snake-like Locomotor and Manipulator). Oxford University Press, Oxford

Howe RD, Cutkosky MR (1992) Touch Sensing for Manipulation and Recognition. In: Kathib O, Craig J, Lozano-Pérez T (eds) The Robotics Review 2: 55–112, MIT Press, Cambridge Mass

http://csmt.jpl.nasa.gov/csmtpages/Technologies/mgyro/mgyro.html

Inoue H (2000) HRP: Humanoid Robotics Project of MITI. In: First IEEE-RAS International Conference on Humanoid Robots – Humanoids 2000, Cambridge Mass, September 7–8

INTER Project promoted by the European Commission ("Intelligent Neural Interface" Esprit Basic Research Project #8897)

Johnson MW, Peckham PH, Bhadra N, Kilgore KL, Gazdik MM, Keith MW, Strojnik P (1999) Implantable transducer for two-degree-of-freedom joint angle sensing". IEEE Trans Biomed Eng 7(3): 349–359

Kandel ER, Schwartz JH, Jessel TM (1991) Principles of Neural Science, 3rd edition. Elsevier Science, New York

Langer R, Vacanti J (1993) Tissue engineering. Science 260: 920–926

Lazzarini R, Magni R, Dario P (1995) A tactile array sensor layered in an artificial skin. In: Proceedings of the 1995 IEEE/RSJ International Conference on Intelligent Robots and Systems, Iros '95, 3: 114–119

Madou M (1997) Fundamentals of Microfabrication. CRC Press, Boca Raton, New York

Mannion P (1999) Integration and Inductive Sensing Combine to Improve Automotive/Industrial Sensing. Electronic Design

Mehlhorn T (1992) CMOS-compatible capacitive silicon pressure sensors. Micro System Technologies 92: 277–285

Menzel P, D'Aluisio F, Mann CC (2000) RoboSapiens. MIT Press, Cambridge Mass

Muller RS, Howe RT, Senturia SD, Smith RL, White RM (1991) Microsensors, IEEE Press, New York

Najafi K, Wise KD, Mochizuki T (1985) A High-Yield IC- Compatible Multichannel Recording Array. IEEE Trans Electronic Devices ED-32: 1206–1211

Nicholls HR, Lee MH (1989) A survey of robot tactile sensing technology. Int J Robotics Research 8: 3–30

Nicholls HR, Lee MH (1999) Tactile sensing for mechatronics – a state of the art survey. Mechatronics 9: 1–32

Russel RA (1990) Robot Tactile Sensing. Prentice Hall Ltd, Australia

Schuettler M, Stieglitz T, Meyer J-U (1999) A multipolar precision hybrid cuff electrode for FES on large peripheral nerves. 21st Annual International Conference of the IEEE Engineering in Medicine and Biology Society, October 13–16, Atlanta, USA, 1999

Shibata T, Tanie K (1999) Creation of subjective value through physical interaction between human and machine. In: 4th International Symposium on Artificial Life and Robotics, January 19–22, Oita, Japan

Sinkjaer T, Haugland M, Struijk J, Riso RR (1999) Long-term cuff electrode recordings from peripheral nerves in animals and humans. In: Windhorst U, Johansson H (eds) Modern Techniques in Neuroscience Research. Springer-Verlag, New York

Stanford University, http://www-cdr.stanford.edu

Tilmans H, Bouwstra S (1993) A novel design of a highly sensitive low differential-pressure sensor using built-in resonant strain gauges. J Micromech Microeng 3: 198–202

Vecchi F, Freschi C, Micera S, Sabatini AM, Dario P (2000) Experimental evaluation of two commercial force sensors for applications in biomechanics and motor control. 5th Annual Conference of the International Functional Electrical Stimulation Sociaty (IFESS 2000), June 17–24, 2000, Aalborg

Webster JG (1988) Tactile Sensors for Robotics and Medicine. John Wiley & Sons, New York

B. Force and Motion

9. The Physics of Arthropod Medium-Flow Sensitive Hairs: Biological Models for Artificial Sensors

Joseph A.C. Humphrey, Friedrich G. Barth,
Michael Reed, and Aaron Spak

Abstract

Theoretical analysis is applied at two levels to the mechanical properties of medium motion-sensing filiform hairs of arthropods. Emphasis is on the hairs of terrestrial animals, namely the spider *Cupiennius salei* and the cricket *Gryllus bimaculatus*, for which experimental data exist. A physically-exact analysis of hair motion yields the work and far field medium velocity required to attain an imposed threshold angular displacement. Far field velocity decreases and work increases with increasing hair length, with spider hairs requiring slightly smaller far field velocities and less work $(2.5 \times 10^{-20} - 1.5 \times 10^{-19}$ Joules) than cricket hairs $(9 \times 10^{-20} - 8.4 \times 10^{-19}$ Joules) to attain the same threshold displacement. These values of energy compare to that of a single photon $(10^{-18} - 10^{-19}$ Joules) for light in the visible spectrum. When the fluid medium dominates viscous damping, both the diameter, d, and density, ρ_h, of a hair are important in air but not in water, and increasing either of these two parameters works to increase the maximum displacement angle and velocity at resonance frequency while decreasing the corresponding resonance frequencies themselves. For hairs in air, the lengths of hairs with greatest sensitivity to changes in medium motion scale with the hair substrate boundary layer thickness. Present findings further suggest that filiform hair motion sensors may have evolved over geological time scales to perform optimally in a specific range of temperatures because of the dependence of medium viscosity on temperature. Considering the design and fabrication of corresponding artificial sensors, the challenge is to: a) fabricate a micro-electro-mechanical system consisting on an array of $N \times N$ artificial hairs analogous to the filiform hairs; and, b) select an appropriate mechanism to transduce mechanical motion into a useful, electrically measurable signal.

I. Introduction

Hair-like sensory structures exist in almost all animal phyla living in both air and water. These include arthropods (insects, arachnids and crustaceans), annelids (leeches), chaetognaths, mollusks, fish, amphibians and mammals (the verterbrate inner ear). In particular, the filiform hairs of terrestrial arthropods serve as uniquely sensitive medium motion sensors that have been refined by natural selection pressures over hundreds of millions of years. The filiform hairs on the cerci of the cricket *Gryllus bimaculatus* and on the legs of the spider *Cupiennius salei* can detect air motions as weak as $1\,\text{mm/s}$ induced by prey or predators. The maximum hair deflection threshold required to elicit a nervous response ranges between 0.001 and 0.1 degrees. The filiform hairs on *Cupiennius* and other spiders, also referred to as trichobothria, appear in spatially distributed clusters or arrays capable of discriminating the magnitude, direction and frequency of air-borne signals as well as their variation in space and time over the animal's body (Barth 2002).

Given their remarkable performance characteristics, filiform hairs provide an especially insightful biological model for the conceptualization and fabrication of artificial medium motion sensors. Such sensors would be invaluable to industry; for example, to resolve the time and spatial characteristics of near wall turbulent boundary layer flows. Not surprisingly, however, experimental and theoretical (analytical and/or numerical) investigations of the performance characteristics of filiform hairs are rendered especially complex by the highly interdisciplinary nature of the problem which involves major facets of biology, fluid and solid mechanics, dynamics, and applied mathematics.

Our work has two major objectives. The first is to uncover and understand the basic "design" principles underpinning the performance characteristics of filiform hairs. The second is to derive and implement a realistic physical-mathematical model for these exquisitely sensitive natural sensors that renders predictable their performance characteristics. With this knowledge in hand, we can hope to address two highly complex questions with mutually informing answers: a) How has physics impacted the sensory ecology and adaptive evolution of the natural motion sensors? b) How might we design and fabricate artificial sensors of similar function and characteristics? The physical-mathematical model underpins and extends the laboratory model, by allowing the systematic analysis of all major system variables in a mere fraction of the time required by experimentation. The tradeoff, of course, is the unavoidable (but often quantifiable) uncertainty embodied in the model due to necessary physical approximations. In contrast, calculation accuracy is not a concern.

In this chapter we first summarize the physical-mathematical model that describes the motion of a single hair. Of particular interest is to use a physically-approximate version of the model to illustrate the effects on hair maximum angular deflection and velocity, and on their preferred (resonance) frequencies, of varying quantities such as the density, viscosity, velocity amplitude and frequency of the oscillating medium, as well as the diameter, length, torsional restoring constant and damping constant of the hair. We consider the constraints and limitations imposed by physics on the evolution of filiform hairs and how, notwithstanding, biology may work to relieve such constraints and limitations to evolve increasingly better adapted motion sensing systems. This is followed by a discussion of the energetics of hair motion, where we seek to bound the minimum amount of work required to detect the weakest meaningful signal. The chapter concludes by positing the ideal artificial motion sensor and considering briefly how it might be realized using microfabrication techniques. Where possible in the presentation, results for the filiform hairs of spiders are compared with corresponding results for crickets using the same physical-mathematical model. In this regard, see also chapter 10, this volume by Shimozawa et al.

II. The Physically-Exact and Physically-Approximate Mathematical Models

Several authors have derived and applied physical-mathematical models to describe the behavior of filiform cuticular hairs responding to oscillatory movements of the surrounding fluid medium in both terrestrial and aquatic arthropods (Fletcher 1978, Tautz 1979, Shimozawa and Kanou 1984, Barth et al. 1993, 1995, Humphrey et al. 1993, 1997, 1998, Devarakonda et al. 1996, Shimozawa et al. 1998, Barth 2000). The numerical model developed by Humphrey et al. (1993, 1997, 1998) solves a rigorously derived form of the equation for the conservation of angular momentum for a single hair. However, the use of this model requires the judicious calculation and plotting of numerous individual cases in order to elucidate the effects on hair performance characteristics due to changes in the hair or fluid medium properties. Thus, in spite of its predictive power and accuracy, the numerical solution approach does not provide easily usable and interpretable analytical formulae from which to derive generally applicable and insightful conclusions.

In a recent study, Humphrey et al. (2001) have derived an analytically-based mathematical model to describe the time-periodic medium-driven motion of single filiform hairs. Throughout this chapter we will refer to this model as the

physically-exact (P-E) model. As in all previous modeling efforts, the P-E analysis neglects flow-mediated viscous interactions among hairs in a cluster but the model yields calculated results for single hairs in good agreement with experimental measurements. A simpler version of the P-E model allows the authors to further explain the complex relationships that exist among some of the major variables affecting hair motion. We will refer to this second model as the physically-approximate (P-A) model. The conclusion in Humphrey et al. (2001) is that all major quantitative aspects of hair behavior are correctly captured by the P-E model while the relations derived from the P-A model provide additional and invaluable qualitative insight. The P-E and P-A models provide the basis for the present study.

A. Equations of the Physically-Exact Model

Attention is restricted here to the final forms of the relevant equations. For details concerning their derivation the reader should consult Humphrey et al. (2001). Except where otherwise indicated, quantities in all equations, tables and plots are expressed in SI units. The P-E model equations for the time-periodic (sinusoidal) angular displacement, θ (rad), velocity, $\dot{\theta}$ (rad s^{-1}), and acceleration $\ddot{\theta}$ (rad s^{-2}), of a hair undergoing relatively small ($\theta < 0.17$ rad or $10°$) displacements in an oscillating fluid medium are given by:

$$\theta = C_1 \cos(\omega t) + C_2 \sin(\omega t) \qquad (1)$$

$$\dot{\theta} = -C_1 \omega \sin(\omega t) + C_2 \omega \cos(\omega t) \qquad (2)$$

$$\ddot{\theta} = -C_1 \omega^2 \cos(\omega t) - C_2 \omega^2 \sin(\omega t). \qquad (3)$$

In these equations the quantities C_1 and C_2 are complex functions of the general form:

$$C_1 = f_1(d, L, \rho_h, S, I, I_\rho, I_\mu,$$
$$R, R_\mu, \rho, \mu, U_0, \omega) \qquad (4a)$$

$$C_2 = g_1(d, L, \rho_h, S, I, I_\rho, I_\mu,$$
$$R, R_\mu, \rho, \mu, U_0, \omega). \qquad (4b)$$

In spite of their algebraic complexity, Eqs. (1–4) are easy to evaluate numerically. The equations apply to a hair attached to a flat substrate and immersed in a fluid oscillating at frequency ω ($= 2\pi f$) (rad s^{-1}) with far field velocity amplitude U_0 (m s^{-1}). They are written for a straight, cylindrically-shaped (rod-like) hair of length L (m), effective diameter d (m) and density ρ_h (kg m^{-3}) immersed in a fluid medium of density ρ (kg m^{-3}) and dynamic viscosity μ (kg m^{-1} s^{-1}). The quantities I (N m s^2 rad^{-1}), R (N m s rad^{-1}) and S (N m rad^{-1}) are, respectively, the moment of inertia, the damping constant and the torsional restoring constant of the hair. These are mechanical properties inherent to the hair; the first can be determined from the hair shaft geometry but the latter two must be determined experimentally, and all three are taken as being constant for a given hair. The quantities I_ρ, I_μ and R_μ denote additional contributions to the moment of inertia and the damping constant of the hair associated with the fluid medium density and viscosity, respectively. The exact forms of C_1 and C_2 as well as the definitions for I, I_ρ, I_μ and R_μ are provided in Humphrey et al. (2001; see also Humphrey et al. 1993, 1997, 1998). In particular, it can be shown that the viscous damping R_μ is proportional to μL^3 ($R_\mu \propto \mu L^3$).

The derivation of Eqs. (1–4) assumes that the velocity of the fluid medium

driving hair motion is given by Stokes' 1851 analytical expression

$$V_F = U_0(-\cos(\omega t) + \cos(\omega t - \beta y)e^{-\beta y}). \qquad (5)$$

This equation describes the velocity at time t and position y of a fluid oscillating at frequency ω and far field amplitude U_0 parallel to a fixed flat substrate supporting the hair. The quantity $\beta = (\omega\rho/2\mu)^{1/2}$ is related to the oscillating flow boundary layer thickness, δ, according to

$$\delta = \frac{4.5}{\beta} = 6.4\left(\frac{\mu}{\rho\omega}\right)^{1/2}. \qquad (6)$$

With V_F given by Eq. (5), it is also possible to directly calculate the total torque, T, and hence the work, W, to drive hair motion. The total torque is composed of two contributions. One, which we label T_D, is proportional to the relative velocity, V_r, between the hair and the fluid medium. The other, which we label T_{VM}, is proportional to the relative acceleration, $\dot{V}_r$. Analytical expressions for T_D and T_{VM} are provided in Appendix 1 of Humphrey et al. (1993; see also Humphrey et al. 1997, 1998). The total torque, T ($= T_D + T_{VM}$), varies sinusoidally with time and the work, W, required to drive hair motion during one oscillation cycle is given by

$$W = 2\int_{-\theta_{max}}^{\theta_{max}} T \, d\theta = \int_0^{2\pi} T\frac{\dot{\theta}}{\omega}d(\omega t) \qquad (7)$$

where θ_{max} is the maximum angular displacement of a hair during an oscillation cycle in a fluid medium oscillating at fixed frequency ω.

Equations (1–7) constitute the P-E model. They describe the behavior of a forced, damped, harmonic rod-like oscillator including the "added" or "virtual" mass, VM, of fluid (air or water) immediately around and moving with it. Their solutions are very powerful because they accurately quantify the dependence of hair deflection angle, velocity, acceleration and work on all the physical parameters that affect these four quantities.

B. Equations of the Physically-Approximate Model

Additional analysis yields physically-approximate but even more insightful explicit results. The details of the analysis are given in Humphrey et al. (2001) who obtain expressions for the resonance frequencies, $\omega_{res(\theta)}$ and $\omega_{res(V)}$, at which hair displacement and velocity maximize (as a function of frequency), and for the corresponding values of the maxima, θ_{res}, and $V_{res}(\equiv \dot\theta_{res})$, at these frequencies. The expressions are:

$$\omega_{res(\theta)} = \left[\frac{S}{I_t} - \frac{R_t^2}{2I_t^2} \right]^{1/2} \quad (8a)$$

$$\theta_{res} = \frac{2I_t}{R_t} \left[\frac{P^2 + Q^2}{4I_t S - R_t^2} \right]^{1/2} \quad (8b)$$

$$\omega_{res(V)} = \left[\frac{S}{I_t} \right]^{1/2} \quad (9a)$$

$$V_{res} = \frac{1}{R_t}[P^2 + Q^2]^{1/2} \quad (9b)$$

where, $I_t = I + I_\rho + I_\mu$ and $R_t = R + R_\mu$ are the total inertia and total damping, and

P and Q are complex but readily calculable functions of the form

$$P = f_2(d, L, \rho, \mu, U_0, \omega) \quad (10a)$$

$$Q = g_2(d, L, \rho, \mu, U_0, \omega). \quad (10b)$$

Equations (8–10) constitute the P-A model.

III. Results and Discussion

A. Observations Based on the Physically-Approximate Model

1. Parameter Dependence of Hair Motion

The accuracy and range of applicability of the P-A expressions given by Eqs. (8–10) has been verified by Humphrey et al. (2001) by reference to the work of Barth et al. (1993) for the trichobothria of the spider *Cupiennius salei* under controlled oscillating flow conditions. A further scaling analysis that retains only the dominant terms contributing to these equations yields qualitative expressions that explain the parameter dependence of hair motion (Humphrey et al. 2001). Table 1 summarizes the results obtained for hairs in air and in water, respectively. Relations are provided for the power dependence of the physical variables that affect hair maximum deflection angle (θ_{res}) and velocity (V_{res}) and their respective resonance frequencies ($\omega_{res(\theta)}$ and $\omega_{res(V)}$).

Table 1. Approximate functional dependencies (given as products of the relevant physical parameters raised to their respective powers) of hair maximum deflection angle, θ_{res}, and maximum velocity, V_{res}, and of their associated resonance frequencies $\omega_{res(\theta)}$ and $\omega_{res(V)}$. Numerical coefficients affecting the tabulated relations (ranging in value between 0.1 and 10, approximately) are omitted for clarity (Humphrey et al. 2001)

	Air	Water
$\theta_{res} \propto$	$\rho_h^{3/4} d^{3/2} L^{13/4} S^{-3/4} \mu^{3/2} \rho^{-1/2} U_0 / [R + \mu L^3]$	$L^{11/2} S^{-3/2} \mu^3 \rho^{-1/2} U_0 / [R + \mu L^3]$
$V_{res} \propto$	$\rho_h^{1/4} d^{1/2} L^{7/4} S^{-1/4} \mu^{3/2} \rho^{-1/2} U_0 / [R + \mu L^3]$	$L^{5/2} S^{-1/2} \mu^2 \rho^{-1/2} U_0 / [R + \mu L^3]$
$\omega_{res(\theta)} \propto$ and $\omega_{res(V)} \propto$	$\rho_h^{-1/2} d^{-1} L^{-3/2} S^{1/2}; (R \ll \mu L^3 \text{ for } \omega_{res(\theta)})$	$L^{-3} S \mu^{-1}; (R \ll \mu L^3 \text{ for } \omega_{res(\theta)})$

The P-A equations (8–10) are used by Humphrey et al. (2001) to directly calculate the relative sensitivities of hairs in air and water to variations in the physical parameters that affect hair motion. However, the qualitative nature of these dependencies is readily illustrated by differentiation of the expressions in Table 1. For the case of viscous-dominated damping ($R \ll \mu L^3$) in air:

$$\frac{d\theta_{res}}{\theta_{res}} \propto \frac{3}{4}\frac{d\rho_h}{\rho_h} + \frac{3}{2}\frac{dd}{d} + \frac{1}{4}\frac{dL}{L} - \frac{3}{4}\frac{dS}{S}$$

$$+ \frac{1}{2}\frac{d\mu}{\mu} - \frac{1}{2}\frac{d\rho}{\rho} + \frac{1}{1}\frac{dU_0}{U_0} \qquad (11a)$$

$$\frac{dV_{res}}{V_{res}} \propto \frac{1}{4}\frac{d\rho_h}{\rho_h} + \frac{1}{2}\frac{dd}{d} + \frac{5}{4}\frac{dL}{L} - \frac{1}{4}\frac{dS}{S}$$

$$+ \frac{1}{2}\frac{d\mu}{\mu} - \frac{1}{2}\frac{d\rho}{\rho} + \frac{1}{1}\frac{dU_0}{U_0} \qquad (11b)$$

$$\frac{d\omega_{res(\theta)}}{\omega_{res(\theta)}} \approx \frac{d\omega_{res(V)}}{\omega_{res(V)}} \propto -\frac{1}{2}\frac{d\rho_h}{\rho_h} - \frac{1}{1}\frac{dd}{d}$$

$$- \frac{3}{2}\frac{dL}{L} + \frac{1}{2}\frac{dS}{S} ; \qquad (11c)$$

while for viscous-dominated damping in water:

$$\frac{d\theta_{res}}{\theta_{res}} \propto \frac{5}{2}\frac{dL}{L} - \frac{3}{2}\frac{dS}{S} + \frac{2}{1}\frac{d\mu}{\mu} - \frac{1}{2}\frac{d\rho}{\rho} + \frac{1}{1}\frac{dU_0}{U_0}$$

$$(12a)$$

$$\frac{dV_{res}}{V_{res}} \propto -\frac{1}{2}\frac{dL}{L} - \frac{1}{2}\frac{dS}{S} + \frac{1}{1}\frac{d\mu}{\mu}$$

$$-\frac{1}{2}\frac{d\rho}{\rho} + \frac{1}{1}\frac{dU_0}{U_0} \qquad (12b)$$

$$\frac{d\omega_{res(\theta)}}{\omega_{res(\theta)}} \approx \frac{d\omega_{res(V)}}{\omega_{res(V)}} \propto -\frac{3}{1}\frac{dL}{L} + \frac{1}{1}\frac{dS}{S} - \frac{1}{1}\frac{d\mu}{\mu}$$

$$(12c)$$

In these expressions and the table, "$\propto$" denotes "varying according to", while "$\approx$" denotes "approximately equal to."

To within the accuracy of the analysis, Eqs. (11–12) give the relative sensitivity of θ_{res}, V_{res}, $\omega_{res(\theta)}$, and $\omega_{res(V)}$ to relative changes in the parameters that affect these four quantities. The three equations are all of the general form

$$\frac{dY}{Y} \propto \Sigma a_i \frac{dX_i}{X_i} \qquad (13)$$

where the a_i are sensitivity (or multiplier) coefficients denoting the positive or negative dependence of dY/Y on the corresponding relative changes dX_i/X_i. The equations show that both the diameter, d, and density, ρ_h, of a hair are important in air but not in water, and that increasing either of these two parameters works to increase θ_{res} and V_{res} while decreasing $\omega_{res(\theta)}$ and $\omega_{res(V)}$. In contrast, numerical solutions of Eqs. (8–10) for the case when $R \approx \mu L^3$ show that increasing d works to increase $\omega_{res(\theta)}$ in both air and water (Humphrey et al. 2001). Thus the entries in Table 1 or any results derived from them must be used paying special attention to the magnitude of the $R/\mu L^3$ ratio. The above analytical findings concerning the parameter dependence of hair motion are in broad agreement with our earlier work based on detailed numerical solutions of the complete physical model (Devarakonda et al. 1996).

2. Influence of Physical Parameters on Evolution and Adaptation

Notwithstanding their quantitative limitations, Eqs. (11–12) suggest some intriguing thoughts concerning the role of the medium viscosity, μ. A relative increase in this parameter is seen to increase θ_{res} and V_{res} in both air and water, and to decrease $\omega_{res(\theta)}$ and $\omega_{res(V)}$ in water. More accurate calculations (Devarakonda et al. 1996, Humphrey et al. 2001) confirm these trends. Depending on geographical and environmental factors, in the course of a diurnal cycle changes in air temperature can be fairly large. For example, in the Sonora desert the temperature of air can drop from 45°C during the day to 25°C at night inducing a corresponding decrease in viscosity of about 5%. In still water, such as a pond exposed to direct solar radiation, the temperature of the surface layer can range from 15°C in the early morning to 30°C in the evening, corresponding to a decrease in viscosity of about 21%. From the entries in Table 1 it follows that, for viscous-dominated damping ($R \ll \mu L^3$), a 5% decrease in air viscosity translates into a 2.5% decrease in θ_{res} and V_{res}, while a 20% decrease in water viscosity translates into a 36% decrease in θ_{res}, a 20% decrease in V_{res}, and a 25% increase in $\omega_{res(\theta)}$ and $\omega_{res(V)}$.

The interesting possibility arises that filiform hair motion sensors perform optimally, and have evolved to perform optimally, in a specific range of temperatures because of the dependence of medium viscosity on temperature. The temperature dependence of viscosity has been illustrated in the previous paragraph for changes in temperature during a diurnal cycle. However, changes in temperature over geological time scales overlapping with evolutionary time scales, meaning tens to hundreds of millions of years, may have also worked to periodically evolve the shapes and forms, material properties, and performance of these sensors by making incremental tradeoffs over long periods of time among the parameters affecting sensor response to medium motion. Thus, for example, by reference to Eq. (12c) above or Fig. 7 in Humphrey et al. (2001), climatological increases (decreases) in water viscosity could have been offset by corresponding evolutionary decreases (increases) in hair length or increases (decreases) in the hair torsional restoring constant in order to avoid drifts in hair resonance frequency. In terms of numbers (and accounting for the signs of the terms in Eq. (12c)!) a gradual 20% increase (decrease) in water viscosity would require a gradual 6.7% decrease (increase) in hair length, or a gradual 20% increase (decrease) in the hair torsional restoring constant (or some combination of the two), to avoid a drift in hair resonance frequency due to the drop in viscosity. In the same way, if the sensor's target frequency were itself to evolve over evolutionary time scales, it is conceivable that the changing frequency could be "tracked" by incremental (positive or negative), genetically transmitted adaptive changes in any one (or combinations) of the several parameters ρ_h, d, L, R and S specific to the hair.

3. Hair Sensitivity to Changes in Fluid Medium Velocity

It is of interest to consider the variation of $L dV_{res}/dU_0$ with respect to $L/\delta_{res(V)}$. The former is a dimensionless measure of hair sensitivity; it characterizes the rate of change of hair tip velocity with respect to a change in the amplitude of the far field fluid medium velocity. The latter is a dimensionless measure of hair length, where $\delta_{res(V)} \sim (\mu/\rho\omega_{res(V)})^{1/2}$ is an estimate of the boundary layer thickness based on the velocity resonance frequency. The qualitative variation of $L dV_{res}/dU_0$ with $L/\delta_{res(V)}$ can be derived from the relations in Table 1 (Humphrey et al. 2001). Alternatively, the variation of sensitivity with hair length can be calculated using the more accurate relations given by the P-E analytical solution, Eqs. (1–4). Figure 1 provides a comparison between these two approaches for calculating sensitivity with frequency set to the hair velocity resonance frequency, $\omega_{res(V)}$. All the values in the figure are for straight cylindrically-shaped (rod-like) hairs oscillating in air at 27°C ($\rho_h = 1100\,\mathrm{kg\,m^{-3}}$, $\rho = 1.1614\,\mathrm{kg\,m^{-3}}$,

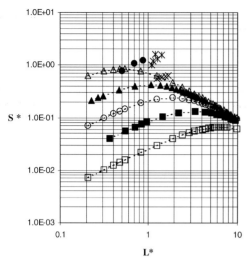

Fig. 1. Variation of hair sensitivity S^* ($=LdV_{res}/dU_0$) with respect to hair length L^* ($=L/\delta_{res(V)}$). Continuous profiles are obtained using the qualitative relations in Table 1. They correspond to: $[S, 0.1R + \mu L^3]$, open triangles; $[S, R + \mu L^3]$, open circles; $[S, 10R + \mu L^3]$, open squares; $[0.1S, R + \mu L^3]$, filled triangles; $[10S, R + \mu L^3]$, filled squares. Data point clusters are obtained using the P-E calculation approach. Points correspond to: MeD1 hairs, filled circles; TiDA1 hairs, asteriks; cricket hairs, crosses; see Table 2

$\mu = 1.85\,10^{-5}\,\mathrm{kg\,m^{-1}\,s^{-1}}$). Note that in Figs. 1–3 and the discussion below, MeD1 refers to hair group D1 on the metatarsus of the spider leg and TiDA1 refers to hair group DA1 on the tibia of the spider leg.

The continuous profiles in Fig. 1 correspond to the qualitative approach, based on the entries to Table 1. The data point clusters correspond to the more accurate P-E solution approach. The profile labeled $[S, R + \mu L^3]$ represents a nominal, "reference" profile (open circles). It is obtained using values for d, L, S, and R corresponding to the MeD1 group of hairs on *Cupiennius* (Barth et al. 1993). For this, we use the fits of $d(L)$, $S(L)$, $R(L)$ to L derived by Humphrey et al. (2001);

$d = 6.343 \times 10^{-5}\,L^{0.3063}$, $S = 1.272 \times 10^{-5}$ $L^{2.03}$, $R = 2.031 \times 10^{-9}\,L^{1.909}$ with all quantities in SI units. With respect to this reference profile, two more profiles are obtained by shifting the value of $R(L)$ in the total damping, $R(L) + \mu L^3$. These profiles are labeled $[S(L),\ 0.1R(L) + \mu L^3]$ (open triangles) and $[S(L),\ 10R(L) + \mu L^3]$ (open squares), meaning that the damping constant, $R(L)$, used in the total damping in one case is 10 times smaller and in the other case is 10 times larger than the nominal, reference value. The other pair of profiles is obtained by shifting the value of $S(L)$. These profiles are labeled $[0.1S(L),\ R(L) + \mu L^3]$ (filled triangles) and $[10S(L),\ R + \mu L^3]$ (filled squares), meaning that the torsional restoring constant, $S(L)$, used in one case is 10 times smaller and in the other case is 10 times larger than the nominal, reference value.

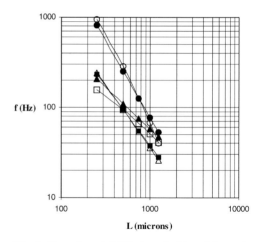

Fig. 2. Experimental (filled symbols) and theoretical (empty symbols) plots of filiform hair resonance frequency as a function of hair length for spider MeD1 hairs (squares) and TiDA1 hairs (circles), and for cricket hairs (triangles). The experimental values plotted are obtained from the correlations provided in the text. These correlations are obtained from the data in Barth et al. (1993) and in Kumagai et al. (1998). The theoretical values are obtained from the P-E analytical solution

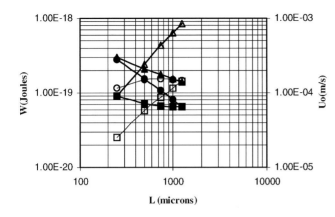

Fig. 3. Work (open symbols) and velocity (filled symbols) to displace spider MeD1 hairs (squares) and TiDA1 hairs (circles), and cricket hairs (triangles), to a threshold of 0.01 degrees at resonance frequency. Energy values are plotted for one complete oscillation cycle

The data clusters in the figure, based on the more accurate P-E approach, correspond to: a) the MeD1 hairs mentioned above (filled circles); b) the TiDA1groups of hairs on *Cupiennius* (Barth et al. 1993) (asterisks); c) the filiform hairs of crickets (Shimozawa et al. 1998) (crosses). For the TiDA1 hairs we take $d(L)$ as for the MeD1 hairs and $S = 1.068 \times 10^{-9}$ $L^{0.631}$, $R = 7.895 \times 10^{-9}$ $L^{2.109}$. For the cricket hairs we take an effective average diameter, obtained by integrating the hair shape along its length, given by $d = 5.487 \times 10^{-4}$

$L^{0.67}$, and $S = 1.944 \times 10^{-6}$ $L^{1.67}$, $R = 1.88 \times 10^{-6}$ $L^{2.77}$. (Note that for the cricket hairs this is not an entirely consistent approach since the values of S and R were obtained by Shimozawa et al. 1998 assuming paraboloid hair shapes.) Table 2 provides the numerical values of the data point clusters in Fig. 1.

Although the dimensionless results in Fig. 1 are expected, they are remarkable in several ways. i) The qualitative profiles based on the relation in Table 1 show that

Table 2. Numerical values of the data point clusters plotted in Fig. 1. Values have been calculated using the P-E model, Eqs. (1–4) in the text. MeD1 and TiDA1, groups of trichobothria on the spider leg metatarsus and tibia, respectively

	Hair length (µm)	$L^* = L/\delta_{\text{res(V)}}$	$S^* = LdV_{\text{res}}/dU_0$
MeD1	250	0.50	0.77
	500	0.70	1.08
	750	0.87	1.22
TiDA1	400	1.10	1.00
	500	1.15	1.20
	650	1.22	1.60
	700	1.27	1.30
	1150	1.44	1.55
Cricket	384	0.93	0.57
	486	1.00	0.59
	640	1.10	0.60
	900	1.21	0.62
	1172	1.32	0.64

increasing the contribution of the damping constant, R, to the total damping $R(L) + \mu L^3$, significantly reduces hair sensitivity while shifting the maximum sensitivity to larger values of hair length. ii) The same observation applies, although the reduction and shift in sensitivity are smaller, when the torsional restoring constant, S, is increased. iii) For the conditions examined, the data clusters based on the P-E calculations suggest that spider and cricket hairs have similar sensitivities and that both experience a viscous-dominated regime $(R(L) < \mu L^3)$. iv) All the data in the clusters fall between $L/\delta_{\mathrm{res}(V)} = 0.5$ and 1.5, approximately, and in this range both the spider and cricket hairs show increasing sensitivity with increasing hair length. v) Such is the result expected from the qualitative profile corresponding to $[S, \ R + \mu L^3]$, based on the physical properties of MeD1 hairs. By comparing this profile with the P-E data cluster for MeD1 hairs we get an idea of the uncertainty in the qualitative profiles. (Note that the accuracy of the entries in Table 1 can be improved by including the values of pre-multiplier numerical coefficients, ranging in magnitude from 0.1 to 10, neglected in the derivation.) This further emphasizes the point that the qualitative information provided in Table 1 and derived from it, such as Eqs. (11–12) and the continuous profiles in Fig. 1, is primarily intended to provide relative trends.

B. Observations Based on the Physically-Exact Model: Energetics of Hair Motion

The work and far field velocity required to displace a filiform hair to a particular maximum threshold angle, θ_{max}, during a single oscillation cycle of a developed, steady periodic flow are obtained from the P-E model using Eqs. (1–7). In particular,

Eq. (7) is integrated numerically using the expressions for torque provided in Appendix 1 of Humphrey et al. (1993, 1997, 1998). The expressions for torque follow exactly from the P-E analysis outlined above and discussed in detail in Humphrey et al. (1993, 2001). Calculations are performed for the spider MeD1 and TiDA1 hairs and for the cricket hairs at their respective resonance frequencies. For values of $d(L)$, $S(L)$ and $R(L)$ we use the experimental fits given above. However, to maintain consistency in the theoretical approach, hair resonance frequencies are evaluated from the P-E equations, Eqs. (1–4), rather than from fits to the experimental data. The plots in Fig. 2 show that the P-E approach yields resonance frequencies in fairly good agreement with the experimental fits. The experimental fits in the figure are given by: MeD1 hairs: $f_{\mathrm{res}(\theta)} = 4.25 \times 10^{-3} \ L^{-1.316}$; TiDA1 hairs: $f_{\mathrm{res}(\theta)} = 6.35 \times 10^{-4} \ L^{-1.695}$; cricket hairs: $f_{\mathrm{res}(\theta)} = 0.1008 \times L^{-0.92}$; where $f_{\mathrm{res}(\theta)}$ is in Hz and L is in m.

Calculations of work and far field velocity are plotted in Fig. 3 for a threshold displacement angle $\theta_{\mathrm{max}} = 0.01$ degrees. Velocity is observed to decrease, and work to increase, with increasing hair length. The first result is in agreement with the entry for θ_{res} for air in Table 1. If all quantities except L and U_0 are kept constant in the entry, for θ_{res} to remain constant increasing L requires U_0 to decrease. Thus, for $R \ll \mu L^3$ we find $U_0 \propto \theta_{\mathrm{res}}/L^{1/4}$, in agreement with the form of the variation observed. Similarly, the second result is expected from inspection of the expressions for torque in Humphrey et al. (1993, 1997), which show $T \propto L^3$.

The calculations suggest that, in the range of frequencies analyzed, spider hairs require smaller medium velocities than cricket hairs to attain the same threshold displacement angle. We also find that the total work required by cricket and spider hairs is similar, with the work for the spider hairs being a little less $(2.5\,10^{-20} - 1.5\,10^{-19}$ Joules) than for the cricket $(9\,10^{-20} - 8.4\,10^{-19}$ Joules). In chapter 10, this volume, Shimozawa et al. report a value of 4×10^{-21} Joules for the minimum work required over an oscillation cycle of frequency 5 Hz to deflect a cricket filiform hair to its maximum deflection at sensory threshold. Using the expressions for torque given in Appendix 1 of Humphrey et al. (1993, see, also, Humphrey et al. 1997, 1998), from Eq. (7) it is possible to show that, to first order, the work over an oscillation cycle varies according to $W \propto \omega$. Thus, for the frequencies explored in this study relative to 5 Hz we expect larger values of work, ranging from 20 to 200 times larger.

IV. Artificial Hair Arrays to Sense Fluid Motion

A. The Application Envisioned

Whether analytical or numerical, the availability of physical-mathematical models to predict the performance of single filiform hairs as medium motion sensors makes it possible to explore the suitability of artificial hair arrays to detect and measure the spatial and temporal characteristics of wall-bounded flows. Especially challenging to characterize are the highly unsteady, multi-scale, three-dimensional vortical structures embedded in the near wall region of turbulent boundary layer flows. The resolution of the time and space characteristics of these structures remains a major unsolved problem in fluid mechanics because of various limitations in the sensors currently available. For example, the measurement volume of a laser-Doppler velocimeter (LDV), a non-invasive flow measurement technique, is shaped as an ellipsoid of revolution and has typical dimensions $25\,\mu m \times 25\,\mu m \times 100\,\mu m$. Because the LDV technique relies on light scattered from small particles suspended in the flow, it yields discontinuous records of velocity. Thus, to obtain spatial correlations revealing the characteristics of flow structures would require relatively high particle concentrations as well as varying the relative locations of two control volumes simultaneously present in the flow. The smallest resolvable flow scale would be $25\,\mu m$, approximately, and the measurements would be restricted to a relatively small region of flow. Also, they would have to be obtained sequentially over relatively long periods of time. The availability of a flow-sensing micro-device with an array of $N \times N$ hair-like sensors on it would circumvent these limitations.

For the purposes of this section, we assume that hair-hair interactions are negligible in the artificial microsensor array imagined, and that geometrical and flow conditions are such that the P-E model applies to any hair. (Actually, the vector equivalent of the P-E model is required in a rigorous analysis.) The micro-electro-mechanical device in mind consists of an array of $N \times N$ hairs all of which are free to deflect rigidly in any direction in the x-y plane of the flat platform supporting them. The values of L, d, ρ_h, R and S are presumed known for each hair as a result of the materials used and the fabrication process employed. Each hair is electromechanically connected at its base from where an electrical or optical signal originates that informs on hair deflection angle and direction as a function of time. In the application, the hairs move in response to the passage of unsteady flow structures and, using image analysis techniques, the size, speed and direction of the structures deflecting the hairs are quantified. From measurements of hair relative displacement, velocity and acceleration it is possible to obtain the total torque acting on any hair. By making measurements and/or informed physical assumptions of the values of the fluid medium velocity and acceleration it is possible, in principle, to separate the viscous drag contribution to torque from that due to the virtual mass. From this information it should then be possible to determine the average force acting on each hair in the array and, from this, the average shear stress of the fluid acting on the wall.

B. Practical Considerations

The fabrication of an artificial sensor mimicking the structure and function of an arthropod filiform hair involves two engineering challenges. The first is to fabricate a mechanical structure analogous to the hair itself that will move in response to the local fluid medium parameters. The second challenge is to select an appropriate transduction mechanism that will transform the mechanical motion into a useful signal, and to interface this signal to electronic circuitry.

Representative dimensions of an arthropod hair are $10\,\mu m$ in diameter and $10\,\mu m$ to $1000\,\mu m$ out of the surface of the organism. An artificial "hair" will likely have different dimensions owing to the differences in material mechanical properties. In microfabrication we typically use stiff materials (silicon, SiO_2, metals) which have a higher Young's modulus than the organic building blocks used by nature. The diameter of the artificial hair can easily be controlled lithographically so that a range of device dimensions could be employed. This would allow for sensing of different strata within the boundary layer, and for tuning the resonant behavior of the microstructures.

Controlling the height out of the plane is not so straightforward. Microfabrication processes employ lithographic pattern transfer tools and techniques which were largely developed for the microelectronics industry (Reed and Fedder 1998). In these applications, the objective is to reduce device scale in all three dimensions in order to improve performance. In particular, electronic devices are very shallow, and the optical tools used to fabricate them have a limited depth of focus, less than $1\,\mu m$. Other tools (e.g., thin film deposition, oxidation furnaces, chemical-mechanical polishing) are also optimized for maintaining a flat profile. In contrast, the present microsystem requires microstructures tens to hundreds of μm out of the plane of the substrate. This difficulty is frequently encountered in microsystem design, and a host of new fabrication techniques are under development to overcome this challenge.

One possibility is to use reactive ion etching to make high aspect ratio pillars. Figure 4 illustrates high aspect ratio structures in silicon approximately $1\,\mu m$ in diameter and $6\,\mu m$ high (Jansen et al. 1995). This technique could be extended

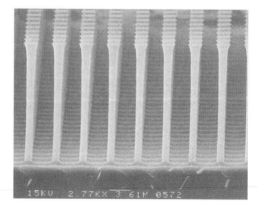

Fig. 4. High aspect ratio microstructures fabricated from a silicon wafer using reactive ion etching (Jansen et al. 1995). In this process, chemical etching is made anisotropic by a directional flux of energetic ions generated in a glow discharge plasma

to structures perhaps 15 to 20 µm out of the plane, but is not practical for larger heights. Another option is to lithographically define tall narrow holes in a thick photoreactive polymer, and selectively fill the holes with metal by electroplating (Frazier and Allen 1993). The sacrificial polymer is removed after plating by an oxygen plasma, leaving behind the desired metal pillars.

Another high aspect ratio fabrication technique takes advantage of the highly ordered structure of porous anodic alumina. When aluminum (and other metals) are anodized, a quasi-regular pore structure results. The arrangement of pores can be controlled with a high degree of precision by lithographically or mechanically deforming the metal surface prior to anodization (Masuda et al. 1997). Figure 5 (left) shows a top view of anodic alumina under natural and patterned anodization conditions. The pores are continuous through the oxide layer, and are oriented perpendicular to

the metal surface, as shown in the right half of Fig. 5. This micrograph shows a porous film approximately 8 µm high, but anodic films can be grown hundreds of microns thick. High aspect ratio microstructures result when these films are lithographically patterned and etched as shown in Fig. 6 (Tan et al. 1995).

In addition to the artificial hair, a transduction technology must be integrated into the device to output the motion sensed. Candidate methods include piezoelectric thin films to transform the stress into an electrical signal; porous membranes where the stress modulates the transverse ionic conductivity; a tunneling current device where the mechanical deflection exponentially controls the tip current; micromirror devices which deflect a probe beam in response to the induced strain. In each of these the integration of the sensing mechanism with sensitive, low-noise electronics is a formidable but tractable problem.

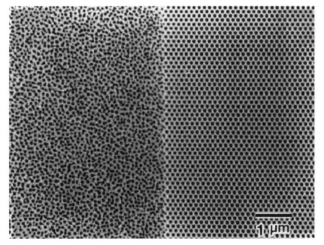

Fig. 5. *Left:* Top view of the porous structure of anodic alumina (Masuda et al. 1997). When aluminum (and other metals) are anodized, a quasi-regular pore structure results. The arrangement of pores can be controlled by lithographically etching or mechanically deforming the metal surface prior to anodization. *Right:* Cross-sectional view of anodic alumina, approximately 8 µm thick (Masuda et al. 1997). Anodization of mechanically deformed aluminum results in a highly regular porous oxide, where the pores are continuous through the oxide layer

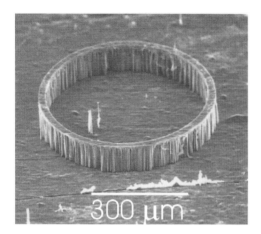

Fig. 6. High aspect ratio microstructures formed by etching of lithographically patterned porous anodic alumina (Tan et al. 1995). Unlike conventional microfabrication techniques, this method can be used to fabricate structures with large out-of-plane dimensions (tens to hundreds of micrometers)

Piezoelectric thin films, such as ZnO and $PbZr_xTi_{1-x}O_3$, have been successfully deposited on silicon substrates, and could conceivably be used to transform the mechanical motion of an artificial hair into a voltage. A likely problem is that most highly piezoelectric materials are extremely stiff, so that the small stresses induced by motion of the hair will not couple into a significant strain in the transduction layer. More compliant piezoelectric materials, such as the polymeric polyvinylidenedifloride (PVDF) have very low piezoelectric coefficients, which will again produce only a small response.

A porous membrane whose ionic conductivity is controlled by the hair stresses is an attractive possibility. Technologies for fabricating thin nanoporous membranes exist and can be integrated into sensor structures. Sensors incorporating modulation of tunneling current have been demonstrated. In both of these methods, noise performance is a concern; because the transduction mechanism is extraordinarily sensitive to minute mechanical motions, other sources of displacement not related to the fluid/hair interaction will introduce significant noise.

A micromirror device has a useful feature: the sensor can be interrogated optically instead of electronically. Also, small angular deflections of a mirror arising from the hair motion can be amplified by placing the optical sensor detecting the motion at a distance from the mirror. Similar devices are employed in video projection systems.

V. Conclusions

The physics of the medium motion-sensing filiform hairs of spiders and crickets is amenable to theoretical analysis. Quantitative results can be obtained using a physically-exact (P-E) approach and useful qualitative observations can be made using a physically-approximate (P-A) approach. Both approaches neglect possible viscous coupling among hairs and the goodness of the results obtained relative to the experimental data available justifies this. Both approaches also assume that the flow driving hair motion takes place along a flat substrate, thus simplifying considerably the specification of the velocity distribution inducing drag.

Physically-exact calculations of the work required to drive a hair over one oscillation cycle yield values ranging from 2.5×10^{-20} to 1.5×10^{-19} Joules for spider hairs, and from 9×10^{-20} to 8.4×10^{-19} Joules for cricket hairs. These values compare to the energy of a single photon (10^{-18}–10^{-19} Joules for light in the visible spectrum) and even smaller values, of order 10^{-21} Joules, have been determined by Shimozawa et al. (see chapter 10, this volume) for the minimum work required over an oscillation cycle of frequency 5 Hz to deflect a cricket filiform hair to its maximum deflection at sensory threshold.

The physically-approximate analysis yields relations valid at resonance frequency from which it is possible to estimate the dependence of hair motion on relative changes in the physical parameters affecting this quantity. These equations show that both the diameter, d, and density, ρ_h, of a hair are important in air but not in water, and that increasing either of these two parameters works to increase the maximum displacement and velocity, θ_{res} and V_{res}, at resonance frequency while decreasing the resonance frequencies, $\omega_{res(\theta)}$ and $\omega_{res(V)}$, themselves. The same relations make it possible to estimate the variation of dimensionless hair sensitivity as a function of dimensionless hair length. A comparison between the P-A calculations and experimental results shows that both spider and cricket hairs experience viscous-dominated damping conditions. Also, the hair length at which sensitivity is a maximum scales with the thickness of the oscillating flow boundary layer, suggesting that this is an optimal hair length.

The P-A relations reveal a significant dependence of hair displacement and velocity on medium viscosity, and this is twice as large in water compared to air. Because viscosity depends on temperature, this suggests the intriguing possibility that hair geometry and material properties may have evolved over evolutionary time scales to perform optimally under gradually changing temperature conditions.

Materials and methods exist with which to microfabricate artificial medium motion sensors modeled on their biological analogs. New technologies for fabricating high-aspect-ratio microstuctures with significant dimensions out of the surface plane must be perfected to build these biomimetic systems. Signal transduction mechanisms which offer integration with electronic circuitry, and can be scaled into large arrays, are necessary to convert the mechanical motions into usable signals.

References

Barth FG (2000) How to catch the wind: Spider hairs specialized for sensing the movement of air. Naturwissenschaften 87: 51–58

Barth FG (2002) A Spider's World. Senses and Behavior. Springer, Berlin Heidelberg New York

Barth FG, Humphrey JAC, Wastl U, Halbritter J, Brittinger W (1995) Dynamics of arthropod filiform hairs. III. Flow patterns related to air movement detection in a spider (*Cupiennius salei* Keys.). Phil Trans R Soc Lond B 347: 397–412

Barth FG, Wastl U, Humphrey JAC, Devarakonda R (1993) Dynamics of arthropod filiform hairs. II. Mechanical properties of spider trichobothria (*Cupiennius salei* Keys.). Phil Trans R Soc Lond B 340: 445–461

Devarakonda R, Barth FG, Humphrey JAC (1996) Dynamics of arthropod filiform hairs. IV. Hair motion in air and water. Phil Trans R Soc Lond B 351: 933–946

Fletcher NH (1978) Acoustical response of hair receptors in insects. J Comp Physiol 127: 185–189

Frazier AB, Allen MG (1993) Metallic microstructures fabricated using photosensitive polyimide electroplating molds. J Microelectromech Systems 2(2): 87–94

Humphrey JAC, Devarakonda R, Iglesias I, Barth FG (1993) Dynamics of arthropod filiform hairs. I. Mathematical modelling of the hair and air motions. Phil Trans R Soc Lond B 340: 423–444

Humphrey JAC, Devarakonda R, Iglesias I, Barth FG (1997, 1998) Errata for Dynamics of arthropod filiform hairs. I. Mathematical modelling of the hair and air motions. Phil Trans R Soc Lond B 352: 1995 and B 353: 2163

Humphrey JAC, Barth FG, Voss K (2001) The motion-sensing hairs of arthropods: Using physics to understand sensory ecology and adaptive evolution. In: Barth FG, Schmid A (eds) Ecology of Sensing, Springer, Berlin Heidelberg etc. pp 105–126

Jansen H, de Boer M, Otter B, Elwenspoek M (1995) The black silicon method IV: the fabrication of three-dimensional structures in silicon with high aspect ratios for scanning probe microscopy and other applications. Proc Eighth International Workshop on Micro Electro Mechanical Systems (MEMS-95), Amsterdam, January 1995, pp 88–93

Kumagai T, Shimozawa T, Baba Y (1998) Mobilities of the cercal wind-receptor hairs of the cricket, *Gryllus bimaculatus*. J Comp Physiol A 183: 7–21

Masuda H, Yamada H, Satoh M, Asoh H, Nakao M, Tamamura T (1997) Highly ordered nanochannel-array architecture in anodic alumina. Appl Phys Lett 71(19): 2770–2772

Reed ML, Fedder GK (1998) Photolithographic Microfabrication. In: Fukuda T, Menz W (eds) Micro Mechanical Systems: Principles and Technology, ISBN 0444-82363-8 Elsevier Science B.V., Amsterdam

Shimozawa T, Kanou M (1984) The aerodynamics and sensory physiology of range fractionation in the cercal filiform sensilla of the cricket *Gryllus bimaculatus*. J Comp Physiol A155: 495–505

Shimozawa T, Kumagai T, Baba Y (1998) Structural scaling and functional design of the cercal wind-receptor hairs of a cricket. J Comp Physiol A 183: 171–186

Shimozawa T, Murakami J, Kumagai T (2002) Cricket wind receptors: Thermal noise for the highest sensitivity known. In: Barth FG, Humphrey JAC, Secomb TW (eds) Sensors and Sensing in Biology and Engineering. Springer-Vienna New York, pp 145–157

Stokes GG (1851) On the effect of the internal friction of fluids on the motion of pendulums. Trans Camb Phil Soc 9: 8ff. (Reprinted in Mathematical and physical papers, vol. III, 1–141. Cambridge University Press, 1901)

Tan SS, Reed ML, Han H, Boudreau R (1995) High aspect ratio microstructures on porous anodic aluminum oxide. Proc Eighth International Workshop on Micro ElectroMechanical Systems (MEMS-95). Amsterdam, January 1995, pp 267–272

Tautz J (1979) Reception of particle oscillation in a medium – an unorthodox sensory capacity. Naturwissenschaften 66: 452–461

10. Cricket Wind Receptors: Thermal Noise for the Highest Sensitivity Known

Tateo Shimozawa, Jun Murakami, and Tsuneko Kumagai

Abstract

The minimum amount of mechanical energy necessary to elicit a neuronal spike in the wind receptor cell of a cricket is determined to be in the order of $k_B T$ (4×10^{-21} Joules at 300°K). Insect mechanoreceptors are therefore bordering the range of thermal noise due to Brownian motion when working near to threshold. Evolution, however, has achieved a paradoxical solution for sensory signal transmission in the presence of thermal noise.

The estimation of mechanical energy is based on three measurements: deflection sensitivity of hair to air motion; sensory threshold of receptor cell in terms of air velocity; and mechanical resistance of hair support. The deflection sensitivity to air motion was measured by laser-Doppler velo-cimetry and Gaussian white noise analysis. Three mechanical parameters, i.e. the moment of inertia of hair shaft, the spring stiffness of hair support, and the torsional resistance within the support were estimated by applying Stokes' theory for viscous force to the data of deflection sensitivity.

The mechanical energy consumed by the resistance provides a maximum estimate of the energy available to the receptor cell for stimulus transduction. The estimated energy threshold of the mechanoreceptor, is far below that of photoreceptors which can detect a single photon (ca. 3×10^{-19} Joules). The mechanoreceptor is 100 times more sensitive than a photoreceptor.

The spike train of the wind receptor cell fluctuates with time, when responding to weak stimuli near threshold. Simultaneous double recordings from two cells revealed that the fluctuations are uncorrelated between cells. An array of mechanoreceptors exposed to thermal noise is able to detect weaker signals, below the threshold, by a paradoxical use of the thermal noise as the seed of uncorrelated randomness for stochastic sampling.

Biological systems evolved under the inevitable presence of thermal noise teach us a design principle for future information systems that will be faced with the same noise. Future technology needs to be informed by the wide and sound bases of the natural sciences, not only those of physics, chemistry and mathematics, but also of biology.

I. Introduction

What is the limit of a biological sensor's sensitivity? Photoreceptors are sensitive to a single photon (Laughlin 1981), and are said to be the most elaborated case of evolutionary improvement. In terms of energy, however, mechano-, chemo- or electro-receptors ought to be much more sensitive than photoreceptors, because stimuli of the former have no quantum structure (Thurm 1983).

The sensitivities of mechanoreceptors have often been described in terms of minimum force or displacement necessary to cause a cellular response (Autrum 1984). The displacement thresholds of mechanoreceptors in a variety of insects and vertebrates are said to lie within the range of 10^{-10} to 10^{-11} m, a fraction of the size of the hydrogen atom (Autrum 1984, Narins and Lewis 1984). This kind of description for mechanical sensitivity is, however, misleading because it ignores the mechanical properties of the receptors. Mechanical impedance is the ratio of force divided by velocity, representing how hard it is to displace or deform an object. When a force causes a displacement, mechanical work (energy), defined by force × displacement (= time integral of impedance × velocity^2 or of force^2/impedance), is done. Any physical measurement requires a certain amount of energy. Violation to this energy requirement will result in the same inconsistency as Maxwell's demon in thermodynamics (Feynman et al. 1963). Thus, the sensory threshold of the mechanoreceptor ought to be expressed in terms of mechanical energy, rather than in displacement alone or force alone. Mechanical impedance must be measured in addition to displacement or force threshold, in order for us to judge the energy sensitivity of a biological sensor.

Crickets and other orthopteroid insects have hundreds of wind sensitive hairs on a pair of appendages named cerci at the abdominal end (Fig. 1). The hairs, under a dissection microscope, are seen to deflect when exposed to the slightest motion of medium air. The hair deflected by air

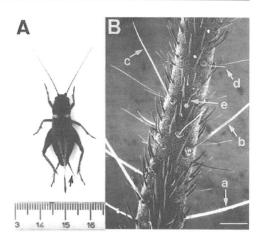

Fig. 1. A: Female of the cricket, *Gryllus bimaculatus*, with scale in cm. Arrow head: cercus. B: Scanning electron microgram of cercal surface, Scale bar: 100 µm; a–e: Wind receptor hairs with wide variety of size

motion gives a mechanical stress to a sensory receptor cell under the hair base. The receptor cell generates nerve impulses and sends them out to the central nervous system along its axon. Animals equipped with sensory capacity for air motion can detect a moving object, which drags surrounding air, without relying on vision. In contrast to the scalar quantity of pressure in a sound wave, the velocity of air motion is a vector quantity by which the central nervous system of the insects can localize the moving object. The sensing of air motion is essential for predator avoidance and other forms of behaviour (Gnatzy and Heusslein 1986; Baba and Shimozawa 1997). The hair is a mechanical lever that is deflected by moving air and transmits a force to a receptor cell under the hair base (Gnatzy and Tautz 1980, Humphrey et al. 1993, Keil 1998, Kumagai et al. 1998a, b).

Arachnids, particularly spiders, have hairs called thrichobothria very similar to the wind sensing filiform hairs on the cricket cerci (Barth and Blickhan 1984, Barth 2002). Mathematical analysis on the trichobothria revealed that their mechanical design and sensitivity are comparable to those of the cricket cercal hairs (Barth et al. 1993, Humphrey et al. 1993, Humphrey et al. in chapter 9, this volume).

II. Mechanical Design of Wind-receptor Hair of Cricket

The cercal wind-receptor hair is a mechanical interface which, on being subjected to a minute amount of viscous force from moving air, transmits its deflection

to a sensory cell beneath the cuticular armor. The size of the cricket wind-receptor hair varies widely between 30–1500 µm in length and 1–9 µm in diameter at the base. The sensory afferents of the cercal wind-receptor sensilla show a length-dependent range fractionation in frequency domain characteristics (Fig. 2): long hairs are velocity sensitive and short ones are acceleration sensitive in the frequency range of 2–100 Hz (Shimozawa and Kanou 1984a).

The cercal wind-receptor hair is essentially an inverted pendulum with a solid

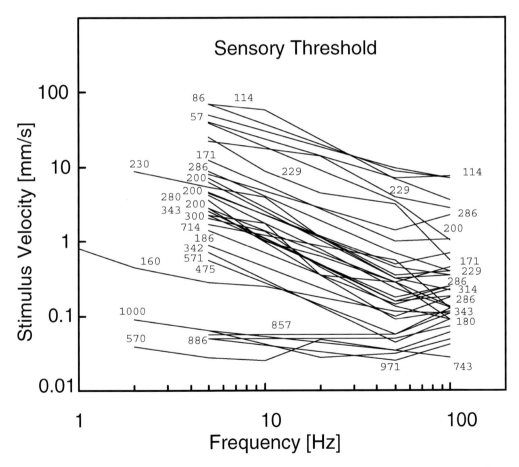

Fig. 2. Sensory thresholds of wind receptors in terms of peak velocity of sinusoidal air motion. Numbers beside lines: hair length in µm

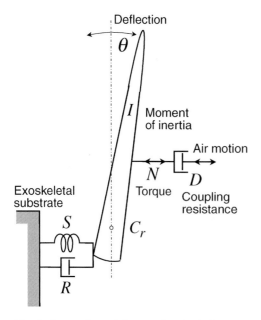

Fig. 3. Inverted pendulum model of wind receptor hair. D Coupling resistance between hair shaft and air, Cr Center of rotation. See text for further explanation

shaft supported by a spring at the base on the cuticular substrate (Fig. 3). The pendulum shaft protrudes into the air and is driven by the viscous force of moving air. The inverted pendulum is a typical second order mechanical system, which consists of three intrinsic parameters, the moment of inertia that represents the mass distribution along the hair shaft (I), the spring stiffness that provides the restoring torque towards a resting position (S), and the torsional resistance within the hair base (R). The pendulum becomes immobile during high frequency air motions due to the moment of inertia.

Neuronal information signaled by the sensory afferents to interneurons in the central nervous system largely depends on the physical structures of the mechanical interface. As a crucial part of the extraction mechanisms of biologically relevant signals from air motion, we must

clarify, 1) the extent of mechanical filtering at the wind-receptor hairs of different lengths, 2) the shape of the hairs, and 3) the mechanical design of the hairs whose length spans a 50 fold range.

A. Deflection Sensitivity of Hair to Air Motion

Deflection sensitivities of the cercal filiform hairs of a cricket were determined by a direct measurement (Kumagai et al. 1998a). The tangential velocities of deflecting hair shafts in response to stimulus air motion were measured *in situ* by a laser-Doppler velocimeter using light scattering from the shaft surface. The stimulus air motion velocity generated in a small wind tunnel was calibrated by the same velocimeter with smoke from a joss stick. The deflection sensitivity of a hair was obtained from the former measurement with reference to the latter calibration of the same apparatus. A Gaussian white noise signal with bandwidth of 500 Hz was employed as the stimulus waveform. Cross-correlation between a measured response and the stimulus waveform gave a time-domain transfer function from the stimulus to the response. The frequency domain transfer functions were obtained by the Fourier transform of the time-domain functions. This method of Gaussian white noise and cross-correlation provides greater precision and wider frequency for a longer period of measurement. The mobility of a hair was expressed in deflection amplitudes and phase shifts with reference to the velocity sinusoid of a stimulus at various frequencies.

The mobilities for the frequency range of 2–500 Hz were measured for a total of 28 hairs whose length ranged over 160–1500 μm (Fig. 4). All wind-receptor hairs were mechanical band-pass filters. Their gain increases with frequency along the slope of 10 ~ 20 dB/decade until the best frequency, and decreases after the best

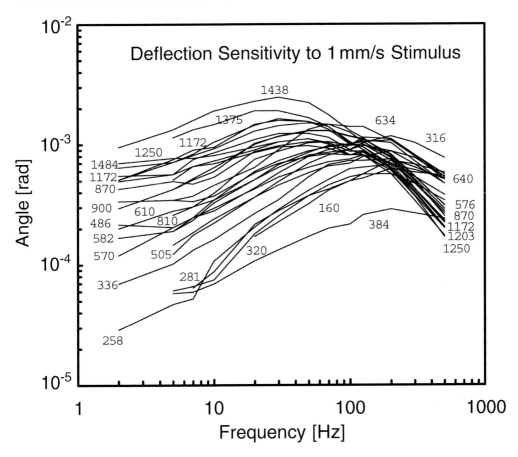

Fig. 4. Deflection sensitivities of cricket wind receptor hairs to 1 mm/s sinusoid stimulus. Numbers beside lines are hair length in µm. Labels at left side tend to be in order of hair length from bottom to top, whereas those at right-hand side are in opposite order

frequency along the slope of −20 ∼ −10 dB/decade. The best frequency varied with hair length, in shorter hairs it was higher. No sharp peaks were seen around the best frequency. This suggests that the wind-receptor hair is a strongly damped mechanical system. The phase of hair deflection lags with frequency along a sigmoidal curve. The rate of increase in the center of the phase shift curve was about $\pi/2$ per decade change of frequency in all hairs (Kumagai et al. 1998a). This also indicates that this system is strongly damped, twice the optimal damping.

The mobility measurements established the following conclusions. The wind receptor hairs comprise an array of mechanical band-pass filters whose best frequencies are inversely proportional to the length. The motion dynamics of the wind-receptor hairs have strong damping.

B. Shape of Wind Receptor Hair

We examined the exact shapes of the thread-like wind receptor hairs in the cricket (Kumagai et al. 1998b). The diameters of the hair shaft at various distances from the hair tip were measured by scanning electron microscopy. The exact shape of a hair is essential for theoretical estimation of the drag force acting on each portion of the hair shaft and for the geometrical estimation of the moment of inertia of the hair.

The measurements revealed hairs in the shape of extremely elongated paraboloids. We had expected, a priori, that the shape of the hair would be a slender linearly-tapered cone. The diameter of the wind-receptor hairs varies with the square root of the distance from the hair tip, i.e., the diameter rapidly increases with the distance from the tip and is asymptotic to the base diameter. The base diameter varies with an allometry to the power of 0.67 on the hair length. Establishing the exact shape and the base diameter-length allometry enables us to make an accurate calculation of the moment of inertia from the hair length alone. The shape of the hair largely constrains the mobility of the wind receptor hair, because the drag force caused by moving air and the moment of inertia of motion dynamics are functions of shaft diameter. The shape of the hair is therefore a crucial biological trait for the sensory information transmitted to the central nervous system, and is certainly subjected to selection pressure.

C. Dynamic Estimation
of Mechanical Parameters

The estimation of I, R, and S was based on the method of least squares fitted to the equation of motion (Shimozawa et al. 1998). The equation of motion of the inverted pendulum model (Fig. 3) is

$$I \cdot d^2\theta(t)/dt^2 + R \cdot d\theta(t)/dt + S \cdot \theta(t) = N(t),$$

where $\theta(t)$, $N(t)$, and I, R, S are respectively, hair deflection, deflecting torque acting on hair shaft, and the three me-

chanical parameters to be determined. Replacing $\theta(t)$ and its time derivatives with the measured values shown in Fig. 4 and $N(t)$ with theoretical values of torque, at different frequencies and instances of time, we obtained 50–70 equations for the three unknowns of each hair. The least square analysis on the set of simultaneous equations provides the most reasonable values for I, R, and S. The theoretical torque is given by Stokes' theory of mechanical impedance of an oscillating cylinder in viscous fluid (Stokes 1851, Humphrey et al. 1993). The effect of the boundary layer in which air is stagnating on the substrate surface is also taken into account. The values of three intrinsic parameters obtained by the dynamic estimation were compared with those estimated in other ways. The moment of inertia was compared with a geometrical estimation based on the hair shape. The spring stiffness was compared with a previous static measurement in which a tiny plastic sphere was loaded on a hair shaft to give a static gravity torque (Shimozawa and Kanou 1984b).

The moment of inertia obtained by the dynamic estimation showed a clear length dependency to the power of 4.32 of the hair length (Fig. 5). The dynamic estimation nicely fits with those obtained by geometrical estimation based on the hair shape of square root cone, for the range of 160–1500 μm in hair length (Shimozawa et al. 1998). As the moment of inertia of a cone shaped hair is proportional to both the square of the base diameter and the cube of hair length, the geometrical estimation should vary as a power function of hair length with an exponent of 4.34, very close to 4.32. This agreement of the two independent estimations strongly supports the validity of the dynamic estimation of the mechanical parameters.

The dynamic estimation also showed clear length dependencies of the stiffness of hair supporting spring and the

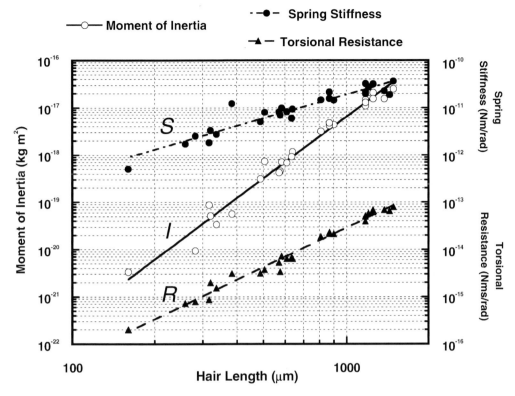

Fig. 5. Three mechanical parameters of cricket wind receptor hairs plotted in relation to hair length

torsional resistance within the hair base, to the power of 1.67 and 2.77 respectively. The slope of length dependency of the spring stiffness also fits with the previous static measurement of the exponent of 1.54, although the absolute value of the dynamic estimation is two times larger than that obtained by the static method. The estimated values of torsional resistance within the hair base are so large that the hairs are strongly damped non-oscillatory second order systems. Because the internal resistance can only be estimated by the dynamic measurements, there is no comparable measurement. The large internal resistance is consistent with the over-all sigmoid shape of the phase shift curve (Shimozawa et al. 1998).

D. Impedance Matching

The length dependency of the internal resistance R was estimated to the power of 2.77. This is fairly close to the theoretical expectation of a power of 3 for the length dependency of the frictional resistance D (see Fig. 3) at the site of air-hair contact (Shimozawa et al. 1998). The coupling resistance D is given by an integral of the real part of Stokes' impedance for an infinitesimally thin cylinder along its length. This large length dependence of D is a physical constraint for efficient energy transmission to the sensory cell. The power supply to R is maximum when R is the same as D. Biological design seems to have attempted to match R (2.77 power of length) with D, although exact matching has not yet been achieved. The internal resistance seems to be in the impedance matching with the frictional resistance at the site of air-hair contact, through which mechanical energy is transferred from air to hair. The energy transmission from the moving air to the sensory cell is maximized by the impedance matching.

The cricket wind-receptor hairs with different sizes compose an array of band-pass filters with different best frequencies. The estimated length dependency of best frequency is to the power of -0.92, and that of best sensitivity is -0.55. The wind-receptor hairs are designed so as to shift the frequency band two decades

while keeping a similar sensitivity. They exhibit a combination of allometric scaling of structural elements: the diameter of hair base, the spring stiffness, and the internal resistance. The impedance matching, which provides a maximum energy transmission from moving air to sensory cell, seems to be a biological adaptation for higher energy sensitivity.

III. Energy Available to the Receptor Neuron

The amount of energy consumed by the resistance R during one period of hair

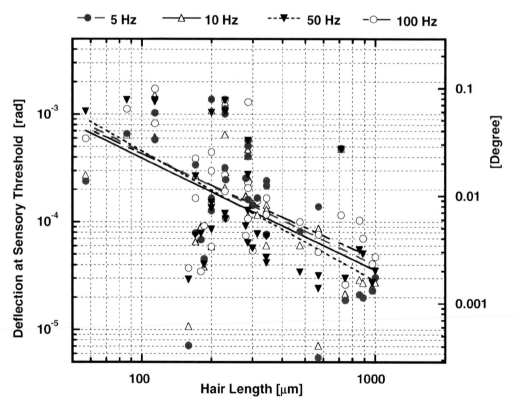

Fig. 6. Hair deflection at sensory threshold. Lines: regressions at four different frequencies

motion is given by $R \cdot (2\pi f \cdot \Theta_t)^2/(2f)$, where Θ_t and f are the deflection and frequency of hair motion. The peak amounts of kinetic and potential energy stored in and released from the moment of inertia and the spring stiffness during hair motion were similarly calculated as $(1/2) \cdot I \cdot (2\pi f \cdot \Theta_t)^2$ and $(1/2) \cdot S \cdot \Theta_t^2$, respectively. The moment of inertia of the air mass stagnating on the hair shaft due to viscosity was taken into account for calculation of the kinetic energy (Shimozawa et al. 1998).

For the estimation of the amount of mechanical energy available to the receptor cell at sensory threshold, the deflection angle of the hair at sensory threshold is necessary in addition to the mechanical parameters shown above. We determined the deflection at sensory threshold (Θ_t, Fig. 6) by combining two measurements, the sensory threshold in terms of wind velocity (Fig. 2; Shimozawa and Kanou 1984a) and the deflection sensitivity measured by the laser-Doppler velocimeter

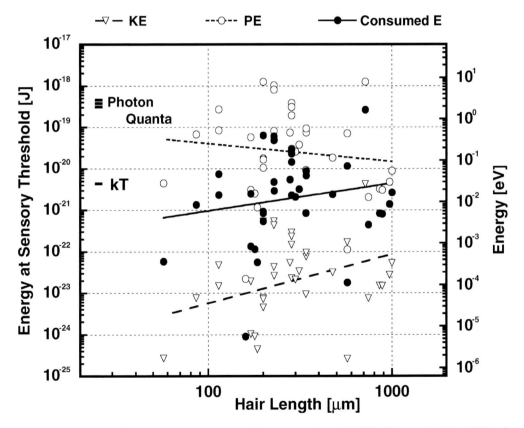

Energy Threshold at 5 Hz

Fig. 7. Mechanical energy available to receptor cell at sensory threshold (filled dots and thick solid lines) at 5 Hz. Open circles (PE) and triangles (KE): Potential and kinetic energies stored in and released from spring stiffness and moment of inertia. Photon quanta: Energies of blue, green and red light photons from top to bottom. kT: Thermal noise level at 300°K (4×10^{-21} J or 25 meV)

(Fig. 4). Deflections at the sensory threshold are frequency independent (Fig. 6).
The receptor cells of a 1000 μm hair fire
a spike when the hair is deflected by ca.
4×10^{-5} radians, regardless of the frequency of hair motion, whereas those of
a 100 μm hair fire when deflected by ca.
4×10^{-4} radians.

The amount of mechanical energy consumed by the torsional resistance within
one cycle of hair motion was calculated
for the deflections at sensory threshold
(Fig. 7, filled dots). The amount of energy
consumed by the resistance provides the
maximum estimate for the energy available to the receptor cell for activating a
biological process of stimulus transduction. The energy thresholds estimated are
around $k_B T$ of 300°K (4×10^{-21} Joules),
the thermal fluctuation level. This mechanoreceptor is therefore 100 times more
sensitive than photoreceptors, which respond to a single photon quantum (ca.
3×10^{-19} Joules). The cercal wind receptor of the cricket is the most sensitive cell
ever reported. If any fraction of the measured torsional resistance is truly frictional, the amount of energy available to
the receptor cell becomes much smaller.

In chapter 9, this volume, the total
work done by air required to displace a
hair by 0.01 degrees during one oscillation cycle at resonant frequency is estimated. Figure 6 indicates that the length
of a cricket cercal hair whose deflection
threshold is 0.01 degrees is about 250 μm.
According to figures 2 and 3 of chapter
9 (Humphrey et al. 2002), the resonant frequency and the work required to
deflect a 250 μm hair are estimated to be
about 200 Hz and 10^{-19} Joules, respectively. Because the work done on a linear
resistance Z during one cycle of motion is
proportional to frequency when deflec-
tion amplitude is constant ($Z \cdot (\dot{\theta})_{max}^2 \cdot$
$(1/f) \propto Z \cdot (\theta_{max})^2 \cdot f$), the work at 5 Hz becomes 1/40 of that at 200 Hz. The similar-
ity of mechanical properties of hairs between cricket and spider may suggest a
comparable energy sensitivity of sensory
cells.

IV. Cell-intrinsic Noise
in Neuronal Spike Train

Because the energy thresholds of the receptor cells are around $k_B T$, we must expect that the cells are exposed to the thermal noise of Brownian motion when
working near threshold. When stimulated
by a pure sinusoid near threshold, the receptor cells actually show considerable
fluctuations in spike timing around a
mean firing phase, suggesting the presence of thermal noise in the transduction
process. In order to discriminate whether
the fluctuation is truly due to the cell-
intrinsic thermal noise or an atmospheric
or substrate born external noise, we made
intracellular recordings of spike responses from two receptor cells simultaneously and examined temporal correlations between the two spike trains.

First, the raw joint probability density
for a pair of spikes within one cycle of stimulus actually fired at time (τ_A, τ_B) from
the mean firing phase of each neuron was
constructed in a form of a two dimensional
contour map. Then, the direct product of
the fluctuation probability densities of
each neuron (null-hypothesis for indepen-
dence) was subtracted from the raw joint
probability (Aertsen et al. 1989). No statistical co-variance remained after the subtraction (Fig. 8a), i.e. spike fluctuations
in the cells under the double recordings
were independent. The apparent temporal
correlation by stimulus driven component
shown in the diagonal projection also dis-
appeared after subtraction. A clear correlation was seen, however, if a common external noise of similar intensity was superimposed on the sinusoid (Fig. 8b). The
receptor cell therefore has an internal
noise whose magnitude is comparable to

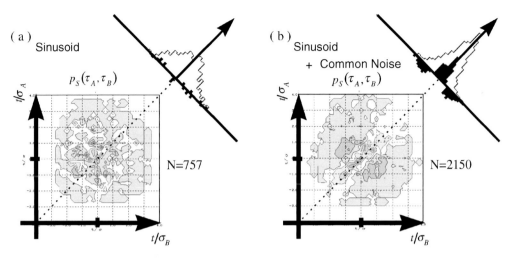

Fig. 8. Joint probability of firing timing of two receptor cells (A and B). Two cells fluctuate in timing, but practically no spike-to-spike covariance remains after subtraction of stimulus driven component (a). A common noise superimposed to sinusoid stimulus produces a diagonal ridge in the covariance (b)

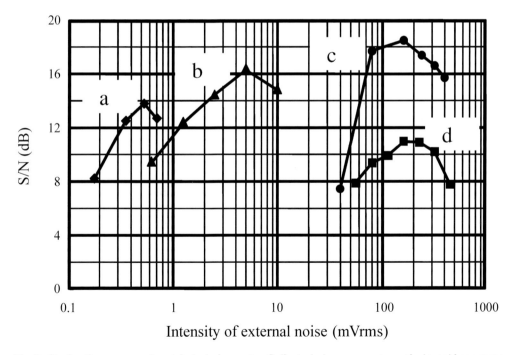

Fig. 9. Stochastic resonance in cricket wind receptor. Spike train in response to weak sinusoid can carry larger amount of signal component, if external noise of higher intensity is superimposed, though the range is limited. a–d: receptor cells with different sensitivity

that of cellular response to the stimulus at sensory threshold. Taking into account that the amount of energy available at the sensory threshold, is around k_BT at $300°K$, the internal noise must be of thermal origin. Although the thermal noise is inevitable for living cells, the common component in the external air motion can still be encoded in the form of covariance of spike timing between receptor cells.

V. Enhancement of Signal Detection Helped by Noise

The wind receptor cell elicits no spikes to a very weak, sub-threshold, sinusoid stimulus. The same receptor, however, elicits a spike train with irregular missing, if an external noise of similar intensity is superimposed to the sub-threshold sinusoid. Within a certain range, larger noise can elicit spike trains that have a better signal to noise ratio (Fig. 9). This enhancement of signal detection by the help of noise is known as stochastic resonance (Douglass et al. 1993). The intensity of the external noise which provides the best enhancement corresponds to the sensory threshold of the receptor. Because the magnitude of the cell-intrinsic noise is estimated to be at threshold, the receptor cells are already under the condition for the enhancement by the intrinsic noise. The cell-intrinsic thermal noise is thus working to trigger the stochastic sampling and essential for the enhancement of signal detection.

VI. Conclusions with Biological Insight into Engineering

We conclude that the cricket wind receptor has improved the mechanical sensitivity up to the ultimate physical limit of thermal noise in the process of evolution. Although the receptor cells set their firing threshold above the k_BT to avoid the meaningless response to the thermal noise, they are still

able to detect weaker signals below the threshold. The insect mechanoreceptor utilizes the cell-intrinsic thermal noise for stochastic sampling as the source of non-correlated noise and contributes to signal detection. An ensemble of spiking sensory cells can transmit sensory information and reconstruct the stimulus wave form in high fidelity, if each member of the ensemble has an independent noise source (Collins et al. 1995, Shimokawa et al. 1999). Other mechanoreceptors including the vertebrate hair cells would have reached similar sensitivity at the ceiling of the thermal noise limit (Denk and Webb 1989), because mechanical energy is transferred without any quantum structure in the frequency range of biological relevance (Thurm 1983).

Social demands on electronic devices are: faster in operation speed, smaller in size and lighter in power consumption. In order to meet the demands, new devices such as single electron transistors, molecular or quantum devices are being developed. All these devices, however, are essentially constrained by noise or the stochastic nature of quantum mechanics. We must tame or cope with the noise, instead of trying to avoid it.

Biological systems evolved under the inevitable thermal noise can teach us a design principle for future information systems that will be faced with the same noise. Future technology needs to be informed by the wide and sound bases of the natural sciences, not only physics, chemistry and mathematics, but also biology.

References

Aertsen AMHJ, Gerstein GL, Habib MK, Palm G (1989) Dynamics of neuronal firing correlation: Modulation of "effective connectivity". J Neurophysiol 61: 900–917

Autrum H (1984) Comparative physiology of invertebrates: Hearing and vision. In: Dawson WW, Enoch JM (eds) Foundations of Sensory Science. Springer-Verlag Berlin, pp 1–23

Baba Y, Shimozawa T (1997) Diversity of motor responses initiated by a wind stimulus in the freely moving cricket, *Gryllus bimaculatus*. Zool Sci 14: 587–594

Barth FG (2002) A Spider's World. Senses and Behavior. Springer, Berlin Heidelberg New York

Barth FG, Blickhan R (1984) Mechanoreception. In: Bereiter-Hahn J, Matoltsy AG, Richards KS (eds) Biology of the Integument, Vol. 1 (Invertebrates), VIII (Arthropoda), Springer-Verlag, Berlin, pp 554–582

Barth FG, Wastl U, Humphrey JAC, Devarakonda R (1993) Dynamics of arthropod filiform hairs. II. Mechanical properties of spider trichobothria (*Cupiennius salei* Keys.). Phil Trans R Soc Lond B 340: 445–461

Collins JJ, Chow CC, Inhoff TT (1995) Stochastic resonance without tuning. Nature 376: 236–238

Denk W, Webb WW (1989) Thermal-noise-limited transduction observed in mechanosensory receptors of the inner ear. Phys Rev Lett 63: 207–210

Douglass JK, Wilkens L, Pantazelou E, Moss F (1993) Noise enhancement of information transfer in crayfish mechanoreceptors by stochastic resonance. Nature 365: 337–340

Feynman R, Leighton R, Sands M (1963) Ratchet and pawl. In: The Feynman Lectures on Physics, Vol. I, Addison-Wesley, Reading, MA, pp 46-1/46-9

Gnatzy W, Heusslein R (1986) Digger wasp against crickets. I. Receptor involved in the antipredator strategies of the prey. Naturwissenschaften 73: 212

Gnatzy W, Tautz J (1980) Ultrastructure and mechanical properties of an insect mechanoreceptor: Stimulus-transmitting structures and sensory apparatus of the cercal filiform hairs of *Gryllus*. Cell Tissue Res 213: 441–463

Humphrey JAC, Devarakonda R, Iglesias I, Barth FG (1993) Dynamics of arthropod filiform hairs. I. Mathematical modeling of the hair and air motion. Phil Trans R Soc Lond B 340: 423–444; Errata added in ditto (1997) 352: 1995

Humphrey JAC, Barth FG, Reed M, Spak A (2002) The physics of arthropod medium-flow sensitive hairs: Biological models for artificial sensors. In: Barth FG, Humphrey JAC, Secomb

TW (eds) Sensors and Sensing in Biology and Engineering. Springer, Wien New York

Keil TA (1998) The structure of integumental mechanoreceptors. In: Harrison FW, Locke M (eds) Microscopic Anatomy of Invertebrates. Vol. 11B: Insecta, Wiley-Liss, New York, pp 385–404

Kumagai T, Shimozawa T, Baba Y (1998a) Mobilities of the cercal wind-receptor hairs of the cricket, *Gryllus bimaculatus*. J Comp Physiol A 183: 7–21

Kumagai T, Shimozawa T, Baba Y (1998b) The shape of wind-receptor hairs of the cricket and cockroach. J Comp Physiol A 183: 187–192

Laughlin S (1981) Neural principles in the peripheral visual systems of Invertebrates. In: Autrum H (ed) Vision in Invertebrates, Handbook of Sensory Physiology VII/6B, Springer-Verlag, Berlin, pp 133–280

Narins PM, Lewis ER (1984) The vertebrate inner ear as an exquisite seismic sensor. J Acoust Soc Amer 76: 1384–1387

Shimozawa T, Kanou M (1984a) Varieties of filiform hairs: range fractionation by sensory afferents and cercal interneurons of a cricket. J Comp Physiol A 155: 485–493

Shimozawa T, Kanou M (1984b) The aerodynamics and sensory physiology of range fractionation in the cercal filiform sensilla of the cricket *Gryllus bimaculatus*. J Comp Physiol A 155: 495–505

Shimozawa T, Kumagai T, Baba Y (1998) Structural scaling and functional design of the cercal wind-receptor hairs of cricket. J Comp Physiol A 183: 171–186

Shimokawa T, Rogel A, Pakdaman K, Sato S (1999) Stochastic resonance and spike-timing precision in an ensemble of leaky integrate and fire neuron models. Phys Rev E 59: 3461–3470

Stokes GG (1851) On the effect of the internal friction of fluid on the motion of pendulums. Trans Cambridge Philos Soc 9: 8ff. (Reprinted in Mathematical and physical papers, Vol. III, pp 1–14: Cambridge University Press, 1901)

Thurm U (1983) Biophysics of mechanoreception. In: Hoppe W, Lohman W, Markl H, Ziegler H (eds) Biophysics, Springer-Verlag, Berlin, pp 666–671

11. Arthropod Cuticular Hairs: Tactile Sensors and the Refinement of Stimulus Transformation

Friedrich G. Barth and Hans-Erich Dechant

Abstract

Arthropods are richly supplied with mechanosensitive cuticular hairs. Much of the refinement of these sensors including their specificity and their sometimes exquisite sensitivity lies in the mechanical processes characterizing the uptake and transformation of their adequate stimulus. In order to understand these processes the application of approaches used in engineering is highly rewarding. In this chapter we report on micromechanical measurements and the application of Finite Element analysis to spider tactile hairs after first contrasting these with hairs responding to airflow. The most relevant mechanical difference between these two types of hairlike sensors is the stiffness of their articulation, which is larger by about four powers of ten in tactile hairs. As a consequence tactile hairs are not only deflected but in addition bent by the stimulus. Their actual bending is dominated by the modulus of elasticity of the hair shaft and the second moment of area and its changes along the hair. The spider tactile hairs examined perfectly combine high sensitivity (threshold deflection by torques measuring ca. 5×10^{-10} Nm) with efficient protection from being overloaded. They may be classified as lightweight structures of equal maximum strength, representing sophisticated micro-electromechanical systems.

I. Introduction

One of the main characteristics of arthropods such as insects, spiders, and scorpions is their cuticular exoskeleton. It protects the animals and provides the lever system needed for their locomotion with segmented extremities. In addition it plays a dual sensory role. One of these roles is to simply carry a large variety of sensors which are mostly hair-like and referred to as hair sensilla. The overwhelming majority of these cuticular hairs is mechanosensitive. The second sensory role of the arthropod exoskeleton is more special. It is its stimulus conducting role for the detection of cuticular strains by insect campaniform sensilla and arachnid slit sensilla. These sensilla are not hair like structures but, instead, tiny holes in the exoskeleton which are covered by a thin membrane to which the dendritic end of a sensory cell attaches. Adequate stimulation of the hole sensilla is due to strains in the surrounding cuticle. The strains in turn result from muscular activity during locomotion, from an increase of the hemolymph pressure (used by spiders to hydraulically extend some of the joints of their legs) (Blickhan and Barth 1985) or from vibrations of the substrate (Barth 1998). The slightest deformation of the hole (in the range of a few Ångström units) elicits a nervous response in the sensory cell following the minute deformation of its dendrite tip (Barth 1981, 2002).

This chapter will not deal with these hole – sensilla but with mechanosensitive hair sensilla. It is the displacement of the hair shaft which makes hair sensilla send a nervous signal to the central nervous system. Here we will focus on some details of stimulus uptake by tactile hairs and the way in which the stimulus is transformed on its way from the outside to the sensory cell. It will become clear that even in the case of a seemingly simple tactile hair the underlying mechanical processes are by no means trivial. In order to better illustrate the wide range of sensitivities represented by mechanosensitive cuticular hair sensilla a short paragraph is also dedicated to hair-like air movement detectors. These are also the subject of the chapters by Humphrey et al.(9) and by Shimozawa et al. (10) in this volume. Our efforts to understand the mechanical design of mechanosensitive arthropod hair sensilla may inspire the engineer to take notice of the tricks found in nature when designing a synthetic hair-like micro-mechanosensor.

II. Hairs as Mechano-Sensors

There is great variation among hair-like arthropod sensors with regard to both their structure and their mechanical pro-

perties. Going along with this, one finds substantial differences in a hair's specific sensitivity to a particular stimulus. Much of this variation, however, can be accommodated in a few simplifying schemes which are presented at the outset and point to the main parameters to be considered (Fig. 1).

Figure 1a shows the hair shaft of length L and diameter d, and its suspension in a cuticular socket by a soft joint membrane. S (Nm/rad) and R (Nms/rad) denote the spring stiffness and the damping constant of the articulation, respectively. When the hair shaft is deflected by an external force both the elastic restoring force and the viscous damping resist its deflection.

The force which deflects the hair shaft and thereby elicits nervous impulses in the sensory cells attached to the sensillum may either be a directly coupled tactile force or a viscous force resulting from the medium flow around the hair. The geometrical and mechanical properties of only a few parameters largely determine whether a mechanosensitive cuticular hair is a tactile sensor depending on comparatively large forces to be deflected, or a considerably more sensitive medium flow sensor like an insect filiform hair or an arachnid trichobothrium.

Joint stiffness S is considerably smaller in flow sensitive hairs than in tactile hairs. As a result the forces acting on flow sensors are too small to bend the hair shaft to any relevant degree. This, in turn, implies that in flow sensors mass M and torsional inertia I are the parameters dominating their mechanical behavior (Fig. 1b). In the tactile hairs examined by us the situation is radically different (Fig. 1c). Here the forces needed to overcome joint stiffness S do bend the hair shaft in addition to deflecting it. Due to the relative size of the stimulating force, the forces due to the inertia of mass M of the hair shaft may be neglected whereas Young's modulus E and the second

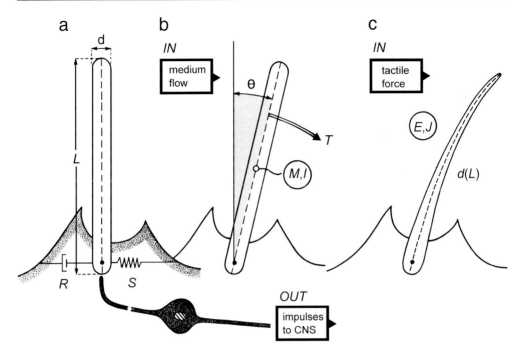

Fig. 1. a Schematic representation of an *arthropod mechanosensory hair* of length L and diameter d and with its articulation characterized by the damping constant or frictional resistance R and the torsional restoring constant or spring stiffness S. **b** *Air flow sensitive hair* deflected (Θ) by viscous forces (T, torque) acting along its length and obeying fluid mechanics. The hair parameters other than length L and diameter d to be mainly considered are its mass M and its torsional inertia I. **c** *Tactile hair*, both deflected and bent by stimulation from above. Here the parameters to be taken into account in an alalysis of stimulus uptake and transformation are the modulus of elasticity E and the second moment of area J along hair length L

moment of area J along the length L of the hair shaft are dominant features.

A. Medium Flow Sensors

Spider filiform hairs or trichobothria are thin hairs (ca. $10\,\mu m$ at their base), only about $0.1\,mm$ to $1.4\,mm$ long and with very small values for R (in the order of $10^{-15}\,Nms/rad$) and S (in the order of $10^{-12}\,Nm/rad$). By contrast the S values of spider tactile hairs (Albert et al. 2001, Barth et al. in prep.) and fly macrochaetae (Theiß 1979) are all in the range of 10^{-9} to $10^{-8}\,Nm/rad$ which is larger by between three and four powers of ten.

The interaction between the moving air and a flow sensor is an intriguing process, which, however, is well understood in terms of the underlying physics (Shimozawa and Kanou 1984a, b, Humphrey et al. 1993, 1998, Barth et al. 1993, Barth 2000; chapters 9 and 10, this volume). In a recent physical and mathematical approach Humphrey et al. (2001) examined the relative importance of the various hair parameters in determining a hair's absolute sensitivity to medium flow as well as its frequency tuning. For the biologist an important question asked at the same time was which effects natural selective pressures and ecology may have had on the adaptive evolution of these parameters. As a result of our computational approach the design principles can now be described even more quantitatively and thereby also rendered more meaningful for an engineer. Only a short summary of the results is given here.

Maximum deflection Θ of a trichobothrium at resonance frequency, ω_{res} (Θ), depends on both the damping constant, R, and the torsional restoring constant, S. The respective effects of R and S on the hair velocity at resonance, V_{res}, and on the maximum velocity resonance frequency, ω_{res} (v), are independent of each other, whereas both ω_{res} (Θ) and ω_{res} (v) depend on S to the 1/2 power and on the hair length, L, to the $-3/2$ power. With increasing farfield velocity amplitude, V_0, of the medium, both the hair displacement at resonance, Θ_{res}, and its velocity at resonance, V_{res}, increase linearly. The effect of air viscosity on the hair's total inertia is much larger than that of its density. Both ω_{res} (Θ) and ω_{res} (v) are independent of air density and viscosity.

When calculating the relative sensitivity of hair-resonance frequency ω_{res} (Θ) for each parameter (hair diameter d, length L, R, S, viscosity and density of the medium) it can be shown that $\Delta L/L \sim -0.50$ $\Delta S/S$, 0.20 $\Delta R/R$, and -0.10 $\Delta d/d$. In more practical terms this means that the fractional change of L is the most effective way to change ω_{res} (Θ). Spider trichobothria do indeed occur in groups of hairs substantially differing in length. A length variation between 100 µm and 1400 µm goes along with best frequency ranges between 40 Hz and 600 Hz (Barth et al. 1993). It can also be shown that in terms of the required mass change it is considerably more economical to vary hair length rather than hair diameter in order to modify the range of best frequencies (Humphrey et al. 2001). Thus predictions on evolutionary changes and the fine tuning of the hairs to the relevant natural stimuli become possible.

The reader interested in more details and their mathematical derivation is referred to chapter 9, this volume, and to Humphrey et al. (2001), where the effects of the various parameters affecting the response of a hair like medium flow detector are also given for hairs in water (see also Devarakonda et al. 1996). It may suffice to add here that boundary layer effects have to be taken into account and that a hair's response to medium flow much depends on how much it is embedded in the boundary layer or poking through it.

The value of the engineering approach to hair sensilla sensitive to medium flow is underlined by the analysis of Shimozawa and colleagues presented in chapter 10 of this volume. In this chapter perfect mechanical impedance matching and the effects of stochastic resonance are related to the outstanding absolute sensitivity seen in the nervous response of cricket filiform hairs which appear to represent the most sensitive biological sensors so far known.

B. Tactile Hairs

A spider has up to about 950 trichobothria (Barth et al. 1993, Barth 2002) sensitive enough to elicit prey capture behavior when stimulated by the air flow produced by a fly passing by in flight at a distance of up to about 30 cm. The same spider, however, may have hundreds of thousands of tactile hairs which at some areas of its body surface form dense coats of up to 400 hairs/mm^2 (Fig. 2a). Apart from being exteroreceptors sensing stimuli (objects) from outside the animal, tactile hairs also form proprioreceptors which sense stimuli produced during locomotion by the animal itself like the contact between neighboring body parts. An example for such proprioreceptive hairs are the rows of hairs seen at the joints between the segments of the leg (Fig. 2a), which are stimulated when the more distal leg segment is flexed against the more proximal one.

Any effort to understand the refinement of the uptake of tactile stimuli depends on an in depth knowledge of the mechanics of the hair shaft and its articulation with the general cuticle. This will be detailed in the following.

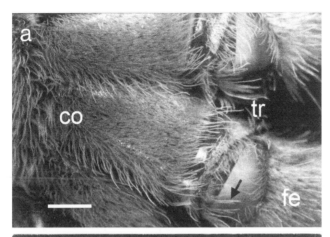

Fig. 2. *Mechanosensitive cuticular hairs* covering the body surface of the spider *Cupiennius salei.* **a** Ventral view of proximal leg segments (*co* coxa, *tr* trochanter, *fe* femur); note proprioreceptive hairs bordering the joints (arrow). **b** The most distal segment of the leg (tarsus); arrow points to tactile hair examined in detail. Scale bars 0.6 mm in **a** and 1 mm in **b** (**b** from Dechant et al. 2001)

III. Mechanics of a Spider Tactile Hair

A. Deformation of the Hair Shaft

A joint membrane divides the hair into an outer lever arm which takes up the incoming mechanical stimulus and a much shorter inner lever arm which transmits the stimulus energy to the dendritic terminals of the sensory cells attached to the hair sensillum. Whereas the hair shaft of medium flow detectors like spider trichobothria may be considered a rigid structure under the conditions of natural stimulation this is different in many tactile hairs as demonstrated for the conspicuously long hairs on the dorsal aspect of the spider leg tarsus and metatarsus seen on Fig. 2b.

The outer hair shaft of these and many similar hairs is a surprisingly complex stimulus transmitting structure (Albert et al. 2001, Dechant et al. 2001). When touched by an object from above as it happens when the night active spider is using its front legs for tactile orientation during its so-called "guide stick walk" in complete darkness (Schmid 1997) the hair shaft is both deflected and bent by direct contact. The formalization and quantification of this process is demanding and all but trivial because of non-linearities. The most obvious of these is the fact that the load on the hair is a function of hair deformation. Therefore linear theory with regard to the equilibrium forces in the unstimulated hair cannot be applied and the hair cannot be adequately modeled analytically. For this reason a numerical model was developed using the finite element method (FEM) and representing the hair shaft by about 50 beam elements (Dechant et al. 2001). For this, the details of the hair's geometry had to be known, hair deflection and hair deformation under load had to be visualized microscopically for precise quantification, and forces in the order of micro-Newtons had to be measured. Forces of that order of magnitude have to be applied to overcome the spring stiffness S of the hair's articulation when the hair is deflected. In addition, at least a good estimate of the material properties is needed.

When the outer hair shaft (length L of the examined tarsal hair measures roughly 2.6 mm, its insertion angle ca. 58°, and its outer diameter d at its base ca. 20 μm) is contacted by a stimulating large area ob-

ject from above the hair is bent towards the leg surface. This can be watched under the microscope and quantified using image analysis. As a consequence of the bending, the contact point of the loading force is shifted towards the hair base with increasing static load and hair deflection (Fig. 3). As a result, the effective lever arm transmitting the contact force toward the socket and the stimulating moment decrease with increasing loading force. Due to its bending, the maximum deflection of the hair shaft at its base never exceeds the remarkably low value of ca. 12° and the bending moment increases much more slowly with large loading forces than with small ones (Fig. 4), approaching a saturation value of ca. 4×10^{-9} Nm.

These properties have two important consequences. (i) The hair is protected against breaking and (ii) the hair's mechanical working range is considerably extended as compared to a hair not bent under the stimulus load. The latter effect goes along with higher mechanical sensitivity for small deflections (force needed to deflect the hair ca. 5×10^{-5} N/°) than for large ones (ca. 1×10^{-4} N/°) and with a corresponding relative decrease of the moments at the socket with increasing stimulus load from above.

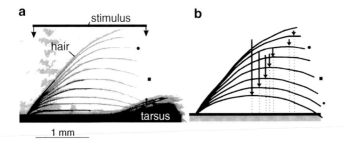

Fig. 3. *Bending* of a spider tactile hair by mechanical stimulation from above. **a** A hair observed under the microscope and increasingly deflected towards the leg surface (tarsus) by the loading force (flat surface). **b** Computer simulation of the situation also shown in **a**. With increasing load (arrows) the effective lever arm represented by the hair shaft shortens. Force vectors are not drawn to scale; in reality the smallest and the largest force differ by two orders of magnitude. Symbols at hair tips mark same status of deflection in the original (**a**) and the simulated (**b**) hair. (Dechant et al. 2001)

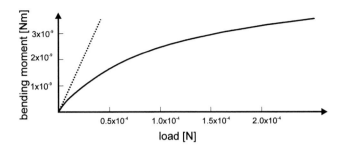

Fig. 4. According to Finite Element simulation the *bending moment* increases more slowly with increasing load (hair stimulated from above as shown in Fig. 3a). The dotted line describes the bending moments to be expected if the lever arm were remaining constant (Dechant et al. 2001)

B. Stresses in the Hair Shaft

Using their FEM model Dechant et al. (2001) calculated the major stress components (the axial stresses due to the bending moment) in the shaft of the spider tactile hair when stimulated/loaded from above. The largest stresses do not occur near the hair base proper where the moment is largest, nor are they found at the point of contact proper, but near to it (about one eighth of the hair length towards the hair base). Stresses reach values of up to about $3.2 \times 10^5 \, N/m^2$. With increasing load the region of maximum stress shifts towards the hair base as does the point of load contact. Remarkably, the maximum stresses remain largely the same although the loads introduced at the different contact points differ considerably during a loading cycle. This finding implies that the hair shaft is a structure of equal maximal strength and that critical stress values are avoided by appropriately varying the hair's cross section along its length. Indeed Dechant et al. (2001) distinguish six regions of the hair shaft each with its specific mechanical properties (second moment of area J) resulting from differences in diameter, wall thickness, and curvature.

C. Restoring Torque

In addition to the non-linear geometrical effects, the force resisting tactile hair deflection increases in a non-linear way with increasing deflection angle. The torsional restoring "constant" S measures about $3 \times 10^{-8} \, Nm/rad$ or 5.2×10^{-10} $Nm/°$ up to a deflection angle of about $10°$. It increases at still larger deflection angles (Albert et al. 2001). These values are larger than those for medium flow sensors (trichobothria; Barth et al. 1993, Barth 2000) by about four powers of ten. Although joint restoring torques do vary among individual tactile hairs of the type examined, maximum stresses and bending do not. Considering the uniformity of hair morphology (Dechant et al. 2001) we suggest that the similarity found among maximal stresses in different individual hairs is a result of variations of Young's modulus along and in accordance with joint stiffness. The constant maximum stress phenomenon works in a narrow range of elasticity values only. A smooth distribution of maximum axial stresses is achieved at a specific value for Young's modulus only (Fig. 6). In our analysis this value was obtained by iterative fitting; it corresponds to the deformation found in experiments with actual hairs.

D. Coupling of Dendrites

The inner lever arm of the spider tactile hair is very short as compared to the outer lever arm, measuring about 3.5 µm only. The axis of rotation was found to divide the hair shaft at a ratio of about 760:1 (Nemeth 2000). This means that the

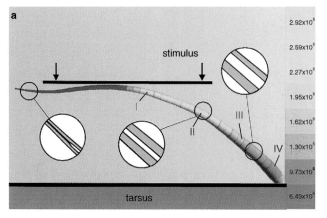

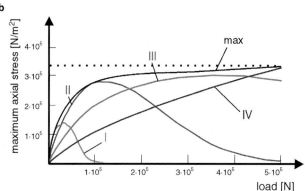

Fig. 5. *Finite Element simulation* of a tarsal tactile hair deflected by a flat surface from above (stimulus). **a** Different colors indicate differences in *maximum axial stresses* at each of the four cross sections of the hair shaft marked by I, II, III, and IV. Whereas red indicates small stresses, yellow indicates large stresses reaching values as high as $3.2 \times 10^5 \, \mathrm{N/m^2}$ (see color scale on right side). **b** The *maximum axial stresses* due to hair bending at the four selected sites marked I to IV in **a** as a function of the load. Black line represents maximum of axial stresses along the hair shaft. (Dechant et al. 2001)

displacement of the hair tip is scaled down by the same ratio and that force increases accordingly. Similar findings are available for insect mechanoreceptive hairs. Examples are the macrochaetae of a fly with a lever ratio of about 400:1 (Keil 1978) and the cricket filiform air movement detectors with a ratio of up to about 1.000:1 (Gnatzy and Tautz 1980). When deflecting the spider tactile hair by 15° the movements of the inner edge of the hair shaft close to the dendritic tip were between 0.7 µm and 0.9 µm (Nemeth 2000). A deflection by 15° is probably never exceeded under natural conditions (see above under hair shaft deformation) except in cases leading to injury. Taking

threshold angular deflections for the elicitation of a physiological response (action potentials) as 1° (Albert et al. 2001), threshold deformation at the dendrite is likely not to be larger than ca. 5 nm. For insect filiform hairs (cricket cercus, *Acheta domesticus*) Thurm (1982) calculated a change of dendrite diameter of 10 to 20 nm under maximally effective stimulation and of only 0.05 nm at threshold when the inner lever arm moves by 0.1 nm only. A similarly small value for the displacement of the inner lever arm at threshold (< 0.1 nm) was calculated by Shimozawa and Kanou (1984a, b). Remember that the cricket filiform hairs are believed to be the most sensitive

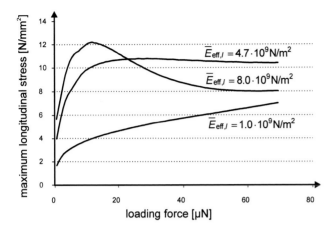

Fig. 6. Calculated distribution of *axial stresses* in the loaded hair shaft *for different Young's moduli (E)* of the material making up the hair shaft. An even stress distribution is only found for the value actually obtained numerically by "best fitting" for the original hair and corresponding to its actual deformation under load ($E = 4.7 \times 10^9$ N/m²). When increasing E the largest stresses are larger than those found in case of the "optimised" value for E (see $E = 8.0 \times 10^9$ N/m²). With decreasing values of E (see $E = 1.0 \times 10^9$ N/m²) the hair shaft becomes too flexible, bending away from the stimulus too easily. (Dechant 2001)

receptors so far known (see chapter 10). For insect tactile hairs such as the fly macrochaetae the displacement of the inner lever arm at the site of the dendritic end is given as about 0.2 µm per degree of hair shaft deflection (Keil 1978).

Clearly the deformation at the dendritic terminal is very small. Contrasting the insect hairs so far described (Gaffal et al. 1975, Gaffal and Theiß 1978, Thurm 1982, Keil 1997), there is no direct interaction between a cuticular structure of the hair base and the sensory cell dendrite in the spider tactile hair. The dendritic tip ends directly beneath the axis of rotation of the hair shaft, i.e. in an area of minimal displacement. It is separated from the hair shaft by material which looks amorphous under the transmission electron microscope (Nemeth 2000). We conclude that the sensillum is not "designed" for large displacements in the region of stimulus transduction but instead for very small ones. The mechanical properties of the outer lever arm of the hair

shaft, the shortness of the inner lever arm, and the location of the dendritic tip so close to the axis of rotation all contribute to this effect. To put it differently the mechanosensitivity of the dendritic membrane transducing the stimulus into a nervous signal must be correspondingly high.

E. Electrophysiological Properties

After having passed through all of the complex mechanical transformations so far described the stimulus leads to changes of the permeability of the sensory cell membrane. An increase of ion currents through specific membrane channels will entail a sensory signal which is sent to the spider central nervous system (Fig. 7). Several properties of the information encoding process demonstrate that the filtering still continues. The central nervous system is informed only about some of the properties of a tactile

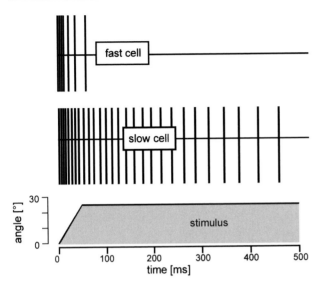

Fig. 7. Digitized action potentials recorded from the slow sensory cell and from the fast sensory cell associated with a spider tactile hair when deflected by a ramp and hold stimulus. Note difference between response patterns (Albert 2001)

stimulus. These are considered the biologically relevant ones.

Judging from the physiological response (tactile hair on spider tarsus) to behaviorally meaningful stimulus velocities of up to 11 cm/s the tactile hair's main function is to signal the mere presence of a stimulus, not however its time – course and its orientation in space (Albert et al. 2001). There are many arguments in favor of this interpretation. (i) The sensory cells are movement detectors; their maximum impulse rate is correlated with the angular velocity of the hair's deflection, not however with the deflection angle as such. (ii) Depending on the sensory cell considered (spider tactile hairs are innervated by three sensory neurons) thresholds of deflection velocity for the firing of impulses are $30 \pm 9°/s$ and $< 0.1°/s$. (iii) Remarkably, the responses to stimulus velocities higher than only 2.5 cm/s are already saturated. Whereas this implies a small working range with regard to the resolution of stimulus velocity, threshold velocity is exceeded by the slightest initial touch. (iv) The static torques leading to supra – threshold deflection measure less than 5×10^{-10} Nm. (v) In addition the nervous responses are largely independent of the azimuth of the hair shaft's deflection (Albert et al. 2001).

From the specializations of a single tactile hair it follows that a large number of neighbouring hairs (indeed found on the spider exoskeleton) is needed to resolve the two-dimensional spatial and the temporal pattern of a stimulus, the shape and texture of an object. The next step of the analysis then is the study of tactile sensing as a spatially distributed process.

IV. Discussion and Conclusions

The mechanoreceptive hairs of arthropods exhibit a wide spectrum of morphological, mechanical, and physiological properties. Hairs extremely sensitive to air flow represent one end of this spectrum, tactile hairs the other. The mechanical properties of these hairs differ greatly from each other reflecting differences in the physics of their respective adequate stimuli. The tactile hairs introduced here may be classified as lightweight structures. When exposed to a tactile stimulus from above the resulting main stresses are distributed evenly pointing to a structure of equal maximum strength. The properties mainly responsible for this are (i) the hair's diameter which considerably increases from its tip towards its base, (ii) the way the hair bends away from the stimulus thus protecting weaker sections of the hair shaft from being loaded excessively, and finally (iii) the "correct" combination of joint stiffness and the hair shaft's Young's modulus.

Several different principles have been used to build force and motion sensors for applications in robotics, prosthetics, and in a wealth of mechatronic devices. Examples are piezoresistive and capacitive pressure sensors, resonance sensors, inductive position sensors, and others (Lee and Nicholls 1999, Dario et al. 2000). The relevance of tactile sensing and the necessity of the corresponding sensory feedback as part of finely tuned manipulation in robots is obvious. In arthropods, unlike in mammals with soft skin such as our own, the arrangement of tactile sensors beneath the body surface, that is beneath the hard exoskeleton, would imply extreme insensitivity. Hairs as a form of short range proximity sensors are the solution of this particular problem. Similarly, in robots whiskers have been used to detect the presence or absence of objects, to protect the robot against damaging collisions, and to search for small parts to be assembled. Such synthetic whiskers, however, seem to be far from the sophistication found in the spider tactile hairs (e.g. Russell 1990).

A recently developed hair-like technical tactile sensor (Schmidt 1999) does not make explicit use of any of the three properties listed above for the spider tactile hair either. Its purpose is to get tactile feedback from an anthropomorphic robot hand, the longterm goal being to improve grip technology of robots and also to assist robot vision by tactile information. The task of measuring static touch has been solved in devices such as computer touch pads a while ago. The problem always has been to somehow imitate the ability of the human skin to sense the dynamic component of a tactile stimulus. The hairlike tactile sensors (Schmidt 1999) were indeed designed to represent the dynamic sensors of the human skin, that is the rapidly adapting Meissner corpuscles and the vibration (acceleration) sensitive Vater-Pacini corpuscles. Hairs taken from cow ears were used to couple the tactile object to the flexible membrane of a capacitive sensor. Upon contact of a hair's tip with an object the distance between the condensor plates (diameter 2 mm, distance at rest 0.04 mm) changes. This change is used to monitor dynamic tactile stimuli. As in the case of the spider tactile hair, the measurement of purely static stimuli is not possible because of the hair's flexibility. Vibrations as such can be detected, not however their frequencies.

The sensitivity of the device is high (< 10 mN). It permits the measurement of the object's velocity of approach up to about 5 cm/s (as compared to about 12 cm/s for the human rapidly adapting tactile sensors). The spatial resolution of such a technical tactile sensing device can be varied by the number of the sensors used and the density of their arrangement. Schmidt (1999) gives 5 mm as the minimum distance of the sensors in an array. This is considerably less than in spiders where one finds up to 400 hairs on one square millimeter.

Acknowledgements

Supported by grant P 12192-BIO of the Austrian Science Foundation to FGB.

References

Albert J, Friedrich O, Dechant H-E, Barth FG (2001) Arthropod touch reception. I. Spider hair sensilla as rapid touch detectors. J Comp Physiol A 187: 303–312

Barth FG (1981) Strain detection in the arthropod exoskeleton. In: Laverack MS, Cosens D (eds) The Sense Organs. Blacky, Glasgow, pp 112–141

Barth FG (1998) The vibrational sense of spiders. In: Hoy RR, Popper AN, Fay RR (eds) Springer Handbook of Auditory Research. Comparative Hearing: Insects. Springer, New York, pp 228–278

Barth FG (2000) How to catch the wind: Spider hairs specialized for sensing the movement of air. Naturwissenschaften 87: 51–58

Barth FG (2002) A Spider's World. Senses and Behavior. Springer, Berlin Heidelberg New York Tokyo, 394pp

Barth FG, Wastl U, Humphrey JAC, Devarakonda R (1993) Dynamics of arthropod filiform hairs. II. Mechanical properties of spider trichobothria (Cupiennius salei Keys.). Phil Trans R Soc Lond B 340: 445–461

Blickhan R, Barth FG (1985) Strains in the exoskeleton of spiders. J Comp Physiol A 157: 115–147

Dario P, Laschi C, Micera S, Vecchi F, Zecca M, Menciassi A, Mazzolai B, Carrozza MC (2000) Biologically inspired microfabricated force and position mechano-sensors. In: Sensors and Sensing in the Natural and Fabricated Worlds. 2nd Internat'l Symp on the Mechanics of Plants, Animals and their Environment. Il Ciocco, Italy, United Engineering Foundation. 19pp

Dechant H-E (2001) Mechanical properties and finite element simulation of spider tactile hairs. Doctoral thesis, Univ of Technology, Vienna

Dechant H-E, Rammerstorfer FG, Barth FG (2001) Arthropod touch reception: stimulus transformation and finite element model of spider tactile hairs. J Comp Physiol A 187: 313–322; see also Erratum J Comp Physiol A 187: 851

Devarakonda R, Barth FG, Humphrey JAC (1996) Dynamics of arthropod filiform hairs. IV. Hair motion in air and water. Phil Trans R Soc Lond B 351: 933–946

Gaffal KP, Theiß J (1978) The tibial thread-hairs of Acheta domesticus L. (Saltatoria, Gryllidae). The dependence of stimulus transmission and mechanical properties on the anatomical characteristics of the socket apparatus. Zoomorphologie 90: 41–51

Gaffal KP, Tichy H, Theiß J, Seelinger G (1975) Structural polarities in mechanosensitive sensilla and their influence on stimulus transmission (Arthropoda). Zoomorphologie 82: 79–103

Gnatzy W, Tautz J (1980) Ultrastructure and mechanical properties of an insect mechanoreceptor: Stimulus-transmitting structures and sensory apparatus of the cercal filiform hairs of Gryllus. Cell Tissue Res 213: 441–463

Humphrey JAC, Devarakonda R, Iglesias I, Barth FG (1993) Dynamics of arthropod filiform hairs. I. Mathematical modeling of the hair and air motions. Phil Trans R Soc Lond B 340: 423–444

Humphrey JAC, Devarakonda R, Iglesias I, Barth FG (1998) Errata re. Humphrey et al. (1993). Phil Trans R Soc Lond B 352: 1995

Humphrey JAC, Barth FG, Voss K (2001) The motion sensing hairs of arthropods: using physics to understand sensory ecology and adaptive evolution. In: Barth FG, Schmid A (eds) Ecology of Sensing. Springer, Berlin Heidelberg New York, pp 105–125

Keil T (1978) Die Makrochaeten auf dem Thorax von Calliphora vicina Robineau – Desvoidy (Calliphoridae, Diptera). Feinstruktur und Morphogenese eines epidermalen Insekten-Mechanoreceptors. Zoomorphologie 90: 151–180

Keil T (1997) Comparative morphogenesis of sensilla: a review. Int J Insect Morphol & Embryol 26: 151–160

Lee MH, Nicholls HR (1999) Tactile sensing for mechatronics – a state of the art survey. Mechatronics 9: 1–31

Nemeth SS (2000) Zum Berührungssinn von Spinnen: Feinstruktur des reiztransformierenden Apparates von tarsalen Haarsensillen bei Cupiennius salei Keys (Ctenidae). Diploma thesis, University of Vienna

Russell RA (1990) Robot Tactile Sensing. Prentice Hall of Australia Pty Ltd, 174pp

Schmid A (1997) A visually induced switch in mode of locomotion of a spider. Z Naturforsch 52c: 1422–1428

Schmidt P (1999) Entwicklung und Aufbau von taktiler Sensorik für eine Roboterhand. Internal Report 99-05, Institut für Neuroinformatik, Ruhr-Universität Bochum, JSSN 0943-2752

Shimozawa T, Kanou M (1984a) Varieties of filiform hairs: range fractionation by sensory afferents and cercal interneurons of a cricket. J Comp Physiol A 155: 485–493

Shimozawa T, Kanou M (1984b) The aerodynamics and sensory physiology of range fractionation in the cercal filiform sensilla of the cricket *Gryllus bimaculatus*. J Comp Physiol A 155: 495–505

Theiß J (1979) Mechanoreceptive bristles on the head of the blowfly: mechanics and electrophysiology of the macrochaetae. J Comp Physiol 32: 55–68

Thurm U (1982) Grundzüge der Transduktionsmechanismen in Sinneszellen. Mechanoelektrische Transduktion. In: Hoppe W, Lohmann W, Markl H, Ziegler H (eds) Biophysik. Springer, Berlin, pp 681–696

12. The Fish Lateral Line: How to Detect Hydrodynamic Stimuli

Joachim Mogdans, Jacob Engelmann, Wolf Hanke, and Sophia Kröther

Abstract

The lateral line is a hydrodynamic receptor system that enables fishes to detect minute water motions generated by conspecifics, predators or prey. The sensory units of the lateral line are the neuromasts which are dispersed over large portions of the body surface. Whereas superficial neuromasts are freestanding on the surface of the skin and sensitive to water velocity, canal neuromasts are embedded in lateral line canals and sensitive to pressure gradients between canal pores. *Superficial and canal neuromasts* are innervated by distinct populations of nerve fibers. When goldfish are exposed to a constant background water flow, responses of fibers innervating *superficial neuromasts* to superimposed hydrodynamic stimuli are masked due to the continuous stimulation of the *neuromasts* by the running water. In contrast, responses of fibers innervating trunk *canal neuromasts* are hardly affected by background water flow due to the filter properties of *lateral line* canals. These findings are evidence for a strong form-function relationship in the sensory periphery of the fish lateral line system. A functional subdivision similar to that in the periphery can be found in the brainstem suggesting that to a large degree information from *superficial* and *canal neuromasts*, respectively, is processed separately at least at the first stage of central nervous integration.

I. Introduction

Water motions provide important sensory information in the aquatic world. Thus aquatic animals have evolved highly sophisticated hydrodynamic receptor systems (review: Bleckmann 1994). The hydrodynamic receptor system of fishes is the lateral line, a sensory system that was first described by Stenon in the 17th century (Parker 1904). The primary function of the lateral line as a receptor for hydrodynamic stimuli was uncovered by Hofer in 1908. Dijkgraaf (1963) was the first to suggest that the lateral line responds to water displacements produced by objects moving in the water. This chapter describes the organization and biomechanical properties of the lateral line periphery and illustrates how hydrodynamic information is represented by the neural activity of primary afferent nerve fibers and by nerve cells at the first site of integration in the brain. For detailed information about the anatomy and physiology of higher lateral line brain centers the reader is refered to the literature (e.g., Bleckmann 1994, Coombs and Montgomery 1999, Mogdans and Bleckmann 2001).

II. Morphology and Biomechanical Properties of the Lateral Line Periphery

The sensory units of the lateral line are the neuromasts (Fig. 1A) which consist of sensory hair-cells, supporting cells and mantle cells (Münz 1979). Lateral line hair-cells are similar in function and morphology to those in the auditory and vestibular system of vertebrates (Roberts et al. 1988). The ciliary bundles project into a gelatinous cupula which covers the entire neuromast and connects the ciliary bundles with the surrounding water.

Neuromasts occur in lines above and below the eyes, which join behind them to continue as the main trunk line. Several additional lines may lie dorsally on the head, on the cheek and on the lower jaw (Coombs et al. 1988, Northcutt 1989). Two types of neuromasts can be distinguished: (i) superficial neuromasts which occur freestanding on the skin, in pits, or on pedestals raised above the skin, and (ii) canal neuromasts which are located in subepidermal canals that are in contact with the water through a series of pores. Typically a canal neuromast is located halfway between two adjacent canal pores.

Lateral line neuromasts have a circular or elliptical shape. The maximal diameter of the sensory macula of a neuromast can be less than 100 µm or more than 600 µm. Consequently the number of hair-cells within a neuromast ranges from very few (e.g. Münz 1979, Song and Northcutt 1991) up to at least 3000 (Müller 1984). In general superficial neuromasts are smaller and have fewer hair-cells than canal neuromasts (Münz 1989). Each neuromast contains two populations of hair-cells with oppositely oriented ciliary bundles (Flock and Wersäll 1962). Afferent nerve fibers innervate only hair-cells of identical orientation (Görner 1963). Consequently, lateral line neuromasts are bidirectionally sensitive due to the separate innervation of unidirectionally polarized hair-cells (Flock 1965, Kroese and Netten 1989).

The adequate stimulus for a hair-cell are shearing movements of the ciliary bundle. Input-output functions are such that within a certain range of amplitudes the membrane potential of a hair-cell changes in a linear fashion with the displacement of the ciliary bundle (e.g. Hudspeth and Corey 1977). For lateral line hair-cells, displacements of less than one nm are sufficient to cause a neural response, displacements greater 100 nm cause saturation (Kroese and Netten 1989).

The adequate stimulus for a lateral line neuromast are fluid movements along the cupula. These movements cause the cupula to slide over the sensory epithelium (Netten and Kroese 1987) resulting in a shearing of the ciliary bundles. The stiffness of the coupling of the cupula to the sensory macula depends on cupula size and on the pivoting stiffness of the ciliary bundles of the hair-cells and thus also on the number of hair-cells in a neuromast (Denton and Gray 1989, Netten et al. 1990). The fluid forces driving the cupula consist of both viscous and inertial components. The relative contribution of these two components varies with frequency and is related to the frequency-dependent boundary layer around the cupula (Netten and Kroese 1987). Within the frequency range relevant for the lateral line, small diameter cupulae will be more affected by viscous forces whereas large diameter cupulae will be more affected by inertial forces (Netten and Kroese 1989). Consequently, cupulae of small diameter mainly function as velocity detectors whereas cupulae of large diameter mainly detect the acceleration of the fluid flow around the cupula.

Most superficial neuromasts are small and thus velocity detectors. They are driven primarily by the viscous drag forces which are proportional to the velocity of the water flowing along their sides (Kalmijn 1989). In contrast, canal neuromasts are pressure gradient detectors because fluid flow within lateral line canals occurs only as a consequence of a pressure gradient between canal pores. Outside the canal, the pressure gradient is proportional to the acceleration of water particles. Thus, canal neuromasts may be considered acceleration detectors with respect to the water motions outside the canal (Kalmijn 1989). Canal neuromasts are therefore driven more effectively with increasing frequency of an external water displacement. In other words, lateral line canals represent high-pass filters for hydrodynamic stimuli (Bleckmann and Münz 1988, Denton and Gray 1988).

Various morphological patterns of the peripheral lateral line are expressed in different fish species (e.g. Coombs et al. 1988, Webb 1989). These patterns may at least in part represent adaptations to the hydrodynamic conditions prevailing in the habitat of a species. There are several possibilities by which the lateral line can be adjusted to the sensory needs of an animal. For instance, with increasing width of a canal, its high-pass characteristic is shifted towards lower frequencies (Denton and Gray 1989). In some species, a membrane-like covering on a lateral line canal adds resonance to the system

which enhances sensitivity to a particular frequency band (Denton and Gray 1989). Furthermore, the sensitivity of the lateral line can be modified by altering the number and size of canal pores (Bleckmann and Münz 1990). Other ways to modify the hydrodynamic information reaching the brain include the variation of the number and placement of superficial neuromasts and/or the number and design of lateral line canals.

III. Technical Devices for the Measurement of Hydrodynamic Stimuli

A number of devices are available for the measurement of subsurface hydrodynamic stimuli. Pressure waves can be recorded with hydrophones which are available in different sizes and with different frequency characteristics. However, hydrophones are generally too big to measure small scale changes in pressure (Mogdans and Bleckmann 1998) and are not well suited to measure laminar (DC) flow. Water velocity can be measured either with constant-temperature (hot-film) anemometers (Coombs et al. 1989) or laser-doppler anemometers (Blickhan et al. 1992). Both devices yield information about temporal changes in flow and about the frequency content of a hydrodynamic event. However, they measure flow velocity only at a single point in space. Visualization of spatial patterns of water flow in two dimensions can be achieved with particle image velocimetry (PIV). In PIV, a narrow sheet of light is used to illuminate a plane of water and the movements of small, neutrally buoyant particles are filmed with a video camera. Vector analysis of particle motion yields information about flow direction and velocity and can be used to calculate vorticity (Hanke et al. 2000). Measuring devices based on the principles used by the fish lateral line are not available.

IV. Reception of Hydrodynamic Stimuli by the Lateral Line Periphery

A. Sine Wave Stimuli in Still Water

The physiology of the peripheral lateral line has been studied in different fish species using single-frequency sinusoidal water wave stimuli generated by a small sphere placed at various locations along the side of the fish (e.g. Münz 1985, Coombs et al. 1996, Mogdans and Bleckmann 1999). The hydrodynamic stimulus generated by a sphere vibrating in still water is well defined (Harris and van Bergeijk 1962). To monitor neuromast function, in most studies the activity of the afferent nerve fibers innervating the lateral line hair-cells was recorded.

Lateral line afferents are spontaneously active. A constant-amplitude sinusoidal stimulus causes a sustained increase in firing rate (e.g. Mogdans and Bleckmann 1999, Fig. 1B). Fibers respond to frequencies of less than 1 Hz up to about 150 Hz. A peak-to-peak displacement of 0.01 µm at the surface of the skin may be sufficient to cause a neural response. Lateral line afferents phase couple their activity to a sinusoidal wave stimulus (Mogdans and Bleckmann 1999, Fig. 1B). Phase-coupling is weak at low stimulus amplitudes and increases with increasing amplitude until it reaches saturation. Thus, the amplitude of a sine wave stimulus is encoded by the degree of phase-coupling. In addition, stimulus amplitude is encoded by firing rate which also increases with increasing stimulus amplitude. However, firing rates in general increase beyond ongoing discharge rates at stimulus levels about 10 dB higher than those causing significant phase-locking (Mogdans and Bleckmann 1999).

Two types of afferent fibers can be distinguished. One type has an amplitude slope of about 6 dB/octave (Fig. 1C) and a constant phase lead with respect to sphere displacement of about 90 degrees for frequencies ≤ 80 Hz (e.g. Münz 1985, Kroese and Schellart 1992, Montgomery and Coombs 1992, Montgomery et al. 1994). These values are in agreement with the theoretical values for a velocity detector (Kalmijn 1989). A second type of afferents shows the characteristics of an acceleration detector, i.e., an amplitude slope of 12 dB/octave (Fig. 1C) and about 180 degrees phase lead (Kroese and Schellart 1992). Based on the theoretical (Kalmijn 1989) and biomechanical (Netten and Kroese 1987) considerations discussed above, responses to water accelerations originate from fibers which innervate canal neuromasts and responses to water velocity originate from fibers innervating superficial neuromasts.

Canal neuromasts can also be considered pressure gradient receptors. In fact, the receptive field organization of primary lateral line afferents can be predicted from the pressure gradient field around a sinusoidally vibrating sphere (Coombs et al. 1996, Coombs and Conley 1997). The exact shape of the receptive field, however, depends on the relative orientations of the axis of the sphere vibration and of the neuromast (Coombs et al. 1996).

B. Complex Wave Stimuli Generated by Objects Passing by

Peripheral lateral line function has also been studied by moving small objects along the side of a fish (e.g., Bleckmann and Zelick 1993, Mogdans and Bleckmann 1998). A moving object causes large changes in water velocity that consist of an initial short transient and predictable component followed by an ill defined long

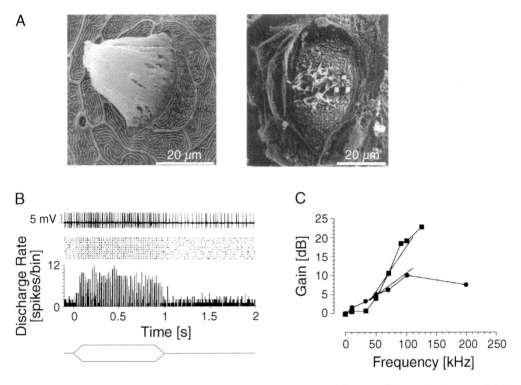

Fig. 1. Morphology and physiology of the lateral line periphery. **A** Electron micrograph of a superficial neuromast of a goldfish. Left: intact cupula. Right: cupula removed to expose the ciliary bundles of the sensory hair-cells. **B** Raster diagram and peri-stimulus-time histogram (binwidth 2 ms) of the responses of a goldfish posterior lateral line nerve (PLLN) fiber to ten repetitions of a 50 Hz wave stimulus generated by a stationary vibrating sphere (vibration amplitude 4 μm, sphere diameter 8 mm). Top trace: original recording; bottom trace: stimulus. **C** Gain of two PLLN fibers plotted as a function of frequency. Note that one fiber (circles) had a gain of about 6 dB/octave, whereas it was 12 dB/octave in the other fiber (squares)

lasting wake (Fig. 2A). These water motions are associated with changes in pressure. However, pressure changes and corresponding pressure gradients are prominent only during the initial transient component and are negligible in the object's wake (Fig. 2B). Measurements with a particle imaging system revealed that the water motions generated by a moving object are changing both in time and space resulting in a complex pattern of water flow across the fish surface (Hanke and Bleckmann 1999, Mogdans et al. 1999).

Afferent fibers in the posterior lateral line nerve of goldfish respond to a moving object with a characteristic discharge pattern which consists of excitation followed by inhibition or vice versa (Mogdans and Bleckmann 1998, Fig. 2). The sequence of excitation and inhibition inverses when the direction of object motion is reversed. This can be predicted from the intrinsic directional sensitivity of the hair-cells within a neuromast. About 70% of the nerve fibers discharged numerous bursts of spikes after an object has passed the fish

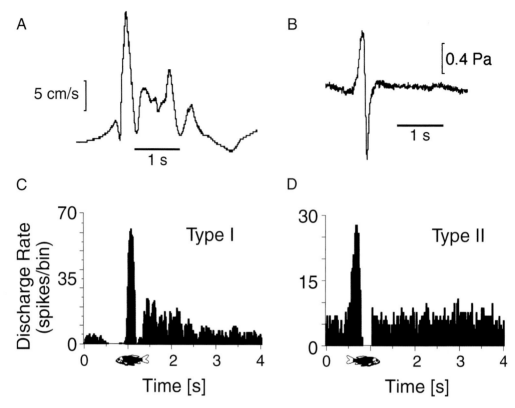

Fig. 2. Peripheral lateral line responses to a moving object. **A** Velocity of the water motions generated by a plexiglass rod (square cross section of 1×1 cm) moving with a speed of 15 cm/s. Ten consecutive measurements made with a constant-temperature anemometer were averaged. **B** Pressure gradient generated by the same object. Ten consecutive pressure wave measurements made with a hydrophone were averaged and the pressure gradient was calculated assuming an average distance between pores of the goldfish trunk lateral line canal of 2 mm (Coombs et al. 1996). **C, D** Peri-stimulus-time histograms (binwidth 2 ms) of the responses to the moving object of two afferent fibers in the goldfish posterior lateral line nerve. Fish symbols represents location, orientation and size of the fish relative to the path of the moving object. Note that the neurophysiological data were not recorded simultaneously with the wave measurements. **C** Response of a type I unit to the object moving from anterior to posterior. The unit responded with inhibition followed by excitation and again inhibition at about the time when the object was closest to the fish. It continued to fire unpredictable bursts of spikes after the object had passed along the side of the fish. **D** Response of a type II unit to the object moving from posterior to anterior. The unit responded with excitation followed by inhibition but barely responded after the object had passed along the side of the fish

(Fig. 2C). These unpredictable bursts are probably caused by the wake of the moving object. About 27% of the fibers did not continue to respond after the object has passed the fish (Fig. 2D). Afferent fibers of the first type (type I) most likely receive input from superficial neuromasts which are highly sensitive to water velocity (e.g. Kroese and Schellart 1992). Canal neuromasts, in contrast, are more sensitive to water acceleration which is proportional to pressure gradient. Thus, fibers of the second type (type II) most likely receive input from canal neuromasts.

C. Hydrodynamic Stimuli with Background Water Flow

Natural hydrodynamic stimuli are often embedded in background noise produced e.g., by water currents, which are a dominant and pervasive feature of many aquatic environments. The responses of afferent fibers in the goldfish posterior lateral line nerve to a background flow have been recorded in a flow tank in which a fish was exposed to a constant water flow (Engelmann et al. 2000, Voigt et al. 2000). Analysis of the water flow using particle image velocimetry showed that the flow was fairly laminar along large portions of the fish surface (Fig. 3A). Neurophysiological data again revealed two types of afferents: type I fibers responded to the flow with an increase in discharge rate for as long as the water flow was maintained, whereas type II fibers did not change their discharge rate in reponse to the water flow (Fig. 3B). The continuous firing of type I units suggests that they were innervating superficial neuromasts which are continuously stimulated by the background flow. In contrast, type II fibers most likely were innervating canal neuromasts which are unresponsive to background flow due to the filter characteristics of the lateral line canals.

When the lateral line was stimulated with sinusoidal water motions in still water both type I and type II afferents exhibited sustained and phase-locked responses which could hardly be distinguished from each other. However, when exposed to a 10 cm/s background water flow a clear difference between the

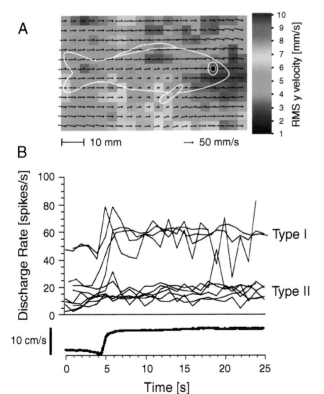

Fig. 3. Peripheral lateral line responses to running water. **A** Vector diagram of water velocity and color-coded plot of water turbulence measured with particle image velocimetry at a distance of about 1 mm from the fish (free-field flow velocity 10 cm/s). Flags represent direction and velocity of water particles in the X-direction. RMS water velocity in the Y-direction was used as an indicator of turbulence. **B** Responses of goldfish type I and type II posterior lateral line nerve fibers to running water (velocity 10 cm/s). Bottom, stimulus trace

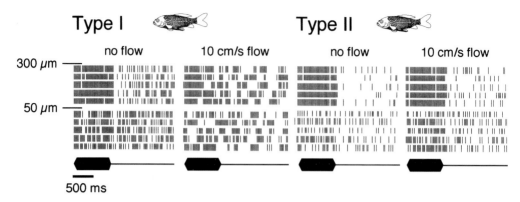

Fig. 4. Responses of goldfish type I and type II posterior lateral line nerve fibers to a 50 Hz sine wave stimulus generated by a vibrating sphere of 10 mm diameter. Raster diagrams of the responses to five stimulus repetitions are shown for two peak-to-peak displacements. Data were recorded in still water (no flow) and in running water (velocity 10 cm/s). Flow direction was from anterior to posterior. Note that the background flow masks the responses of the type I fiber but not that of the type II fiber

responses of type I and type II fibers emerged. Whereas the responses of type I fibers to a sinusoidal water motion were masked, the responses of type II fibers were hardly affected by the flow (Fig. 4). Thus there is a clear functional difference between the superficial and the trunk canal neuromast system. In running water superficial neuromasts are immediately driven into saturation. Consequently, type I afferents respond with high sensitivity to a vibrating sphere in still water only, i.e., only if the fish is not exposed to unidirectional water flow. Trunk canal neuromasts, in contrast, are hardly affected by running water and thus type II afferents respond about equally well to a vibrating sphere in still and running water.

V. Integration of Hydrodynamic Information by the Central Nervous System

A. Hydrodynamic Stimuli in Still Water

Afferent fibers travel to the brain in at least three distinct lateral line nerves which innervate neuromasts on the fish head and trunk (e.g., Northcutt 1989, Puzdrowski 1989, Song and Northcutt 1991). The first site of sensory integration is the medial octavolateralis nucleus (MON) in the fish brainstem (e.g. McCormick and Hernandez 1996, New et al. 1996). Compared to afferent fibers, MON units have lower spontaneous and evoked rates of activity. The responses to sine wave stimuli exhibit greater degrees of adaptation and greater heterogeneity both in terms of the response patterns and in terms of phase-coupling. Moreover, MON units are substantially less sensitive to sine wave stimuli than primary afferents (e.g., Paul and Roberts 1977, Caird 1978, Wubbels et al. 1993, Montgomery et al. 1996, Coombs et al. 1998). In goldfish, about 30% of the units in the MON did not respond to a stationary vibrating sphere, even when tested with displacement amplitudes of up to 800 μm (Mogdans and Goenechea 2000). Such displacement amplitudes are substantially higher than those causing rate saturation in lateral line afferents. However, many of these seemingly insensitive units readily respond to the water motions generated by a moving sphere.

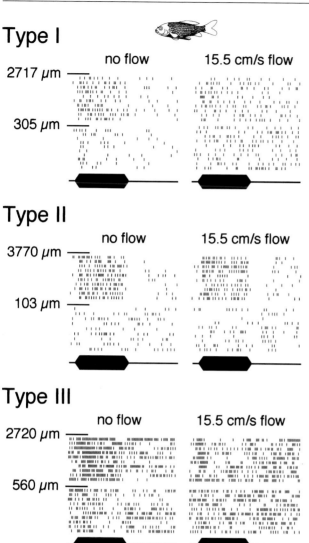

Fig. 5. Responses of goldfish type I, type II, and type III cells in the medial octavolateralis nucleus to a 50 Hz sine wave stimulus generated by a vibrating sphere of 8 mm diameter. Raster diagrams of the responses to ten stimulus repetitions are shown for two peak-to-peak displacements. Data were recorded in still water (no flow) and in running water (velocity 15.5 cm/s). Flow direction was from anterior to posterior. Note that the background flow masks the responses of the type I and type III cells but not that of the type II cell

The responses of MON units to moving object stimuli are very diverse (Mogdans et al. 1997). Nevertheless, two response types can be distinguished. Like type I afferents, many MON units respond to the passing object and to the water motions generated in the wake of it. These units probably receive input from superficial neuromasts. Other MON units, like type II afferents, respond with a transient increase in discharge rate only.

B. Hydrodynamic Stimuli with Background Water Flow

When goldfish MON units are stimulated using a vibrating sphere in the presence of background flow, at least three types can be distinguished (Kröther et al. 2002, Fig. 5). Type I MON units responded to running water with a change in discharge rate. In running water, the responses of these units to a sine wave stimulus showed a decrease of either the response rates or the degree of phase-coupling or both. Thus these units most likely received input from type I afferents, i.e. from superficial neuromasts. Type II MON units hardly responded to running water. Moreover, in these units the responses to sine wave stimuli were not altered under running water conditions. Therefore type II MON units most likely received input from canal neuromasts. Finally, type III MON units also did not respond to running water but their responses to a vibrating sphere in background flow were significantly reduced either in terms of discharge rate or in terms of phase-coupling to the stimulus. Most likely type III MON units received input from both, superficial and canal neuromasts.

VI. Discussion

The lateral line consists of two morphologically and physiologically different subsystems, the superficial and the canal neuromast system. Due to different biomechanical properties superficial neuromasts are sensitive to water velocity whereas canal neuromasts respond to water acceleration or pressure gradients, respectively (Kroese and Schellart 1992, Coombs et al. 1996). When stimulated with sinusoidal water motions in still water, both systems perform about equally well. Only when the lateral line is exposed to running water a clear difference becomes apparent: superficial neuromast function is substantially impaired in running water whereas canal neuromast function is not affected (Engelmann

et al. 2000). This finding indicates a form-function relationship for the lateral line that has been predicted about 40 years ago by Dijkgraaf (1963) but only now has been demonstrated experimentally. Many fishes that live in running or turbulent water or are fast swimmers tend to have extended lateral line canals and canal specialization (i.e., secondary and tertiary branching), but few superficial neuromasts (Bleckmann and Münz 1990). In contrast, species that live in still waters and are slow swimmers or have a sedentary behavior often have reduced, simple canals and an increased number (up to several thousand) of superficial neuromasts (Dijkgraaf 1963, Puzdrowski 1989, Bleckmann and Münz 1990, Vischer 1990). In running water and/or in swimming fish water flow permanently stimulates superficial neuromasts. As a consequence superficial neuromasts are less sensitive or even useless for the detection of small and locally presented objects like small oscillating prey items. In running water, superficial neuromasts most likely mediate rheotaxis, i.e., the behavioral orientation to water currents (Montgomery et al. 1997, Baker and Montgomery 1999). Prey detection and its localization in running water, on the other hand, most likely are made possible by canal neuromasts (Coombs et al. 2001) which respond to local hydrodynamic stimuli about equally well in still and running water, respectively.

In contrast to peripheral lateral line function only little is known about the processing of lateral line information in the fish brain. The apparent loss of sensitivity to vibrating sphere stimuli by MON units (e.g., Paul and Roberts 1977, Caird 1978, Wubbels et al. 1993, Montgomery et al. 1996, Coombs et al. 1998), the response of such MON units to a moving object (Mogdans and Goenechea 2000) and the fairly large and complex receptive fields of many MON units (Mogdans and Kröther 2001) suggest that the central lateral line system is not particularly well adapted for the analysis of pure sine wave stimuli. Instead, many MON units appear to be integrating inputs from large parts of the lateral line periphery. Among MON units responding to sine wave stimuli a functional subdivision similar to that in the lateral line periphery can be found. This suggests that to a large degree information from superficial and canal neuromasts is processed separately in the fish brainstem. Type III MON units have response properties intermediate to those of type I and type II units. This may be due to convergence of inputs from the superficial and canal neuromast system at the first site of sensory integration in the ascending lateral line pathway.

VII. Conclusions

Although the biomechanical properties of lateral line neuromasts are well described, little is known about lateral line function under natural hydrodynamic conditions. Studies using sine wave stimuli applied in still water have revealed fundamental functional properties of the peripheral lateral line. Although sine waves may approximate the oscillatory water movements generated for instance by swimming zooplankton or by fish body vibrations, they are rare in nature. During evolution, the lateral line most likely has adapted to more complex water motions, for instance the complicated wave patterns generated by moving objects. Thus, stimulating the lateral line with corresponding water motions is an important additional approach to understand its function. Likewise, studies in which the lateral line is stimulated under background noise conditions, e.g., in a laminar water flow, will contribute to increase our knowledge about the functional limitations and evolutionary adaptations of the lateral line system.

Acknowledgments

We thank H. Bleckmann for comments on the manuscript. The original research reported herein was supported by the German Science Foundation (grants Bl 242/9-1, Bl 242/10-1, Mo 718/2-1) and the Bundesministerium für Forschung und Technologie (Helmholtz-stipend to JM).

References

Baker CF, Montgomery JC (1999) The sensory basis of rheotaxis in the blind Mexican cave fish, *Astyanax fasciatus*. J Comp Physiol A 184: 519–527

Bleckmann H (1994) Reception of hydrodynamic stimuli in aquatic and semiaquatic animals. In: Rathmayer W (ed) Progress in Zoology. Vol. 41 Gustav Fischer, Stuttgart, Jena, New York, pp 1–115

Bleckmann H, Münz H (1988) The anatomy and physiology of lateral line mechanoreceptors in teleosts with multiple lateral lines. In: Barth FG (ed) Verh Dtsch Zool Ges 81, Gustav Fischer, Stuttgart, p 288

Bleckmann H, Münz H (1990) Physiology of lateral-line mechanoreceptors in a teleost with highly branched, multiple lateral lines. Brain Behav Evol 35: 240–250

Bleckmann H, Zelick R (1993) The responses of peripheral and central mechanosensory lateral line units of weakly electric fish to moving objects. J Comp Physiol A 172: 115–128

Blickhan R, Krick C, Breithaupt T, Zehren D, Nachtigall W (1992) Generation of a vortex-chain in the wake of a subundulatory swimmer. Naturwissenschaften 79: 220–221

Caird DM (1978) A simple cerebellar system: the lateral line lobe of the goldfish. J Comp Physiol A 127: 61–74

Coombs S, Conley RA (1997) Dipole source localization by mottled sculpin II. The role of lateral line excitation patterns. J Comp Physiol A 180: 401–416

Coombs S, Montgomery JC (1999) The enigmatic lateral line system. In: Fay RR, Popper AN (eds) Comparative Hearing: Fish and Amphibians. Springer Handbook of Auditory Research. Springer, New York, pp 319–362

Coombs S, Janssen J, Webb JF (1988) Diversity of lateral line systems: evolutionary and functional considerations. In: Atema J, Fay RR, Popper AN, Tavolga WN (eds) Sensory Biology of Aquatic Animals. Springer, New York, pp 553–593

Coombs S, Fay RR, Janssen J (1989) Hot-film anemometry for measuring lateral line stimuli. J Acoust Soc Am 85: 2185–2193

Coombs S, Hastings M, Finneran J (1996) Modeling and measuring lateral line excitation patterns to changing dipole source locations. J Comp Physiol A 178: 359–371

Coombs S, Mogdans J, Halstead M, Montgomery J (1998) Transformation of peripheral inputs by the first-order lateral line brainstem nucleus. J Comp Physiol A 182: 609–626

Coombs S, Braun CB, Donovan B (2001) The orienting response of Lake Michigan mottled sculpin is mediated by canal neuromasts. J Exp Biol 204: 337–348

Denton EJ, Gray JAB (1988) Mechanical factors in the excitation of the lateral lines of fishes. In: Atema J, Fay RR, Popper AN, Tavolga WN (eds) Sensory Biology of Aquatic Animals. Springer, New York, pp 595–617

Denton EJ, Gray JAB (1989) Some observations on the forces acting on neuromasts in fish lateral line canals. In: Coombs S, Görner P, Münz H (eds) The Mechanosensory Lateral Line. Neurobiology and Evolution. Springer, New York, pp 229–246

Dijkgraaf S (1963) The functioning and significance of the lateral line organs. Biol Rev 38: 51–106

Engelmann J, Hanke W, Mogdans J, Bleckmann H (2000) Hydrodynamic stimuli and the fish lateral line. Nature 408: 51–52

Flock Å (1965) Electronmicroscopic and electrophysiological studies on the lateral line canal organ. Acta Otolaryngol 199: 1–90

Flock Å, Wersäll J (1962) A study of the orientation of sensory hairs of the receptor cells in the lateral line organ of a fish with special reference to the function of the receptors. J Cell Biol 15: 19–27

Görner P (1963) Untersuchungen zur Morphologie und Elektrophysiologie des Seitenlinienorgans vom Krallenfrosch (Xenopus laevis Daudin). J Comp Physiol A 47: 316–338

Hanke W, Bleckmann H (1999) Flow visualization and particle image velocimetry with a custom made inexpensive device. In: Zissler D (ed) Verh Dtsch Zool Ges, Gustav Fischer, Stuttgart, p 352

Hanke W, Brücker C, Bleckmann H (2000) The ageing of the low-frequency water disturbances caused by swimming goldfish and its possible relevance to prey detection. J Exp Biol 203: 1193–1200

Harris GG, van Bergeijk WA (1962) Evidence that the lateral line organ responds to near-field displacements of sound sources in water. J Acoust Soc Am 34: 1831–1841

Hofer B (1908) Studien über die Hautsinnesorgane der Fische I. Die Funktion der Seitenorgane bei den Fischen. Ber kgl Bayer biol Versuchsstation München 1: 115–168

Hudspeth AJ, Corey DP (1977) Sensitivity, polarity, and conductance change in the response of vertebrate hair cells to controlled mechanical stimuli. Proc Natl Acad Sci USA 74: 2407–2411

Kalmijn AJ (1989) Functional evolution of lateral line and inner ear sensory systems. In: Coombs S, Görner P, Münz H (eds) The Mechanosensory Lateral Line. Neurobiology and Evolution. Springer, New York, pp 187–216

Kroese ABA, Netten SMv (1989) Sensory transduction in lateral line hair cells. In: Coombs S, Görner P, Münz H (eds) The Mechanosensory Lateral Line. Neurobiology and Evolution. Springer, New York, pp 265–284

Kroese ABA, Schellart NAM (1992) Velocity- and acceleration-sensitive units in the trunk lateral line of the trout. J Neurophysiol 68: 2212–2221

Kröther A, Mogdans J, Bleckmann H (2002) Brainstem lateral line responses to sinusoidal wave stimuli in still and running water. J Exp Biol 205: 1471–1484

McCormick CA, Hernandez DV (1996) Connections of the octaval and lateral line nuclei of the medulla in the goldfish, including the cytoarchitecture of the secondary octaval population in goldfish and catfish. Brain Behav Evol 47: 113–138

Mogdans J, Bleckmann H (1998) Responses of the goldfish trunk lateral line to moving objects. J Comp Physiol A 182: 659–676

Mogdans J, Bleckmann H (1999) Peripheral lateral line responses to amplitude-modulated sinusoidal wave stimuli. J Comp Physiol A 185: 173–180

Mogdans J, Goenechea L (2000) Responses of medullary lateral line units in the goldfish, Carassius auratus, to sinusoidal and complex wave stimuli. Zoology 102: 227–237

Mogdans J, Bleckmann H (2001) The mechanosensory lateral line of jawed fishes. In:

Kapoor BG (ed) Sensory Biology of Jawed Fishes – New Insights. Oxford and IBH Publishing Co Pvt Ltd, New Delhi, pp 181–213

Mogdans J, Kröther S (2001) Brainstem lateral line responses to sinusoidal wave stimuli in the goldfish, Carassius auratus. Zoology 104: 153–166

Mogdans J, Bleckmann H, Menger N (1997) Sensitivity of central units in the goldfish, Carassius auratus, to transient hydrodynamic stimuli. Brain Behav Evol 50: 261–283

Mogdans J, Wojtenek W, Hanke W (1999) The puzzle of hydrodynamic information processing: how are complex water motions analyzed by the lateral line? Europ J Morphol 37: 195–199

Montgomery JC, Coombs S (1992) Physiological characterization of lateral line function in the Antarctic fish (Trematodus bernacchii). Brain Behav Evol 40: 209–216

Montgomery JC, Coombs S, Janssen J (1994) Form-function relationships in lateral line systems: comparative data from six species of antarctic notothenioid fish. Brain Behav Evol 44: 299–306

Montgomery J, Bodznick D, Halstead M (1996) Hindbrain signal processing in the lateral line system of the dwarf scorpionfish Scorpaena papillosus. J Exp Biol 199: 893–899

Montgomery JC, Baker CF, Carton AG (1997) The lateral line can mediate rheotaxis in fish. Nature 389: 960–963

Müller U (1984) Anatomische und physiologische Anpassungen des Seitenliniensystems von Pantodon buchholzi an den Lebensraum Wasseroberfläche. Dissertation, Universität Gießen 1–201

Münz H (1979) Morphology and innervation of the lateral line system in Sarotherodon niloticus L. (Cichlidae, Teleostei). Zoomorphol 93: 73–86

Münz H (1985) Single unit activity in the peripheral lateral line system of the cichlid fish Sarotherodon niloticus L. J Comp Physiol A 157: 555–568

Münz H (1989) Functional organization of the lateral line periphery. In: Coombs S, Görner P, Münz H (eds) The Mechanosensory Lateral Line. Neurobiology and Evolution. Springer, New York, pp 285–298

Netten SMv, Kroese ABA (1987) Laser interferometric measurement on the dynamic behavior of the cupula in the fish lateral line. Hearing Res 29: 55–61

Netten SMv, Kroese ABA (1989) Dynamic behavior and micromechanical properties of the cupula. In: Coombs S, Görner P, Münz H (eds) The Mechanosensory Lateral Line. Neurobiology and Evolution. Springer, New York, pp 247–264

Netten SMv, Kelly JP, Khanna SM (1990) Dynamic responses of the cupula in the fish lateral line are spatially nonuniform. Association for Research in Otolaryngology 341–342

New JG, Coombs S, McCormick CA, Oshel PE (1996) Cytoarchitecture of the medial octavolateralis nucleus in the goldfish, Carassius auratus. J Comp Neurol 366: 534–546

Northcutt RG (1989) The phylogenetic distribution and innervation of craniate mechanoreceptive lateral lines. In: Coombs S, Görner P, Münz H (eds) The Mechanosensory Lateral Line. Neurobiology and Evolution. Springer, New York, pp 17–78

Parker GH (1904) The function of the lateral-line organs in fishes. Bull US Bur Fish 24: 185–207

Paul DH, Roberts BL (1977) Studies on a primitive cerebellar cortex. III. The projections of the anterior lateral-line nerve to the lateral-line lobes of the dogfish brain. Proc R Soc Lond B 195: 479–496

Puzdrowski RL (1989) Peripheral distribution and central projections of the lateral-line nerves in goldfish, Carassius auratus. Brain Behav Evol 34: 110–131

Roberts WM, Howard J, Hudspeth AJ (1988) Hair cells: Transduction, tuning, and transmission in the inner ear. Ann Rev Cell Biol 4: 63–92

Song J, Northcutt RG (1991) Morphology, distribution and innervation of the lateral-line receptors of the Florida gar, Lepisosteus platyrhincus. Brain Behav Evol 37: 10–37

Vischer HA (1990) The morphology of the lateral line system in three species of Pacific cottoid fishes occupying disparate habitats. Experientia 46: 244–250

Voigt R, Carton AG, Montgomery JC (2000) Responses of anterior lateral line afferent neurones to water flow. J Exp Biol 203: 2495–2502

Webb JF (1989) Developmental constraints and evolution of the lateral line system in teleost fishes. In: Coombs S, Görner P, Münz H (eds) The Mechanosensory Lateral Line. Neurobiology and Evolution. Springer, New York, pp 79–98

Wubbels RJ, Kroese ABA, Schellart, NAM (1993) Response properties of lateral line and auditory units in the medulla oblongata of the rainbow trout (Oncorhynchus mykiss). J Exp Biol 179: 77–92

13. The Blood Vasculature as an Adaptive System: Role of Mechanical Sensing

Timothy W. Secomb and Axel R. Pries

Abstract

The vascular system consists of an extensive network of conduits that carry blood to all parts of the body. The metabolic requirements of tissues, including oxygen demand, vary spatially and temporally. In order to meet these varying requirements, the vascular system must have the ability to adjust and control blood flow in space and time. Centrally driven neural and hormonal signals modulate flow at the whole-organ or regional level. Local modulation of blood flow is achieved by responses of individual microvessels to stimuli that they experience. The responses include acute changes of diameter achieved by alterations in the contractile state of smooth muscle in vessel walls (flow regulation), and long-term changes of vascular dimensions achieved by structural alterations in the vessel walls and by addition or loss of vascular segments (structural adaptation). Here, current understanding of these processes is reviewed, with emphasis on the role of vascular responses to mechanical stresses, i.e., wall shear stress resulting from blood flow and circumferential wall stress resulting from intravascular pressure, and the importance of these responses in flow regulation and structural adaptation. It is concluded that the blood vasculature is a sensitive adaptive system, in which mechanical sensing plays an important role in coordinating vascular responses.

I. Introduction

The delivery of an appropriate amount of blood flow to each part of the body is the main function of the circulatory system. The flow must be sufficient to supply nutrients and remove waste products. Blood flows through an intricate network of vessels, with diameters ranging from 1–2 cm to a few μm. The volume flow rate Q in each vascular segment is given to a first approximation by Poiseuille's law (Caro et al. 1978):

$$Q = (\pi D^4 \Delta P)/(128\,L\,\eta) \qquad (1)$$

where D is the diameter, ΔP is the drop in hydrostatic pressure along the segment, L is the length and η is the apparent viscosity of blood in the segment. From an engineering point of view, the vascular system may therefore be regarded as a network of resistive elements, where the resistance of each element is given by

$$R = \Delta P/Q = (128\,L\,\eta)/(\pi D^4). \qquad (2)$$

The inverse fourth-power dependence of resistance on diameter implies that flow resistance, and therefore blood flow rate, is highly sensitive to changes in vessel diameter. Even in vessels where significant deviations from Poiseuille's law occur, as for instance in arteries where effects of unsteady flow and fluid inertia are important, flow resistance is sensitively dependent on diameter. Therefore, relatively precise control of vessel diameters is required if distribution of blood according to demand is to be achieved while avoiding energetically costly oversupply in some areas.

In the circulatory system, two main types of diameter adjustment may be distinguished. Rapid responses to changing demands are achieved by active changes in vessel diameters. The arterioles, which connect the arteries to the capillaries, are most important in this process of flow regulation, and can vary their diameters over a wide range by changes in the degree of contraction (tone) of the smooth muscle in their walls. Over longer time scales, vessels are capable of structural adaptation of internal diameter and wall thickness. These structural changes occur during growth and development, and in response to injury or changes in functional demands.

Central control mechanisms play important roles in the processes of blood flow regulation and structural adaptation. The approximate positions and dimensions of the major arteries and veins are under genetic control. Centrally generated hormonal and neural signals influence blood flow at the whole-body and whole organ levels. However, tissue metabolic requirements vary over length scales ranging from microscopic to macroscopic (Bassingthwaighte et al. 1989). Control of blood flow on small spatial scales requires the ability of blood vessels to sense local conditions and to respond by appropriate alterations in their diameters. Moreover, flow in a given vascular segment depends not only on the flow resistance of that segment but also on the flow resistance of other segments in the vascular network, particularly those in the same flow pathway through the network. The increase in flow that can be achieved by increasing the diameter of that segment is limited by flow resistance in upstream and downstream segments. Thus, control of flow over a wide range in a given vascular segment requires coordinated responses of multiple segments.

From an engineering point of view, the blood vasculature can therefore be considered as a sensitive adaptive system, consisting of an extensive network of interconnected vessel segments, which act as both sensors and actuators. This system is capable of modulating flow on small and large length scales, and on short time scales (by alterations in vessel tone) and long time scales (by alterations in vessel wall structure). The mechanisms by which this is achieved are not fully understood. However, much experimental work on the factors governing the diameters of individual vessels has been carried out, particularly in the last twenty years, providing a basis for attempts to understand blood flow regulation and structural adaptation at a system level. This chapter provides a brief overview of such work, emphasizing the role of vascular responses to mechanical forces. The main forces acting on blood vessel walls are wall shear stress resulting from blood flow over the internal surface of the vessels, and circumferential tension resulting from hydrostatic pressure within the vessels. Vascular responses to these forces are reviewed, and the role of these responses in overall network behavior are discussed.

II. Vascular Responses to Wall Shear Stress and Pressure

A thin layer of endothelial cells forms the inner surfaces of blood vessels. Blood flowing along a vessel produces a shear stress (τ) acting on the endothelial cell surface:

$$\tau = \Delta P \cdot D/(4L) = (32/\pi)\eta Q/D^3 \qquad (3)$$

Typical levels of wall shear stress are in the range 1–100 dyn/cm^2 (Pries et al. 1995a).

The ability of blood vessels to respond actively, by alterations in diameter, to changes in the level of wall shear stress has long been known (Schretzenmayr 1933, Rodbard 1975). Generally, increased shear stress leads to dilation of arteries and arterioles, by relaxation of the smooth muscle that surrounds vessels. This response has been studied extensively in microvessels (Koller and Kaley 1990, Kuo et al. 1990, Kuo and Hein 2001). Analogous structural responses are also observed. Prolonged alteration in wall shear stress has been shown to lead to structural changes in vessel diameter. Thus, a chronic reduction in flow rate leads to a reduction in arterial diameter (Langille and O'Donnell 1986) while increased flow produces the opposite effect (Tulis et al. 1998). The functional significance of these responses to shear stress is discussed in the following sections.

The difference in hydrostatic pressure between blood and the space exterior to blood vessels induces circumferential tension in the vessel wall. For vessels in which the wall thickness w is small compared to the diameter D, the average circumferential tensile stress in the wall (σ) is given approximately by

$$\sigma = DP/(2w) \qquad (4)$$

where P is the transmural pressure difference. The circumferential stress σ may be distributed among all the structural components of the vessel wall. In vessels with smooth muscle tone and capable of significant dilation, a large fraction of the circumferential stress must be supported by the smooth muscle cells. The transmural pressure difference is typically in the range 10–100 mmHg, i.e., $1.3 \times 10^4 - 1.3 \times 10^5$ dyn/cm^2. Typical values of the radius to wall thickness ratio (D/[2w]) are in the range 0.05 to 0.5, and the resulting average circumferential stresses are in the range $10^5 - 1.5 \times 10^6$ dyn/cm^2 (Pries et al. 1999). It is noteworthy that these stresses are much larger than typical wall shear stresses, by a factor of 10^4 or more.

Arteries and arterioles respond actively to changes in intravascular pressure. In this effect, which is known as the myogenic response, an increase of intravascular pressure in vessels with smooth muscle tone generally leads to constriction of the vessel, and vice versa (Bayliss 1902, Folkow 1949, Johnson 1980). As in the case of wall shear stress, analogous structural responses are observed. Thus, increased pressure causes luminal diameter to decrease with an associated increase in wall thickness (Folkow 1987, Bakker et al. 2000).

Despite intensive studies by a number of groups over recent years, the cellular mechanisms by which vessel walls sense changes in wall shear stress and circumferential tension are not well understood. The sensing of wall shear stress has been the subject of a review by Davies (1995) and is discussed by Kuo and Hein in chapter 14 (Kuo and Hein 2002). The surface of endothelial cells adjacent to flowing blood is coated with a glycocalyx or endothelial surface layer consisting of macromolecules bound or adsorbed to the endothelial surface. This layer has been shown to be of order 0.5–1 μm thick and to have a substantial effect on the hemodynamics of flow in microvessels (Desjardins and Duling 1990, Vink and Duling 1996, Pries et al. 2000). Shear stress is transmitted to the endothelial surface mainly via this endothelial surface layer, and the fluid shear stress acting on the actual cell surface is relatively small (Damiano 1998, Secomb et al.

2001). This is consistent with a scheme in which shear stress is transmitted via the attachment points of endothelial surface layer components to the internal cytoskeleton of the cell and to the adhesion points of endothelial cells to the underlying structures. Changes in the stresses acting on the adhesion proteins or other structural molecules may stimulate or inhibit the release of messenger substances produced by endothelial cells that are capable of influencing smooth muscle tone (Kuo and Hein 2002). Mechanosensitive ion channels may be involved in this process, leading to changes in cell membrane potential (Davies 1995). It has also been proposed that the physical properties of the lipid bilayer component of the cell membrane may vary with the level of shear stress, and play a role in regulating smooth muscle tone (Haidekker et al. 2000).

The signaling mechanisms underlying the myogenic response have been reviewed by Davis and Hill (1999). Changes in vascular tone in response to pressure changes are believed to be mediated by alterations in the permeability or transport properties of the smooth muscle cell membrane, or by alterations in the state of membrane-bound enzymes or proteins, leading to changes in the potential difference across the membrane. As in the case of the shear response, the mechanisms by which mechanical forces are transduced into electrical or chemical changes remain to be determined.

III. Regulation of Blood Flow

The term 'regulation' is used to describe the modulation of blood flow in response to acute changes in tissue demands. The main mechanism of flow regulation is active adjustment of the diameters of small arteries and arterioles by changes in smooth muscle tone. Delivery of sufficient oxygen is generally the most critical demand placed on the vascular system, because the amount of molecular oxygen that can be stored in tissue, either dissolved or bound to hemoglobin or myoglobin, is relatively small, and because the maximum distance over which oxygen can be transported by diffusion into oxygen-consuming tissue is very short, typically $100\,\mu m$ or less. Many studies have shown that arterioles constrict in response to increased oxygen levels and dilate when levels decrease, although the mechanism has been elusive (Jackson 1987, 2000). The sensing of alterations in levels of oxygen or other metabolites provides the initial stimulus for regulatory adjustments.

Flow along a given pathway depends on the resistance of all segments in that pathway. Most of the flow resistance resides in the arterioles and small arteries. Therefore, effective regulation of blood flow requires coordinated changes in the diameter of the vessels upstream of the region of altered demand, as illustrated schematically in Fig. 1. Increased oxygen consumption in the region labeled $(^*)$ causes a reduction of oxygen levels, and dilation of arterioles within that region. However, a substantial increase in flow requires dilation of the segments labeled $(+)$, so that the upstream flow resistance is reduced. This implies the existence of a mechanism for information transfer in the upstream direction.

The sensitivity of arterioles to wall shear stress provides one such mechanism ["hemodynamic coupling" (Secomb and Pries 2002)]. If one segment in a network dilates, reducing its flow resistance and causing flow to increase along a flow pathway containing that segment, then the resulting increase in shear stress can cause dilation of other segments along the same flow pathway, allowing flow to increase further. Additionally, dilation of a given segment causes a reduction in the

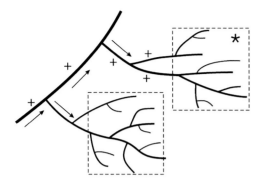

Fig. 1. Schematic representation of arteriolar network supplying a tissue, illustrating necessity for coordinated vascular responses. Arrows denote direction of blood flow. Rectangles denote regions supplied by vessels. If region labeled (*) experiences increased metabolic demand, dilation of vessel segments labeled (+) is required in order to decrease flow resistance along the flow pathway supplying the region

pressure drop across the segment, implying a reduction of pressure in all upstream segments on the same flow pathway. Upstream segments would then dilate in a myogenic response to reduction of pressure (Cornelissen et al. 2000). It should be noted, however, that a second mechanism contributes to this type of information transfer. Vasomotor activity can be conducted along vessel walls by means of electrical coupling between endothelial and smooth muscle cells (Duling and Berne 1970). Recent studies (Segal and Jacobs 2001) on skeletal muscle indicate that this mechanism is responsible for a large part of the dilation of feeding vessels that is observed when muscle is stimulated. However, the relative contributions to vasodilation of conducted and flow-mediated mechanisms remain unclear.

IV. Structural Adaptation

Lengths, diameters and branching patterns of vessel segments in microvascular networks are heterogeneous (Pries et al. 1995b). Even so, a considerable degree of overall organization is apparent in normal tissues, with segment lengths and diameters varying in a systematic way with position in a hierarchical branching system. This organization reflects the growth and adaptation processes governing vascular structure. At the microvascular level, the vast number of vessels and the need for relatively precise control of their diameters imply that local control and feedback mechanisms are largely responsible for setting structural diameters and for dictating the growth or regression of segments. Evidently, organized microvascular structures 'emerge' from the combined effects of local responses by each vessel segment to the stimuli that it receives.

As already discussed, mechanical forces (wall shear stress and intravascular pressure) can stimulate structural changes in vessel walls. Long-term structural responses occur in all types of vessels, including capillaries, venules and veins. For example, elevated pressure leads to transformation of capillaries into arterioles (Skalak and Price 1996). Similarly, a vein grafted into the position of an artery, exposing it to a high-pressure, high-shear environment, undergoes structural adaptation and assumes characteristics of an artery. Returning the grafted segment to the venous circulation reverses some of these changes (Fann et al. 1990).

The structural response to shear stress provides a potential mechanism to ensure an appropriate variation of vessel diameter with flow rate throughout a network of vessels. It follows from Eq. 3, if variations in apparent viscosity η are neglected, that a network in which vessel diameters adjust themselves to reach a target level of wall shear stress would have the property that flow (Q) is proportional to the cube of diameter (D^3). This concept (Kamiya et al. 1984, Rodbard 1975) is appealing because such a scaling corresponds approximately to observations in the arterial tree (Mayrovitz and Roy 1983, LaBarbera 1990). Furthermore, it satisfies the minimum-cost principle known as Murray's law (Murray 1926), which may be stated in the following form: For a given distribution of flow rates in a network, and a given total volume of blood in the network, viscous energy dissipation is minimized if Q is proportional to D^3. Because total flow is conserved in bifurcations, the diameters of the parent vessel (D_0) and the two daughter vessels (D_1 and D_2) in any bifurcation of a network with this property must satisfy a "bifurcation law"

$$D_0^3 = D_1^3 + D_2^3 \qquad (5)$$

and vessel diameters must decrease with each successive bifurcation from the proximal to the distal branches of vascular trees, as illustrated in Fig. 2A.

The dynamics of diameter changes occurring during structural adaptation according to the above scheme can be represented mathematically assuming that the diameter $D(t)$ of each vessel is governed by an equation of the form:

$$\mathrm{d}D/\mathrm{d}t = D \cdot S_{\mathrm{tot}} \text{ where}$$
$$S_{\mathrm{tot}} = f(\tau) - f(\tau_0) \qquad (6)$$

and where S_{tot} represents the net stimulus for diameter growth, f is an increasing function and τ_0 is the target level of wall shear stress. When applied to tree structures with specified flows in the terminal branches, in which the flow in every segment is in effect prescribed, such a system evolves to a stable steady state solution, with diameters consistent with Eq. 5.

However, two deficiencies become apparent when such an approach is applied to microvascular networks including arterioles, capillaries and venules. Firstly, venules are larger than the corresponding

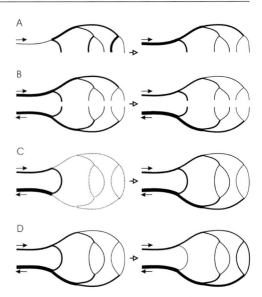

Fig. 2. Schematic representation of roles of adaptive responses in a hypothetical small microvascular network. Solid arrows denote flow direction. Widths of lines represent segment diameter; variations in diameter are exaggerated for clarity. In each case, network characteristics before (left) and after (right) inclusion of specified structural responses are indicated. A: Without response to wall shear stress, diameters are uncoordinated. With response to shear stress, diameters of daughter vessels are less than those of parent vessels. B: Without response to pressure, diameters are identical in corresponding arteriolar and venular segments. With response to pressure, arteriolar segments are narrower than corresponding venular segments. C: Linked arteriolar and venular networks create multiple parallel pathways. Without metabolic response, the network is unstable and decays to a single flow pathway. Lost segments are shown dotted. With metabolic response, parallel flow pathways are stabilized. D: Without information transfer, flow is directed to short pathways at the expense of longer ones. With upstream and downstream information transfer, diameters of longer flow pathways are increased at the expense of short pathways, and flow is balanced between long and short flow pathways

arterioles and are subjected to correspondingly lower shear stress. A fixed set point of wall shear stress cannot therefore be assumed. Based on experimental data derived from intravital microscopy of the mesenteric microcirculation in male Wistar rats, Pries et al. (1995a) proposed that the target level of wall shear stress, $\tau_e(P)$, is an increasing function of intravascular pressure P, so that $S_{tot} = f(\tau) - f(\tau_e(P))$. In fact, such behavior is to be expected because vessels respond to increases in intravascular pressure by structurally decreasing their diameters, as already mentioned. A higher intravascular pressure therefore requires a higher level of wall shear stress to achieve equilibrium between growth and shrinking tendencies. Inclusion of this behavior leads to asymmetric networks, with venous vessels larger than corresponding arterial vessels, and most of the decline in pressure occurring on the arterial side of the network (Fig. 2B). The predicted capillary pressure is substantially lower than the mean of the arterial and venous pressures, which is important with respect to tissue fluid balance, given that the rate of fluid loss from capillaries is dependent on intracapillary pressure.

A second shortcoming of the model shown in Eq. 6 is that, when applied to a network with parallel flow pathways, it leads to instability (Rodbard 1975, Hacking et al. 1996). Such a network would decay to a single flow pathway. To show how this instability may be overcome in real networks, Pries et al. (1998) developed a model including a response to metabolic conditions, by assuming that low-flow pathways experience metabolic deficits which cause a growth stimulus. A sufficiently strong metabolic response was shown to lead to stable network structures (Fig. 2C). Even with this response, however, predicted networks were not consistent with experimentally observed network structures. In particular, unrealistic high-flow 'shunt' pathways between proximal segments of the arteriolar and venular trees were

predicted (Fig. 2D). Therefore, a further term representing propagation of the metabolic stimulus from terminal segments to proximal feeding and draining vessels was included (Pries et al. 1998). Inclusion of this term, together with a constant term ('shrinking tendency') to allow adjustment of the overall level of flow resistance in the network, allowed stable network structures in which flows were balanced between short and long flow pathways (Fig. 2D).

This model was used to predict the distributions of vessel diameters and flow velocities in microvascular networks observed in the rat mesentery (Pries et al. 1998). Three networks were scanned and video-recorded for hematocrit measurement and for determination of flow velocity in every segment using a spatial correlation technique. Vessel diameters, topological network structure and segment lengths between branch points were measured. Mathematical simulations (Pries et al. 1990) were used to estimate pressure, volume flow rate and shear stress in each vessel segment. These simulations took into account non-proportional partition of red cell and plasma flows at diverging bifurcations, and the dependence of effective viscosity of blood flowing through microvessels on diameter and hematocrit (Pries et al. 1994). When applied to the structural data obtained from rat mesentery networks, this model led to predicted vessel diameters and flow velocities in satisfactory agreement with experimentally observed values. In recent work (Pries et al. 2001), the model has been modified to relate its assumptions more closely to known physiological mechanisms. In particular, the metabolic response was based on an explicit consideration of oxygen transport in networks of microvessels, the upstream propagated response was assumed to result from conducted responses as described in the previous section and the downstream propagated response was assumed to result from convective transport of vasoactive metabolites.

In summary, the studies described in this section have explored the role of mechanical sensing in structural adaptation of vessel diameters. Vascular responsiveness to wall shear stress tends to establish the overall relationship between vessel diameter and flow rate, such that flow rate is approximately proportional to diameter cubed. Such a relationship is seen within the arterial and arteriolar systems. This dependence is modulated by the structural response to pressure, which establishes the basic arterio-venous asymmetry of the vasculature, with most of the flow resistance on the arterial side. However, mechanical responses alone are insufficient to establish stable, functionally adequate distributions of diameters. Mechanisms to sense metabolic conditions and to transmit information upstream and downstream along flow pathways are essential in the formation of functionally adequate and efficient vascular networks.

V. Conclusion

The ability of blood vessels to act as mechanical sensors of wall shear stress and of intravascular pressure has long been known, based on their observed active and structural responses to changes in these stresses. In recent years, a considerable amount has been learned about the cellular signaling processes involved in vascular responses to mechanical stress, but the fundamental mechanisms by which mechanical forces are transduced into biochemical signals remain to be elucidated. Although responses to shear and pressure have generally been assumed to be of functional significance, their actual roles in regulation and structural adaptation are only beginning to be assessed quantitatively. In broad terms, the underlying stimuli driving active and structural vascular responses are metabolic and biochemical in nature, including oxygen and other metabolites and a variety of growth factors. Mechanical sensing appears to play an important coordinating role in both flow regulation and structural adaptation. For instance, responses to shear stress govern the overall progression of vascular diameters throughout the vascular system, tend to minimize viscous energy dissipation, and contribute to the coordination of diameter changes in response to changing demands. Responses to pressure are important in establishing the underlying arterio-venous asymmetry of the system. In combination with vascular responses to biochemical signals and other mechanisms of information transfer within networks of vessels, mechanical responses play an important role in allowing the vascular system to function as a sensitive adaptive system, capable of adjusting blood flow according to tissue needs over wide ranges of temporal and spatial scales.

Acknowledgments

Supported by National Institutes of Health grant HL34555 and Deutsche Forschungsgemeinschaft FOR 341, TP1.

References

Bakker EN, Der Meulen ET, Spaan JA, VanBavel E (2000) Organoid culture of cannulated rat resistance arteries: effect of serum factors on vasoactive and remodeling. Am J Physiol Heart Circ Physiol 278: H1233–H1240

Bassingthwaighte JB, King RB, Roger SA (1989) Fractal nature of regional myocardial blood flow heterogeneity. Circ Res 65: 578–590

Bayliss WM (1902) On the local reactions of the arterial wall to changes of internal pressure. J Physiol (Lond) 28: 220–231

Caro CG, Pedley TJ, Schroter RC, Seed WA (1978) The Mechanics of the Circulation. Oxford University Press, Oxford

Cornelissen AJ, Dankelman J, VanBavel E, Stassen HG, Spaan JA (2000) Myogenic reactivity and resistance distribution in the coronary arterial tree: a model study. Am J Physiol Heart Circ Physiol 278: H1490–H1499

Damiano ER (1998) The effect of the endothelial-cell glycocalyx on the motion of red blood cells through capillaries. Microvasc Res 55: 77–91

Davies PF (1995) Flow-mediated endothelial mechanotransduction. Physiol Rev 75: 519–560

Davis MJ, Hill MA (1999) Signaling mechanisms underlying the vascular myogenic response. Physiol Rev 79: 387–423

Desjardins C, Duling BR (1990) Heparinase treatment suggests a role for the endothelial cell glycocalyx in regulation of capillary hematocrit. Am J Physiol 258: H647–H654

Duling BR, Berne RM (1970) Propagated vasodilation in the microcirculation of the hamster cheek pouch. Circ Res 26: 163–170

Fann JI, Sokoloff MH, Sarris GE, Yun KL, Kosek JC, Miller DC (1990) The reversibility of canine vein-graft arterialization. Circulation 82: IV9–18

Folkow B (1949) Intravascular pressure as a factor regulating the tone of the small vessels. Acta Physiol Scand 17: 289–310

Folkow B (1987) Structure and function of the arteries in hypertension. Am Heart J 114: 938–948

Hacking WJG, VanBavel E, Spaan JAE (1996) Shear stress is not sufficient to control growth of vascular networks: a model study. Am J Physiol 270: H364–H375

Haidekker MA, L'Heureux N, Frangos JA (2000) Fluid shear stress increases membrane fluidity in endothelial cells: a study with DCVJ fluorescence. Am J Physiol Heart Circ Physiol 278: H1401–H1406

Jackson WF (1987) Arteriolar oxygen reactivity: where is the sensor? Am J Physiol 253: H1120–H1126

Jackson WF (2000) Hypoxia does not activate ATP-sensitive K+ channels in arteriolar muscle cells. Microcirculation 7: 137–145

Johnson PC (1980) The myogenic response. In: Bohr DF, Somlyo AP, Sparks HV, Jr.: Handbook of Physiology, Section 2, The Cardiovascular System, Vol. II: Vascular Smooth Muscle. American Physiological Society, Bethesda, MD, 409–442

Kamiya A, Bukhari R, Togawa T (1984) Adaptive regulation of wall shear stress optimizing vascular tree function. Bull Math Biol 46: 127–137

Koller A, Kaley G (1990) Endothelium regulates skeletal muscle microcirculation by a blood flow velocity sensing mechanism. Am J Physiol 258: H916–H920

Kuo L, Davis MJ, Chilian WM (1990) Endothelium-dependent, flow-induced dilation of isolated coronary arterioles. Am J Physiol 259: H1063–H1070

Kuo L, Hein TW (2002) Mechanism of shear-stress induced coronary microvascular dilation. In: Barth FG, Humphrey JAC, Secomb TW (eds) Sensors and Sensing in Biology and Engineering. Springer, Wien New York

LaBarbera M (1990) Principles of design of fluid transport systems in zoology. Science 249: 992–1000

Langille BL, O'Donnell F (1986) Reductions in arterial diameter produced by chronic decreases in blood flow are endothelium-dependent. Science 231: 405–407

Mayrovitz HN, Roy J (1983) Microvascular blood flow: evidence indicating a cubic dependence on arteriolar diameter. Am J Physiol 245: H1031–H1038

Murray CD (1926) The physiological principle of minimum work. I. The vascular system and the cost of blood volume. Proc Natl Acad Sci USA 12: 207–214

Pries AR, Reglin B, Secomb TW (2001) Structural adaptation of microvascular networks: functional roles of adaptive responses. Am J Physiol Heart Circ Physiol 281: H1015–H1025

Pries AR, Secomb TW, Gaehtgens P (1995a) Design principles of vascular beds. Circ Res 77: 1017–1023

Pries AR, Secomb TW, Gaehtgens P (1995b) Structure and hemodynamics of microvascular networks: heterogeneity and correlations. Am J Physiol 269: H1713–H1722

Pries AR, Secomb TW, Gaehtgens P (1998) Structural adaptation and stability of microvascular networks: theory and simulations. Am J Physiol 275: H349–H360

Pries AR, Secomb TW, Gaehtgens P (1999) Structural autoregulation of terminal vascular beds: vascular adaptation and development of hypertension. Hypertension 33: 153–161

Pries AR, Secomb TW, Gaehtgens P (2000) The endothelial surface layer. Pflügers Arch 440: 653–666

Pries AR, Secomb TW, Gaehtgens P, Gross JF (1990) Blood flow in microvascular networks. Experiments and simulation. Circ Res 67: 826–834

Pries AR, Secomb TW, Gessner T, Sperandio MB, Gross JF, Gaehtgens P (1994) Resistance to blood flow in microvessels in vivo. Circ Res 75: 904–915

Rodbard S (1975) Vascular caliber. Cardiology 60: 4–49

Schretzenmayr A (1933) Über kreislaufregulatorische Vorgänge an den grossen Arterien bei der Muskelarbeit. Pflügers Arch Ges Physiol 232: 743–748

Secomb TW, Hsu R, Pries AR (2001) Effect of the endothelial surface layer on transmission of fluid shear stress to endothelial cells. Biorheology 38: 143–150

Secomb TW, Pries AR (2002) Information transfer in microvascular networks. Microcirculation (in press)

Segal SS, Jacobs TL (2001) Role of endothelial cell conduction in ascending vasodilation and exercise hyperemia in hamster skeletal muscle. J Physiol 536: 937–946

Skalak TC, Price RJ (1996) The role of mechanical stresses in microvascular remodeling. Microcirculation 3: 143–165

Tulis DA, Unthank JL, Prewitt RL (1998) Flow-induced arterial remodeling in rat mesenteric vasculature. Am J Physiol 274: H874–H882

Vink H, Duling BR (1996) Identification of distinct luminal domains for macromolecules, erythrocytes, and leukocytes within mammalian capillaries. Circ Res 79: 581–589

14. Mechanism of Shear Stress-Induced Coronary Microvascular Dilation

Lih Kuo and Travis W. Hein

Abstract

In the coronary circulation, fluid shear stress acts as an important, moment-to-moment regulator of vascular resistance. Coronary microvessels display profound vasodilation to increased shear stress, a response shown to be mediated by endothelium-dependent release of nitric oxide. However, the sensory transduction mechanism and the intracellular signaling pathway by which shear stress stimulates release of nitric oxide in endothelial cells is not completely understood. In this chapter, the involvement of cytoskeleton, integrin/focal adhesion proteins, protein kinases, membrane potassium channels and calcium mobilization in endothelial activation and vasodilation to elevated shear stress is discussed. The vasomotor regulation by shear stress in the coronary microcirculation is specially emphasized.

I. Introduction

In the circulatory system, the vascular network provides a route for the flowing blood to carry essential nutrients and oxygen to the cells of different tissues. The amount of blood delivered to each tissue is primarily determined and regulated by the activity of arterial microvessels ($<100\,\mu m$ in diameter), which are composed of a single inner layer of endothelium and an outer layer(s) of smooth muscle cells. Changes in vascular tone (or resistance), i.e., constriction or dilation of these microvessels, will decrease or increase blood supply to the tissue, respectively. It is well documented that coronary blood flow is normally closely matched to the metabolic demand of myocardium through a precise regulation of coronary vascular resistance in the microcirculation. Because approximately 90% of total coronary resistance resides in microvessels less than $300\,\mu m$ in diameter, characterizing and assessing microvascular activity in response to physiological disturbances (i.e., hemodynamic and neurohumoral stimulations) are essential to the understanding of coronary flow regulation under physiological and pathophysiological conditions.

Because of the pulsatile flow passing through, as a result of systolic and diastolic actions of the cardiac pump, blood vessels are constantly subjected to cyclic mechanical strain and shear. Blood pressure generates radial and tangential forces that affect all vascular cells. It is well characterized that the systemic (Davis and Hill 1999), as well as the coronary (Kuo et al. 1990a), resistance arterioles can respond to increased and decreased pressure by vasoconstriction and vasodilation, respectively (i.e., myogenic responses). In contrast, fluid shear stress is the drag force derived from the friction of blood against the lumenal surface of endothelial layer and acts in parallel to the vessel wall. In 1933, it was first reported that femoral arteries dilated during increases in blood flow (Schretzenmayr 1933). Mechanistic and quantitative aspects of this phenomenon, termed flow (or shear stress)-induced dilation, have been extensively studied in the isolated vessel and intact organ systems in the past decade. It is generally believed that this vasomotor response is caused by a local vasoregulatory mechanism and not mediated by ascending dilation from the microvasculature (Lie et al. 1970). Adrenergic, cholinergic, and ganglionic blockades have no influence on the flow-induced vasodilation (Lie et al. 1970, Hull et al. 1986). Recent studies, mainly in microvessels, have revealed pronounced flow-dependent changes in vascular diameters. This response clearly may play a significant role in integrative control of tissue blood flow. Although it is well recognized that endothelial mechanotransduction is central for both acute vasoregulation and chronic vascular remodeling, in this chapter the experimental evidence supporting the presence of acute flow-induced vasodilatory responses in the coronary circulation is summarized, with particular emphasis on the cellular mechanism and signal transduction of these responses in resistance microvessels less than 200 μm in diameter. In addition, the physiological and pathophysiological importance of this vasomotor adjustment is discussed.

II. The Role of Endothelium and Nitric Oxide

Integrated vascular responses to flow have been more often studied in the intact organs in the systemic circulation than in the heart, due to the greater practical difficulties inherent in the study of the coronary circulation. Nevertheless, flow-induced dilation was initially documented in large arteries in the coronary circulation in blood-perfused canine hearts (Gerova et al. 1981), and was subsequently demonstrated also in large coronary arteries of conscious dogs (Hintze and Vatner 1984, Holtz et al. 1984). In these vessels, flow-dependent dilation was found to be unaffected by inhibition of prostaglandin synthesis (Holtz et al. 1984), combined α- and β-adrenergic receptor blockade, ganglionic blockade or the adenosine antagonist, aminophylline (Hintze and Vatner 1984). Clinical studies also demonstrated the dilation of normal human epicardial coronary arteries to increased flow (Drexler et al. 1989). Interestingly, selective removal of coronary endothelium abolishes the flow-induced responses (Pohl et al. 1986, Lamping and Dole 1988). Administration of selective compounds that block the release of nitric oxide (NO) from endothelial cells have been shown to cause vasoconstriction of coronary arteries (Amezcua et al. 1989, Chu et al. 1991) and also to inhibit flow-induced dilation of conduit coronary arteries in conscious dogs (Chu et al. 1991). These studies suggest that the release of NO from the endothelium may mediate the coronary arterial dilation to increased flow. Because NO synthesis inhibitors also increased coronary perfusion pressure (i.e., constriction of resistance arterioles) in the intact heart perfused with constant flow (Amezcua et al. 1989), it is suggested that the resting coronary flow can continuously stimulate the release of vasodilator

NO to counterbalance the tonic smooth muscle contraction induced by pressure-associated stretch on the vessel wall.

The first direct evidence that flow-induced responses occurred in the coronary microcirculation was provided by the studies from isolated microvessels (Kuo et al. 1990b). In these studies, coronary arterioles (40 to 80 μm in diameter) were dissected from freshly excised porcine hearts and cannulated with a pair of glass micropipettes. The vessels were perfused with a physiological salt solution over a range of pressure gradients (linearly related to flow rate) at a constant intraluminal pressure by means of a dual reservoir perfusion system. The isolated vessels developed basal tone and exhibited graded dilation to increases in flow rate over the physiological range. Disruption of endothelium abolished the flow-induced dilation, as did treating the intact vessel with an NO synthesis inhibitor, suggesting that this dilation was mediated by NO. Coronary arteriolar dilation to flow was not affected by indomethacin, a cyclooxygenase inhibitor (Kuo et al. 1991), and therefore was not mediated by the release of vasodilator prostaglandins. This finding is important, since prostaglandins appear to mediate flow-induced dilation in a different microvascular bed (Koller and Kaley 1990). A double-microvessel bioassay study confirmed that a transferable substance was responsible for the dilation in response to increased flow, since increasing the flow from the upstream intact arteriole to the downstream denuded arteriole caused dilation of both microvessels, whereas increasing flow in the opposite direction (i.e., denuded vessel at upstream and intact vessel at downstream) led to dilation only of the endothelium-intact arteriole (Kuo et al. 1991). These studies also ruled out the possibility that flow-induced dilation was mediated by electrical conduction mechanisms between endothelial cells

and the underlying smooth muscle. Flow-induced dilation, through an NO-dependent mechanism, can also be observed in the porcine coronary venules (Kuo et al. 1993) and rabbit coronary arterioles (Muller et al. 1999). In a recent human study, isolated coronary arterioles also dilated to increased flow in an endothelium-dependent manner involving the release of NO and the hyperpolarization of vascular smooth muscle (Miura et al. 2001). It appears that NO released from the endothelium is a critical mediator responsible for the dilation of coronary microvessels to increased flow.

Although flow-induced dilation has been reported in both the macro- and microcirculation, there are significant segmental differences in the magnitude of the response. For example, an increase in flow caused a 30% dilation of coronary arterioles with resting diameter of 60–70 μm (Kuo et al. 1990b), whereas a 3% to 10% dilation was found in large conduit coronary arteries (Holtz et al. 1984, Drexler et al. 1989, Chu et al. 1991). These observations suggest that the shear stress influence on vascular tone may be of greater importance in the microcirculation. Further study of flow-induced responses in the coronary microvascular network indicates that large arterioles (80–130 μm) are apparently more responsive to flow than their upstream small arteries (150–300 μm) and the downstream intermediate (50–70 μm) and terminal arterioles (25–45 μm) (Kuo et al. 1995). Interestingly, in contrast to the flow-induced vasodilation, the intermediate and terminal arterioles are more sensitive to pressure-induced myogenic response (Liao and Kuo 1997) and metabolic vasodilation (Kuo et al. 1995), respectively. The heterogeneous distribution of various vasoregulatory responses in the coronary microvascular network is depicted in figure 1. It appears that coronary vascular tone, in different microvascular segments, is

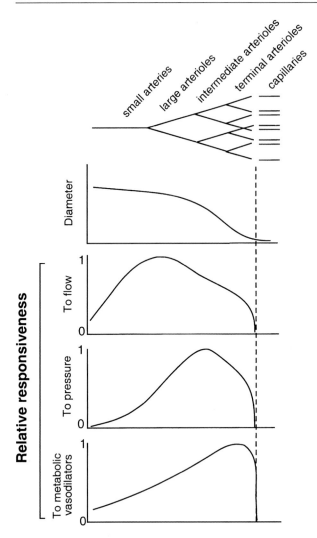

Fig. 1. A scheme for heterogeneous responsiveness of coronary microvessels to pressure (myogenic response), flow (shear stress-induced dilation), and metabolites. Note that each vasoactive mechanism has a dominant site of action in a particular microvascular segment

controlled by different predominating mechanisms, an arrangement that may serve to facilitate the coordination of vascular responses in the intact network during rapid changes in metabolic demand, as discussed later in this chapter.

III. The Role of Integrins and Protein Kinases

Integrins are the transmembrane receptor proteins expressed at the sites of close apposition of the cell surface and the extracellular matrix. They provide linkage between cytoskeletal elements and extracellular matrix and mediate binding and attachment of the cell to the extracellular matrices and the formation of focal adhesions. Although the detailed mechanisms by which endothelium senses shear stress and subsequently transmits the signal for NO synthesis and thus vasodilation are incompletely understood, the involvement of integrins in this mechano-chemical signal transduction has been recently suggested by the following scenario. The shear force generated by viscous fluid flowing in parallel to the surface of anchored cells gives rise to stress that causes cell deformation (strain), which subsequently raises tension and/or deformation of cytoskeletal elements and cell membranes at sites of cell-cell and cell-extracellular matrix adhesion. Therefore, the shear force can be balanced by reaction-tensile forces imposed on cytoskeletal elements and thus consequently transmitted to the anchoring molecules (i.e., focal adhesion molecules and integrins) at the abluminal sides. Because many stretch and stress related signaling pathways are associated with the activation of integrins and focal adhesion-associated protein kinases, the potential pathway for shear stress signals transmission via the integrin-focal adhesion protein cascade for NO release and vasodilation has been extensively examined.

Recent studies in cultured endothelial cells have demonstrated the mechanical connections between integrins and cytoskeletal filaments and the immediate changes in the organization of molecular assemblies in the cytoplasm and nucleus after integrin activation by mechanical stimulation (Maniotis et al. 1997). Focal adhesion sites at abluminal endothelial membrane are also acutely responsive to shear stress stimulation by association/dissociation with the extracellular matrix and redistribution of intracellular stress fibers (Davies et al. 1994). Furthermore, disruption of the cytoskeleton attenuates shear stress-induced release of NO in native endothelial cells (Hutcheson and Griffith 1996). These studies suggest that components of the focal adhesion complex may be mechanically responsive elements coupled to the cytoskeleton for shear stress-induced NO release. Since focal adhesion contacts are integrin-rich complexes, activation of specific integrins may mediate signal transduction elicited by shear stress.

Indeed, integrin activation stimulates many intracellular signal events, several of which resemble those stimulated by shear stress in endothelial cells. For example, both elevation of shear stress (Geiger et al. 1992, Schwarz et al. 1992, Shen et al. 1992, Ando et al. 1993, James et al. 1995, Corson et al. 1996) and activation of specific integrins (Leavesley et al. 1993, Schwartz 1993) in endothelial cells cause a rise in intracellular calcium (Ca^{2+}). In addition, both flow and activation of β_1 integrin produced tyrosine phosphorylation of focal adhesion kinase and stimulated mitogen-activated protein (MAP) kinase in human umbilical vein endothelial cells (Ishida et al. 1996). Recent studies also showed that elevation of shear stress rapidly activates MAP kinase (ERK1/2), possibly via Src signaling pathway, in an integrin-dependent manner (Takahashi and Berk 1996). In mechanotransduction pathways of other cell types, tyrosine kinases are involved at sites of integrin binding to the extracellular matrix, the focal adhesions complex (Burridge et al. 1988, Kornberg et al. 1992). It is possible that tyrosine kinase activation is a subsequent signaling pathway for shear stress-induced NO release and vasodilation. This idea is supported by a cultured human endothelial cell study showing that shear stress activates protein tyrosine kinases and results in enhanced tyrosine phosphorylation of MAP kinase ERK1/2 and several cytoskeletal proteins (Ayajiki et al. 1996). Inhibition of tyrosine kinase activity attenuated the shear stress-induced tyrosine phosphorylation of these cellular proteins (Ayajiki et al. 1996) and abrogated the associated NO production (Ayajiki et al. 1996, Corson et al. 1996).

Although the physiological role of integrin binding, formation of focal adhesion and the tyrosine kinase activation associated with shear stress elevation have been extensively investigated recently, the majority of studies in this nature have been performed using cultured cells, and consequently, do not address the functional significance of these pathways in shear stress-induced vasodilation. The only study to address this issue is that of Muller et al. (1997), who showed that competitive inhibition of integrin binding to extracellular matrix proteins, which contain the Arg-Gly-Asp peptide sequence, blocked shear stress-induced vasodilation of porcine coronary arterioles. The shear stress-induced dilation was also attenuated by a blocking antibody to the integrin β_3 chain, further suggesting the role of integrins in vasodilation to increased shear stress. Moreover, shear stress also significantly increased integrin-sensitive tyrosine kinase activity in the endothelium of these microvessels (Muller et al. 1997). Exposure of these vessels to tyrosine kinase inhibitors attenuated shear stress-induced tyrosine phosphorylation and abolished shear stress-induced vasodilations (Muller et al. 1996). These results indicate that integrin-matrix interactions, possibly at focal adhesions, are of importance in the signaling pathway for shear stress-induced vasodilation of coronary arterioles involving tyrosine kinase activation.

Cumulative evidence indicates that endothelial NO synthase (eNOS) activity can be regulated by the phosphorylation of the enzyme at serine, threonine or tyrosine residues (Fleming and Busse 1999). The rapid (< 30 seconds) phosphorylation of eNOS by shear stress in cultured endothelial cells (Gallis et al. 1999) closely matched the time course of vasodilation initiated by shear stress in intact vessels. Although tyrosine kinase activation is critical for shear stress-induced vasodilation, data suggest that tyrosine kinase may not directly act on the eNOS since the enzyme activity can be negatively controlled by tyrosine phosphorylation (Garcia-Cardena et al. 1996). On the other hand, recent studies demonstrated that eNOS activity could be upregulated by serine phosphorylation through activation of Akt (protein kinase B) (Dimmeler et al. 1999, Fulton et al. 1999). This Akt-mediated NOS activity is involved in the shear stress-induced NO synthesis in cultured endothelial cells since inhibition of phosphatidylinositol 3-kinase (PI 3-kinase), the upstream activator of Akt, inhibited flow-induced eNOS phosphorylation and NO production (Gallis et al. 1999). Interestingly, PI 3-kinase has been shown to be involved in the MAP kinase ERK1/2 activation in endothelial cells subjected to a short-term cyclic strain stimulation (Ikeda et al. 1999) and its level, as well as Akt activity, was rapidly elevated after integrin activation at focal adhesion sites (Khwaja et al. 1997). However, whether shear stress-induced NO release and vasodilation in an intact vessel is mediated by the activation of PI 3-kinase/Akt singling via integrin/cytoskeleton pathway remains to be determined. It is also worth noting that the increased eNOS activity, and thus NO production, through protein kinase phosphorylation does not require an increase in cytosolic Ca^{2+} (Fisslthaler et al. 2000, McCabe et al. 2000), the important signaling molecule which is generally believed to be required for NO synthesis by endothelial cells. This apparent discrepancy, i.e., Ca^{2+}-dependent vs. -independent NOS activation by shear stress is discussed in the later section of this chapter.

IV. Role of Endothelial Potassium Channels and Membrane Hyperpolarization

Enzymatic studies of endothelial particulate and cytosolic proteins indicate that NO production is a Ca^{2+}/calmodulin-dependent process involving the conversion of L-arginine to L-citrulline by NOS (Förstermann et al. 1991). Because Ca^{2+} influx into endothelial cells appears to be linearly related to the electrochemical gradient, hyperpolarization of endothelial membrane by the opened potassium (K^+) channels might contribute to the formation of NO as a result of enhanced Ca^{2+} influx. This K^+ channel-dependent, membrane potential-mediated Ca^{2+} influx for NO release has been demonstrated in cultured endothelial cells in response to pharmacological agonist stimulation (Lückhoff and Busse 1990a, 1990b). Recent electrophysiological studies of cultured endothelial cells implicate that flow/shear stress can activate K^+ channels and lead to membrane hyperpolarization (Nakache and Gaub 1988, Olesen et al. 1988, Jacobs et al. 1995). Furthermore, shear stress-induced endothelial hyperpolarization has been shown to facilitate Ca^{2+} influx by increasing the electrical driving force (Hoyer et al. 1998). However, whether this shear stress-associated membrane potential change contributes to NO production was not determined.

The linkage of K^+ channels to NO release has been demonstrated in cultured bovine aortic endothelial cells exposed to various levels of laminar flow (Ohno et al. 1993). This signal transduction pathway seemingly relies on the activation of G_i- or G_o-proteins since pertussis toxin effectively blocked the shear stress-stimulated NO release (Ohno et al. 1993). The activation of a spectrum of Ca^{2+}-dependent potassium (K_{Ca}) channel subtypes may be involved in the transduction of the pulsatile flow and shear stress to NO production in rabbit abdominal aorta (Hutcheson and Griffith 1994). Likewise, the involvement of endothelial K_{Ca} channels in shear stress-induced, NO-mediated vasodilation of rabbit iliac arteries was also reported (Cooke et al. 1991). However, the activation of endothelial cell inward

rectifier K^+ channels seems to be responsible for the flow-induced dilation of rabbit cerebral arteries (Wellman and Bevan 1995) and facial veins (Xie and Bevan 1998). Although the above intact vessel studies were performed in the large conduit arteries/veins of the same animal species (i.e., rabbit), they show a slight discrepancy in the involvement of a specific K^+ channel in flow-induced vasodilation. Without considering the differences in approaches/preparations used in these studies, it appears that membrane hyperpolarization via opening of K^+ channels is generally involved in the flow-initiated NO release and vasodilation.

In the isolated and pressurized porcine coronary resistance arteries ($<100\,\mu m$), administration of a high concentration of KCl to the lumen of the vessel, without altering resting vascular diameter, abolished the NO-mediated vasodilation in response to increased flow (Kuo and Chancellor 1995, Hein et al. 2000). Since endothelial hyperpolarization was prevented by a high concentration of depolarizing solution KCl, the results suggest that endothelial hyperpolarization is essential for the release of NO and the subsequent vasodilation. Activation of endothelial large-conductance K_{Ca} channels is likely to be responsible for the vasodilation, because administration of a large conductance K_{Ca} channel inhibitor iberiotoxin, rather than of other K^+ channel inhibitors, into the lumen of the vessel abolished the flow-induced dilation (Hein et al. 2000). Interestingly, the dilation of coronary microvessels to an increased flow was potentiated in the presence of a subthreshold concentration of metabolic vasodilator adenosine (Kuo and Chancellor 1995). It has been recently characterized that adenosine, at lower concentrations, preferentially acts on the G-protein coupled adenosine receptors in the endothelium and subsequently leads to NO release through membrane hyperpolarization by opening of ATP-sensitive K^+ channels (Hein et al. 1999). It is believed that increased flow-induced vasodilation by adenosine is a consequence of enhanced NO synthesis by facilitating endothelial cell hyperpolarization during flow. This potentiated effect was abolished when endothelial membrane potential was clamped to a higher level by a high concentration of KCl (Kuo and Chancellor 1995), further emphasizing the cardinal importance of endothelial hyperpolarization in flow-induced vasodilation.

In conjunction with the aforementioned critical role of tyrosine kinase signaling in shear stress-induced vasodilation, the tyrosine kinase Src, which has been shown to be rapidly and transiently activated by the shear stress (Takahashi and Berk 1996), can enhance large conductance K_{Ca} channel activity through direct tyrosine phosphorylation of the channels (Ling et al. 2000). Additionally, inhibition of tyrosine kinase activity significantly reduced K_{Ca} channel current and open probability in a transfected cell system (Prevarskaya et al. 1995). It is possible that activation of tyrosine kinase pathway directly modulates K_{Ca} channel activity in response to shear stress and subsequently leads to Ca^{2+} entry for NO production.

V. Role of Endothelial Cell Calcium

In cultured endothelial cells, increases in fluid shear stress elicit a transient increase in endothelial Ca^{2+} concentration (Geiger et al. 1992, Schwarz et al. 1992, Shen et al. 1992, Ando et al. 1993, James et al. 1995, Corson et al. 1996) and release of NO (Korenaga et al. 1994, Kuchan and Frangos 1994, Ayajiki et al. 1996, Corson et al. 1996). However, the issue whether increased cytosolic Ca^{2+} is

necessary for NO synthesis in shear stress-stimulated endothelial cells is controversial. For example, removal of extracellular Ca^{2+} (Korenaga et al. 1994) or inhibition of Ca^{2+} release from intracellular stores (Hutcheson and Griffith 1997) blocked shear stress-dependent production of NO, but others found that shear stress-induced NO production in cultured endothelial cells can be independent of Ca^{2+} (Ayajiki et al. 1996, Corson et al. 1996, Fleming et al. 1998). The explanation for this discrepancy may be multifactorial and possibly related to the differences in culture medium, source of cell, flow pattern, cell passage, methodology and time course of study.

In a well-defined laminar flow chamber, exposure of primary culture of human umbilical vein endothelial cells to the onset of laminar flow resulted in a rapid (<5 minutes), burst-like stimulation of NO production (Kuchan and Frangos 1994). However, continued exposure (>30 minutes) of the endothelial cells to flow caused a sustained but diminished NO production. In contrast to the NO production during prolonged flow stimulation, the initial NO production in response to a sudden increased flow was inhibited by either Ca^{2+} chelators or calmodulin antagonists, suggesting the Ca^{2+}/calmodulin-dependence of this initial NO release (Kuchan and Frangos 1994). Since the onset latency of flow-induced dilation in small coronary arterioles and large conduit arteries is relatively short, i.e., in a range of 5–10 seconds (Kuo et al. 1990b) and 15–40 seconds (Lie et al. 1970, Gerova et al. 1981, Hull et al. 1986, Pohl et al. 1986), respectively, it is expected that the vasodilation elicited by the initially released NO is dependent upon the increase of endothelial Ca^{2+}. Consistent with this idea, shear stress-induced NO release from a pharmacologically constricted and buffer-perfused rabbit iliac artery segment has been reported to be biphasic and consist of an initial transient (< 20 minutes) Ca^{2+}-dependent phase followed by a Ca^{2+}-independent plateau phase (Ayajiki et al. 1996). Interestingly, stretching the vessels to their *in vivo* length abolished the initial phase of NO release without affecting the later plateau phase (Ayajiki et al. 1996). Since the initial Ca^{2+}-dependent phase was not observed in the stretched vessels, it is thought that this phase may reflect artificial *in vitro* conditions and is irrelevant to the real physiological response. However, stretch of vessels might perturb endothelial and smooth muscle cells leading to changes in basal level of cytosolic Ca^{2+} and activation of stretch-activated ion channels (Naruse and Sokabe 1993, Davis and Hill 1999). Furthermore, the magnitude and characteristics of shear stress-induced changes in endothelial Ca^{2+}, and thus NO release, may be influenced by the vascular smooth muscle Ca^{2+} as proposed recently (Dora et al. 1997). With these considerations in mind, it would be a reasonable approach if the study were performed in the pressurized intact vessels with laminar flow perfusion.

Muller et al. (1999) have recently studied the relationship between endothelial Ca^{2+} and NO-mediated vasodilation of isolated rabbit coronary arterioles in response to increased laminar flow under a constant physiological lumenal pressure. The changes in vessel caliber and endothelial Ca^{2+} concentration in response to an increased shear stress or NO-mediated pharmacological agonists (acetylcholine and substance P) were recorded simultaneously in this study using fluorescence videomicroscopic techniques. It was found that vasodilation in response to acetylcholine, substance P, or shear stress was accompanied by significant increases in endothelial cell Ca^{2+}, although the magnitude of Ca^{2+} changes in response to shear stress was much smaller (Muller et al. 1999). Calcium chelator, BAPTA, eliminated significant changes in endothelial Ca^{2+} and inhibited dilations to acetylcholine and substance P but did not significantly affect shear stress-induced vasodilation. These results suggest that endothelium-dependent, NO-mediated vasodilation of coronary arterioles in response to agonist stimulation is mediated by a rise in endothelial cell Ca^{2+} but that a substantial component of the shear stress-induced response occurs through a Ca^{2+}-insensitive pathway. Although this intact microvessel study does not support the role of increased Ca^{2+} in the short-term regulation of NO-mediated vasodilation to shear stress, recent studies have shown that focal elevations in subplasmalemmal Ca^{2+}, rather than increases in overall Ca^{2+}, trigger NO biosynthesis in endothelial cells (Graier et al. 1998, Paltauf-Doburzynska et al. 1998). This consideration raises the possibility that a small change in subplasmalemmal Ca^{2+} might not be effectively buffered by the Ca^{2+} chelator BAPTA and is likely not captured by the fluorescence videomicroscopic techniques. Since eNOS and many proteins that regulate NOS activity have recently been targeted to specialized cell surface signal-transducing domains termed plasmalemmal caveolae (Shaul and Anderson 1998), the small local increases in subplasmalemmal Ca^{2+} after membrane hyperpolarization may be sufficient to trigger shear stress-induced NO release and the initial vasodilation. Because Akt-mediated NOS phosphorylation has been suggested to reduce the dissociation of calmodulin from activated NOS (McCabe et al. 2000), it is possible that Akt activation, subsequent to shear stress stimulation, can maintain or even further increase NOS activity for sustained NO release without the participation of Ca^{2+}. Further study in this area is apparently needed to unequivocally address the role of endothelial cell Ca^{2+} in shear stress-induced NO release and vasodilation.

VI. Physiological and Patho-physiological Considerations

In the circulatory system, the vasculature constantly encounters two physical forces that interact with blood vessels as regulatory signals mediated through stretch (pressure) and shear stress (flow). We have previously shown that vascular smooth muscle in coronary arterioles can respond directly to changes in intraluminal pressure (i.e., myogenic response) (Kuo et al. 1990a) but that endothelial cells sense changes in shear stress as discussed above. In addition to the hemodynamics-induced vascular responses, coronary microvascular diameters are also regulated by the metabolic vasodilators released from the surrounding tissues (Ishizaka and Kuo 1996, Ishizaka and Kuo 1997, Hein and Kuo 1999) and by neurohumoral substances (Feigl 1983). Many in-vivo and in-vitro studies from our (Kuo et al. 1991, Jones et al. 1995, Kuo and Chancellor 1995, Jones et al. 1996, Liao and Kuo 1997) and other (Feigl 1983, Pohl et al. 1994) laboratories have documented that acute coronary flow regulation does not solely rely on a single mechanism but is orchestrated by the participation of multiple mechanisms involving myogenic, shear stress, and metabolic factors operating in a cooperative and integrative manner.

A hallmark of recent findings in the coronary microcirculation is that this microvascular network exhibits heterogeneity in vasomotor responses to physiological and pharmacological stimuli. The in vitro studies indicate that small terminal arterioles are more sensitive to metabolic vasodilators (e.g., adenosine), and inter-mediate size arterioles are more reactive to pressure changes (Liao and Kuo 1997), but large arterioles are more responsive to shear stress stimulation (Kuo et al. 1995) as shown in figure 2. It is speculated that this site-specific preferential response may play a crucial role in coordinating overall vascular function in the coronary microcirculation. The intact heart studies indicate that vasodilations induced by metabolic activation/stress predominate in the smallest microvessels (Kanatsuka et al. 1989, Chilian and Layne 1990, Jones et al. 1995). If the metabolic demands of the tissue are increased, it is expected that small downstream arterioles would preferentially dilate, presumably as a result of increased production of metabolic vasodilators. Metabolic dilation of distal arterioles would lower intraluminal pressure in intermediate-size arterioles upstream, which possess a strong myogenic response. The myogenic dilation of these vessels would further reduce vascular resistance and subsequently increase flow. The increased flow could initiate flow-induced dilation in larger arterioles upstream, which are characterized by a less potent metabolic and myogenic response but possess a dominant flow-dependent vasodilatory mechanism. Thus, metabolic dilation of small arterioles can potentially recruit dilation of upstream arterioles via myogenic and flow-induced mechanisms, which would serve to lessen overall vascular resistance and improve O_2 delivery to the active tissue. This segmental coordination of microvascular reactions mediated by series coupling of metabolic, myogenic and flow-induced responses is illustrated in figure 2. A recent mathematic network model describing the interaction of metabolic-, pressure-, and shear stress-sensitive vessel segments supports the above integrative flow regulation mechanism (Liao and Kuo 1997). The model indicates that the heterogeneous vascular response

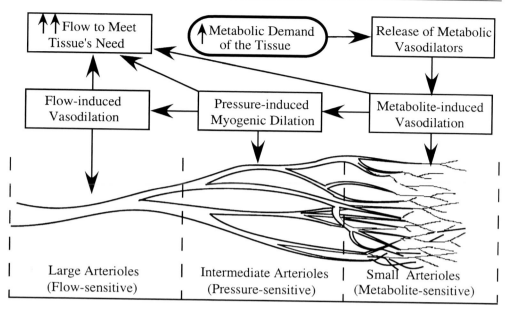

Fig. 2. Integrative regulation of coronary flow by metabolic, myogenic, and shear stress-induced mechanisms in response to a metabolic activation

to metabolic vasodilators and the predominant flow-induced dilation in large arterioles play an important role in overall increase in flow during metabolic activation (Liao and Kuo 1997). Furthermore, shear stress-induced vasodilation in large arterioles can help to stabilize the downstream microvascular pressure (Liao and Kuo 1997), which is essential for the homeostatic movement of fluid and solutes across the capillaries and venules.

Because the shear stress-induced response in the coronary microcirculation is mediated by the intact function of NOS in the endothelium, it is conceivable that dysfunction of NO/endothelium and impairment of shear stress-sensitive mechanism under pathophysiological conditions such as hypercholesterolemia/atherosclerosis (Kuo et al. 1992, Drexler 1999, Hein et al. 2000) and ischemia-reperfusion injury (DeFily 1998, Vinten-Johansen et al. 1999) could potentially reduce tissue perfusion as a result of

losing the counterbalance effect of shear stress on myogenic vasoconstriction. The uncoupling between coronary blood flow and myocardial metabolic demands would aggravate the inadequacy of flow supply to the tissue during intense metabolic activation.

VII. Conclusion

The endothelium is a biological mechano-transducer that senses shear and stretch forces and converts these mechanical stimuli into biochemical signals for various biological functions. Shear stress generated by the flowing fluid on the endothelial layer results in a change in the tension of the endothelial cytoskeleton and in the transmission of this signal throughout the cell, which immediately activates NOS activity. It has been shown that the glycocalyx is involved in sensing shear stress, because its removal by

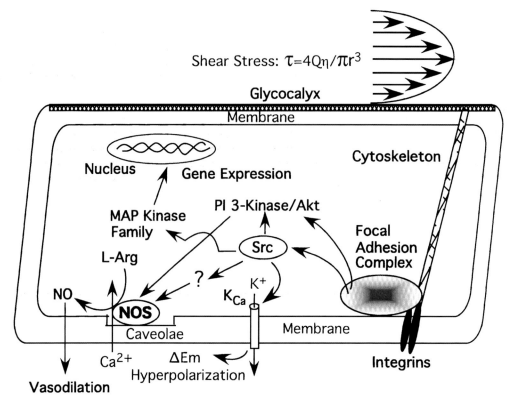

Fig. 3. Putative signaling pathways for shear stress-induced nitric oxide (NO) release from the endothelium. Shear stress (τ) is a function of flow (Q), blood viscosity (η) and the radius of the vessel. Shear stress elicits various signaling pathways for NO release following NO synthase (NOS) activation (see text for details)

neuraminidase blocked flow-dependent dilation (Hecker et al. 1993). Based on the aforementioned signaling pathways, it can be synthesized into a scheme in which shear forces are transmitted from the glycocalyx, through the cytoskeleton, to the focal adhesions of endothelial cells. Accordingly, transmission of shear force to the abluminal side of endothelial cells should initiate shear stress-induced integrin-mediated vasodilation possibly through activation of protein kinase pathways, opening of potassium channels, and phosphorylation of NOS by activated PI 3-kinase/Akt. The acute vasodilation induced by an abrupt increase in shear stress may require a rise in cytosolic Ca^{2+} in endothelial cells, but the prolonged activation of NOS by protein kinase phosphorylation is believed to be independent of Ca^{2+} mobilization (Fig. 3). The basal release of NO in response to resting flow is important in counterbalancing the tonic contraction of smooth muscle associated with pressure stretch, thus preventing vasospasm. The shear stress-induced vascular response participates in many physiological events involving flow recruitments for metabolic needs. The dysfunction of endothelium and impairment of NOS activity during disease states could compromise shear stress-mediated local flow regulation and jeopardize tissue oxygenation and survival.

References

Amezcua JL, Palmer RM, de Souza BM, Moncada S (1989) Nitric oxide synthesized from L-arginine regulates vascular tone in the coronary circulation of the rabbit. Br J Pharmacol 97: 1119–1124

Ando J, Ohtsuka A, Korenaga R, Kawamura T, Kamiya A (1993) Wall shear stress rather than shear rate regulates cytoplasmic Ca^{++} responses to flow in vascular endothelial cells. Biochem Biophys Res Commun 190: 716–723

Ayajiki K, Kindermann M, Hecker M, Fleming I, Busse R (1996) Intracellular pH and tyrosine phosphorylation but not calcium determine shear stress-induced nitric oxide production in native endothelial cells. Circ Res 78: 750–758

Burridge K, Fath K, Kelly T, Nuckolls G, Turner C (1988) Focal adhesions: transmembrane junctions between the extracellular matrix and the cytoskeleton. Annu Rev Cell Biol 4: 487–525

Chilian WM, Layne SM (1990) Coronary microvascular responses to reductions in perfusion pressure. Evidence for persistent arteriolar vasomotor tone during coronary hypoperfusion. Circ Res 66: 1227–1238

Chu A, Chambers DE, Lin C-C, Kuehl WD, Palmer RMJ, Moncada S, Cobb FR (1991) Effects of inhibition of nitric oxide formation on basal vasomotion and endothelium-dependent responses of the coronary arteries in awake dogs. J Clin Invest 87: 1964–1968

Cooke JP, Rossitch E Jr, Andon NA, Loscalzo J, Dzau VJ (1991) Flow activates an endothelial potassium channel to release an endogenous nitrovasodilator. J Clin Invest 88: 1663–1671

Corson MA, James NL, Latta SE, Nerem RM, Berk BC, Harrison DG (1996) Phosphorylation of endothelial nitric oxide synthase in response to fluid shear stress. Circ Res 79: 984–991

Davies PF, Robotewskyj A, Griem ML (1994) Quantitative studies of endothelial cell adhesion. Directional remodeling of focal adhesion sites in response to flow forces. J Clin Invest 93: 2031–2038

Davis MJ, Hill MA (1999) Signaling mechanisms underlying the vascular myogenic response. Physiol Rev 79: 387–423

DeFily DV (1998) Control of microvascular resistance in physiological conditions and reperfusion. J Mol Cell Cardiol 30: 2547–2554

Dimmeler S, Fleming I, Fisslthaler B, Hermann C, Busse R, Zeiher AM (1999) Activation of nitric oxide synthase in endothelial cells by Akt-dependent phosphorylation. Nature 399: 601–605

Dora KA, Doyle MP, Duling BR (1997) Elevation of intracellular calcium in smooth muscle causes endothelial cell generation of NO in arterioles. Proc Natl Acad Sci USA 94: 6529–6534

Drexler H (1999) Nitric oxide and coronary endothelial dysfunction in humans. Cardiovasc Res 43: 572–579

Drexler H, Zeiher AM, Wollschläger H, Meinertz T, Just H, Bonzel T (1989) Flow-dependent coronary artery dilatation in humans. Circulation 80: 466–474.

Feigl EO (1983) Coronary Physiology. Physiol Rev 63: 1–205

Fisslthaler B, Dimmeler S, Hermann C, Busse R, Fleming I (2000) Phosphorylation and activation of the endothelial nitric oxide synthase by fluid shear stress. Acta Physiol Scand 168: 81–88

Fleming I, Bauersachs J, Fisslthaler B, Busse R (1998) Ca^{2+}-independent activation of the endothelial nitric oxide synthase in response to tyrosine phosphatase inhibitors and fluid shear stress. Circ Res 82: 686–695

Fleming I, Busse R (1999) Signal transduction of eNOS activation. Cardiovasc Res 43: 532–541

Förstermann U, Pollock JS, Schmidt HHHW, Heller M, Murad F (1991) Calmodulin-dependent endothelium-derived relaxing factor/nitric oxide synthase activity is present in the particulate and cytosolic fractions of bovine aortic endothelial cells. Proc Natl Acad Sci USA 88: 1788–1792

Fulton D, Gratton JP, McCabe TJ, Fontana J, Fujio Y, Walsh K, Franke TF, Papapetropoulos A, Sessa WC (1999) Regulation of endothelium-derived nitric oxide production by the protein kinase Akt. Nature 399: 597–601

Gallis B, Corthals GL, Goodlett DR, Ueba H, Kim F, Presnell SR, Figeys D, Harrison DG, Berk BC, Aebersold R, Corson MA (1999) Identification of flow-dependent endothelial nitric-oxide synthase phosphorylation sites by mass spectrometry and regulation of phosphorylation and nitric oxide production by the phosphatidylinositol 3-kinase inhibitor LY294002. J Biol Chem 274: 30101–30108

Garcia-Cardena G, Fan R, Stern DF, Liu J, Sessa WC (1996) Endothelial nitric oxide synthase is regulated by tyrosine phosphorylation and interacts with caveolin-1. J Biol Chem 271: 27237–27240

Geiger RV, Berk BC, Alexander RW, Nerem RM (1992) Flow-induced calcium transients in single endothelial cells: spatial and temporal analysis. Am J Physiol 262: C1411–C1417

Gerova M, Gero J, Barta E, Dolezel S, Smiesko V, Levicky V (1981) Neurogenic and myogenic control of conduit coronary a.: a possible interference. Basic Res Cardiol 76: 503–507

Graier WF, Paltauf-Doburzynska J, Hill BJ, Fleischhacker E, Hoebel BG, Kostner GM, Sturek M (1998) Submaximal stimulation of porcine endothelial cells causes focal Ca^{2+} elevation beneath the cell membrane. J Physiol 506: 109–125

Hecker M, Mülsch A, Bassenge E, Busse R (1993) Vasoconstriction and increased flow: two principal mechanisms of shear stress-dependent endothelial autacoid release. Am J Physiol 265: H828–H833

Hein TW, Belardinelli L, Kuo L (1999) Adenosine A_{2A} receptors mediate coronary microvascular dilation to adenosine: role of nitric oxide and ATP-sensitive potassium channels. J Pharmacol Exp Ther 291: 655–664

Hein TW, Kuo L (1999) cAMP-independent dilation of coronary arterioles to adenosine: role of nitric oxide, G proteins, and K_{ATP} channels. Circ Res 85: 634–642

Hein TW, Liao JC, Kuo L (2000) oxLDL specifically impairs endothelium-dependent, NO-mediated dilation of coronary arterioles. Am J Physiol 278: H175–H183

Hintze TH, Vatner SF (1984) Reactive dilation of large coronary arteries in conscious dogs. Circ Res 54: 50–57

Holtz J, Forstermann U, Pohl U, Giesler M, Bassenge E (1984) Flow-dependent, endothelium-mediated dilation of epicardial coronary arteries in conscious dogs: effects of cyclooxygenase inhibition. J Cardiovasc Pharmacol 6: 1161–1169

Hoyer J, Kohler R, Distler A (1998) Mechanosensitive Ca^{2+} oscillations and STOC activation in endothelial cells. FASEB J 12: 359–366

Hull SS Jr, Kaiser L, Jaffe MD, Sparks HV Jr. (1986) Endothelium-dependent flow-induced dilation of canine femoral and saphenous arteries. Blood Vessels 23: 183–198

Hutcheson IR, Griffith TM (1994) Heterogeneous populations of K^+ channels mediate EDRF release to flow but not agonists in rabbit aorta. Am J Physiol 266: H590–H596

Hutcheson IR, Griffith TM (1996) Mechanotransduction through the endothelial cytoskeleton: mediation of flow- but not agonist-induced EDRF release. Br J Pharmacol 118: 720–726

Hutcheson IR, Griffith TM (1997) Central role of intracellular calcium stores in acute flow- and agonist- evoked endothelial nitric oxide release. Br J Pharmacol 122: 117–125

Ikeda M, Kito H, Sumpio BE (1999) Phosphatidylinositol-3 kinase dependent MAP kinase activation via p21ras in endothelial cells exposed to cyclic strain. Biochem Biophys Res Commun 257: 668–671

Ishida T, Peterson TE, Kovach NL, Berk BC (1996) MAP kinase activation by flow in endothelial cells. Role of beta 1 integrins and tyrosine kinases. Circ Res 79: 310–316

Ishizaka H, Kuo L (1996) Acidosis-induced coronary arteriolar dilation is mediated by the ATP-sensitive potassium channels in vascular smooth muscle. Circ Res 78: 50–57

Ishizaka H, Kuo L (1997) Endothelial ATP-sensitive potassium channels mediate coronary microvascular dilation to hyperosmolarity. Am J Physiol 273: H104–H112

Jacobs ER, Cheliakine C, Gebremedhin D, Birks EK, Davies PF, Harder DR (1995) Shear activated channels in cell-attached patches of cultured bovine aortic endothelial cells. Pflügers Arch 431: 129–131

James NL, Harrison DG, Nerem RM (1995) Effects of shear on endothelial cell calcium in the presence and absence of ATP. FASEB J 9: 968–973

Jones CJ, Kuo L, Davis MJ, Chilian WM (1996) In vivo and in vitro vasoactive reactions of coronary arteriolar microvessels to nitroglycerin. Am J Physiol 271: H461–H468

Jones CJH, Kuo L, Davis MJ, DeFily DV, Chilian WM (1995) Role of nitric oxide in the coronary microvascular responses to adenosine and increased metabolic demand. Circulation 91: 1807–1813

Kanatsuka H, Lamping KG, Eastham CL, Dellsperger KC, Marcus ML (1989) Comparison of the effects of increased myocardial oxygen consumption and adenosine on the coronary microvascular resistance. Circ Res 65: 1296–1305

Khwaja A, Rodriguez-Viciana P, Wennstrom S, Warne PH, Downward J (1997) Matrix adhesion and Ras transformation both activate a phosphoinositide 3-OH kinase and protein kinase B/Akt cellular survival pathway. EMBO J 16: 2783–2793

Koller A, Kaley G (1990) Prostaglandins mediate arteriolar dilation to increased blood flow velocity in skeletal muscle microcirculation. Circ Res 67: 529–534

Korenaga R, Ando J, Tsuboi H, Yang W, Sakuma I, Toyo-oka T, Kamiya A (1994) Laminar flow stimulates ATP- and shear stress-dependent nitric oxide production in cultured bovine endothelial cells. Biochem Biophys Res Commun 198: 213–219

Kornberg L, Earp HS, Parsons JT, Schaller M, Juliano RL (1992) Cell adhesion or integrin clustering increases phosphorylation of a focal adhesion-associated tyrosine kinase. J Biol Chem 267: 23439–23442

Kuchan MJ, Frangos JA (1994) Role of calcium and calmodulin in flow-induced nitric oxide production in endothelial cells. Am J Physiol 266: C628–C636

Kuo L, Arko F, Chilian WM, Davis MJ (1993) Coronary venular responses to flow and pressure. Circ Res 72: 607–615

Kuo L, Chancellor JD (1995) Adenosine potentiates flow-induced dilation of coronary arterioles by activating K_{ATP} channels in endothelium. Am J Physiol 269: H541–H549

Kuo L, Chilian WM, Davis MJ (1990a) Coronary arteriolar myogenic response is independent of endothelium. Circ Res 66: 860–866

Kuo L, Chilian WM, Davis MJ (1991) Interaction of pressure- and flow-induced responses in porcine coronary resistance vessels. Am J Physiol 261: H1706–H1715

Kuo L, Davis MJ, Cannon MS, Chilian WM (1992) Pathophysiological consequences of atherosclerosis extend into the coronary microcirculation. Restoration of endothelium-dependent responses by L-arginine. Circ Res 70: 465–476

Kuo L, Davis MJ, Chilian WM (1990b) Endothelium-dependent, flow-induced dilation of isolated coronary arterioles. Am J Physiol 259: H1063–H1070

Kuo L, Davis MJ, Chilian WM (1995) Longitudinal gradient for endothelium-dependent and -independent vascular responses in the coronary microcirculation. Circulation 92: 518–525

Lamping KG, Dole WP (1988) Flow-mediated dilation attenuates constriction of large coronary arteries to serotonin. Am J Physiol 255: H1317–H1324

Leavesley DI, Schwartz MA, Rosenfeld M, Cheresh DA (1993) Integrin beta 1- and beta 3-mediated endothelial cell migration is triggered through distinct signaling mechanisms. J Cell Biol 121: 163–170

Liao JC, Kuo L (1997) Interaction between adenosine and flow-induced dilation in coronary microvascular network. Am J Physiol 272: H1571–H1581

Lie M, Sejersted OM, Kiil F (1970) Local regulation of vascular cross section during changes in femoral arterial blood flow in dogs. Circ Res 27: 727–737

Ling S, Woronuk G, Sy L, Lev S, Braun AP (2000) Enhanced activity of a large conductance, calcium-sensitive K^+ channel in the presence of Src tyrosine kinase. J Biol Chem 275: 30683–30689

Lückhoff A, Busse R (1990a) Activators of potassium channels enhance calcium influx into endothelial cells as a consequence of potassium currents. Naunyn-Schmiedeberg's Arch Pharmacol 342: 94–99

Lückhoff A, Busse R (1990b) Calcium influx into endothelial cells and formation of endothelium-derived relaxing factor is controlled by the membrane potential. Pflügers Arch 416: 305–311

Maniotis AJ, Chen CS, Ingber DE (1997) Demonstration of mechanical connections between integrins, cytoskeletal filaments, and nucleoplasm that stabilize nuclear structure. Proc Natl Acad Sci USA 94: 849–854

McCabe TJ, Fulton D, Roman LJ, Sessa WC (2000) Enhanced electron flux and reduced calmodulin dissociation may explain "calcium-independent" eNOS activation by phosphorylation. J Biol Chem 275: 6123–6128

Miura H, Wachtel RE, Liu Y, Loberiza FR Jr., Saito T, Miura M, Gutterman DD (2001) Flow-induced dilation of human coronary arterioles: important role of Ca^{2+}-activated K^+ channels. Circulation 103: 1992–1998

Muller JM, Chilian WM, Davis MJ (1997) Integrin signaling transduces shear stress-dependent vasodilation of coronary arterioles. Circ Res 80: 320–326

Muller JM, Davis MJ, Chilian WM (1996) Coronary arteriolar flow-induced vasodilation signals through tyrosine kinase. Am J Physiol 270: H1878–H1884

Muller JM, Davis MJ, Kuo L, Chilian WM (1999) Changes in coronary endothelial cell Ca^{2+} concentration during shear stress- and agonist-induced vasodilation. Am J Physiol 276: H1706–H1714

Nakache M, Gaub HE (1988) Hydrodynamic hyperpolarization of endothelial cells. Proc Natl Acad Sci USA 85: 1841–1843

Naruse K, Sokabe M (1993) Involvement of stretch-activated ion channels in Ca^{2+} mobilization to mechanical stretch in endothelial cells. Am J Physiol 264: C1037–C1044

Ohno M, Gibbons GH, Dzau VJ, Cooke JP (1993) Shear stress elevates endothelial cGMP. Role of a potassium channel and G protein coupling. Circulation 88: 193–197

Olesen S-P, Clapham DE, Davies PF (1988) Haemodynamic shear stress activates a K^+ current in vascular endothelial cells. Nature 331: 168–170

Paltauf-Doburzynska J, Posch K, Paltauf G, Graier WF (1998) Stealth ryanodine-sensitive Ca^{2+} release contributes to activity of capacitative Ca^{2+} entry and nitric oxide synthase in bovine endothelial cells. J Physiol 513: 369–379

Pohl U, Holtz J, Busse R, Bassenge E (1986) Crucial role of endothelium in the vasodilator response to increased flow in vivo. Hypertension 8: 37–44

Pohl U, Lamontagne D, Bassenge E, Busse R (1994) Attenuation of coronary autoregulation in the isolated rabbit heart by endothelium derived nitric oxide. Cardiovasc Res 28: 414–419

Prevarskaya NB, Skryma RN, Vacher P, Daniel N, Djiane J, Dufy B (1995) Role of tyrosine phosphorylation in potassium channel activation. Functional association with prolactin receptor and JAK2 tyrosine kinase. J Biol Chem 270: 24292–24299

Schretzenmayr A (1933) Über Kreislaufregulatorische Vorgänge an den groben Arterien bei der Muskelarbeit. Pflügers Arch 232: 743–748

Schwartz MA (1993) Spreading of human endothelial cells on fibronectin or vitronectin triggers elevation of intracellular free calcium. J Cell Biol 120: 1003–1010

Schwarz G, Droogmans G, Nilius B (1992) Shear stress induced membrane currents and calcium transients in human vascular endothelial cells. Pflügers Arch 421: 394–396

Shaul PW, Anderson RG (1998) Role of plasmalemmal caveolae in signal transduction. Am J Physiol 275: L843–L851

Shen J, Luscinskas FW, Connolly A, Dewey CF Jr., Gimbrone MA Jr. (1992) Fluid shear stress modulates cytosolic free calcium in vascular endothelial cells. Am J Physiol 262: C384–C390

Takahashi M, Berk BC (1996) Mitogen-activated protein kinase (ERK1/2) activation by shear stress and adhesion in endothelial cells. Essential role for a herbimycin-sensitive kinase. J Clin Invest 98: 2623–2631

Vinten-Johansen J, Zhao ZQ, Nakamura M, Jordan JE, Ronson RS, Thourani VH, Guyton RA (1999) Nitric oxide and the vascular endothelium in myocardial ischemia-reperfusion injury. Ann N Y Acad Sci 874: 354–370

Wellman GC, Bevan JA (1995) Barium inhibits the endothelium-dependent component of flow but not acetylcholine-induced relaxation in isolated rabbit cerebral arteries. J Pharmacol Exp Ther 274: 47–53

Xie H, Bevan JA (1998) Barium and 4-aminopyridine inhibit flow-initiated endothelium- independent relaxation. J Vasc Res 35: 428–436

15. A Possible Mechanism for Sensing Crop Canopy Ventilation

Tony Farquhar, Jiang Zhou, and Henry W. Haslach, Jr.

Abstract

One approach that may help elucidate certain mechanisms of biological sensing is based on a concept developed by engineers in the 1940's. *Operations research* seeks to describe the *black box* or functional behavior of a complex system without determining its exact relationship to the myriad behaviors and interactions of its many components. For example, using this perspective, it may be possible to identify the physical basis underlying sensory capability before identifying the biological pathway by which the sensing occurs at the cellular level. To illustrate this idea, the present Chapter describes a physical process, which provides filtered information that might allow the members of a plant community to monitor the effects of their collective wind-induced motion. Specifically, the energetics of crop canopy ventilation are studied for a wheat crop excited by a non-steady flow. In the model, the wind gust elicits large scale motions facilitating waste gas clearance out of the canopy and into the overlying airspace. As described below, the volumetric flow per unit time is found to be independent of gust velocity but varies in inverse proportion to stalk flexural stiffness. If waste gas clearance represents an important evolutionary constraint, healthy canopy plants may indeed sense intra-canopy gas concentration and modulate their biomechanical properties accordingly.

I. Introduction

The dominant forces in the environment of terrestrial plants are weight and aerodynamic drag, imposed by gravity and wind respectively (Biddington 1986, Speck et al. 1990, Jaffe and Forbes 1993, Mitchell 1996). Plant cells contain dense starch granules called statoliths, whose sedimentary motions provide the signals that allow a growing plant to align itself with respect to gravity (Barlow 1995). In contrast, plant cells do not appear to have specialized structures for obtaining sensory information about wind-induced motion. However, in the present Chapter, we see that such motions facilitate convective exchange between the plant domain and its exterior. Ensuing changes in the gaseous microenvironment of the plants may provide the feedback information needed to modulate their collective structural frequency response.

The productivity of grain crops like wheat and rice strongly depends on ambient wind conditions (Fischer and Stapper 1987, Evans and Fischer 1999). In general, moderate wind exposure increases crop hardiness and fungal disease resistance (Fritig and Legrand 1993) while severe wind exposure damages leaves, slows vegetative growth, and inhibits reproductive development (Mitchell 1996). As evidenced by dramatic changes in their gross morphology, growing plants respond to wind stress by modulating the rates and patterns of cell elongation. However, it is not clear whether and how plant cells discriminate between different levels of wind intensity, whether and how they communicate this information to other remote cells, and whether and how their local responses are coordinated in a globally proportionate manner.

By what mechanisms could a large group of plants sense and respond to exterior air flow? When a steady flow sweeps over a wheat canopy, individual stalks at the leading edge deflect and exclude most of the mean flow energy from the canopy interior. In contrast, the response to nonsteady flow is very different (Shaw 1985). A short gust-like flow is able to penetrate deep into the canopy, where its energy becomes available to flush heat, particulates, and waste gases out of the intracanopy space (Finnigan 1979, Brunet et al. 1995). In the process, the passage of the gust can elicit coherent wave-like disturbances seen sweeping over the canopy surface (Inoue 1955). Proceeding from the assumption that the associated changes in gas concentrations could provide useful sensory input, the objective of the present study was to quantify the dependence of the air exchange rate on wind gust velocity, and on stalk spacing and stiffness.

II. Model Formulation

A maturing wheat field contains about 400 stalks per square meter, with higher densities possible under supranormal conditions (Bugbee and Salisbury 1988). On a windy day, each of the many stalks is subjected to impulsive excitation by drag forces acting on the apical organs. The resulting motions entail myriad collisions with neighboring stalks and leaves. Precise description of these small-scale

events lies beyond the capabilities of present day theory. Nevertheless, it is instructive to study the behavior of an idealized model canopy, which retains essential features and scaling of the real plant system.

Let us consider the dynamical behavior of a canopy made up of many identical stalks, which have been excited by a short gust traveling along the row direction (Fig. 1). To first approximation, the horizontal gust can be idealized as an x-directed square wave of velocity V_0, where the x-axis coincides with the gust direction and the z-axis is the upward vertical direction. The $x - z$ view can now be understood to show a row of width y_0 measured in the out-of-plane direction. The upper boundary of this slice can be visualized as a flexible virtual surface that passes through the head of every stalk in the row.

Let us assume that the stalks are of height L and are spaced at distance s. Each stalk can be idealized as a slender vertical cantilever with flexural stiffness k^* and supporting a grain load of effective mass m_e. Wind tunnel study has shown that the drag force F on the head is linearly proportional to the wind velocity, that is,

$$F = C_d V_0 \qquad (1)$$

where C_d is an empirical drag coefficient (Farquhar et al. 2000 a). As the gust sweeps down the row, neighboring stalks are successively deflected to a maximum displacement

$$u_{\max} = F/k^* = C_d V_0/k^*, \qquad (2)$$

which occurs when the restoring force due to stalk deflection balances the wind-induced drag force. As a simplification, let us assume that the shape of the deflected stem has a cubic dependence on height z (Zebrowski 1999, Farquhar and Meyer-Phillips 2001), implying that

$$u(z) = u_{\max}[1.5(z/L)^2 - 0.5(z/L)^3]. \qquad (3)$$

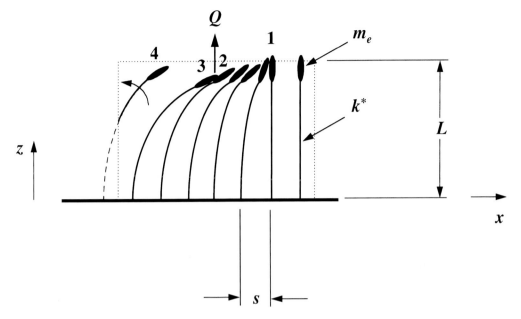

Fig. 1. This schematic of the model canopy shows a row of stalks subjected to a rightwards wind gust. Each stalk has been idealized as a slender vertical cantilever supporting a tip load m_e. As suggested here, the dynamic events include (1) initial contact with the gust, (2) downwind deflection due to it drag, (3) release by the gust, and (4) free underdamped vibration at a specific frequency. The timing of these events can be explicitly related to the gust speed, the stalk height L, the spacing s, and the flexural stiffness k^*. The larger scale response of central interest is the air flux Q between the canopy and its exterior

At some later time, each stalk is released from its deflected position as the trailing edge of the gust moves down the row. When the resulting motions remain confined to the $x-z$ plane (see Farquhar et al. 2000 b), the resonant frequency of stalk vibration is

$$\omega = 2\pi f = (k^*/m_e)^{1/2} \qquad (4)$$

where ω has units of radians per second and f has units of cycles per second (Farquhar and Meyer-Phillips 2001). The sequential release of the stalks sets up a regular pattern of phase-separated vibration. In the model, the relative motion of many stalks is assumed to drive a time-varying flow across the upper boundary. In effect, the gentle but persistent sculling action of many stalks underlies a process of self ventilation.

Referring again to Fig. 1, the distance separating the heads of neighboring stalks is

$$\Delta u = s + u_{max}\{\exp(-\delta\omega t/2\pi)$$
$$[\sin\omega(t+\Delta t) - \sin\omega t]\}, \qquad (5)$$

implying that the volume between the jth stalk-pair is

$$\text{Vol}_j = (0.375)(\Delta u_j)(L)(y_0) \qquad (6)$$

where the area bounded by adjacent stalks has been integrated here by approximation. If the air within the intra-canopy space were incompressible and could only move in the z-direction, the time variation of Vol$_j$ would create a z-directed flux

$$Q_j = \eta \, \Delta\text{Vol}_j \, 2\,\omega \qquad (7)$$

where ΔVol_j is the swept volume per half cycle of vibration and η is an empirical measure of pumping efficiency. Let us assume that $\eta = 1$ while recognizing that the air that is constrained between the stalks is actually compressible and free to move in all directions. Even so, these assumptions establish an upper bound on the volume of air that could be displaced, which becomes increasingly realistic as stalk density increases. On this basis, the vertical flux per unit canopy area per half cycle is

$$Q = Q_j/s, \qquad (8)$$

where Q has already been identified as the dependent variable of particular interest. Using Eq. (5), it can be shown that the condition to avoid inter-stalk collision is

$$u_{\max} < V_0/2\pi\omega. \qquad (9)$$

Due to the coordinated nature of the stalk motions, the upper boundary also exhibits a larger scale motion. As illustrated in Fig. 2, the wavelength of these surface motions is

$$\lambda = V_0/\omega, \qquad (10)$$

where the wavecrest velocity is exactly equal to the gust speed V_0. Guided further by this model, one can show that the elastic work done by the wind on stalk i is

$$W_i = 0.5(k^*)(u_i)^2. \qquad (11)$$

The associated energy is stored in the deflected stalk, released as the gust moves down the row, and then gradually dissipated in a lightly damped cyclic exchange between kinetic and potential energy. In principle, an oscillatory vertical flux continues throughout the latter stage of this process until all plant motions have subsided. The pumping efficiency is the one free parameter in the model, which implies that η can be understood as a measure of damping. The amount of wind energy available to drive the process of self ventilation may be substantial, since the total wind power harnessed per unit row width is

$$P = W/t = V_0W/s^2 \qquad (12)$$

where $W = W_i/s$ is the work per unit canopy area.

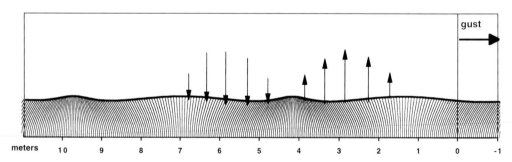

Fig. 2. This snapshot of the deforming canopy at a particular timepoint was obtained by numerical simulation. The gust has swept down the row and its passage has excited the synchronous motions of many stalks. The physical parameters used to predict the plant motions were obtained by experimental study as described in the text. Of central interest, the larger scale canopy motion drives an oscillatory vertical flux, which can continue for many seconds after the gust has passed

III. Results

The model behavior was examined for two varieties of wheat. To facilitate comparisons, all results are for gust speed $V_0 = 5\,\text{m/s}$ and stalk spacing $s = 0.05\,\text{m}$. The stalk stiffness k^* and other plant parameters in the model are in accordance with experimental observation (Farquhar and Meyer-Phillips 2001).

The wheats of interest are nearly isogenic and differ only by the absences of presence of a single dwarfing gene mutation called Rht-1. Global dissemination of this (and other closely related dwarfing genes) began in the mid-1960's and was essentially completed by 1985 (Gent and Kiyomoto 1998). As described earlier, the Rht genes inhibit basal internode extension, and are used to improve standing ability and grain yield (Farquhar et al. 2002).

The two varieties are called *Kauz x rht* (no gene) and *Kauz x Rht* (one allele). The first is a tall research cross developed by Dr. Ravi Singh at the International Center for Maize and Wheat Improvement (CIMMYT) in Mexico. The second is an elite semidwarf variety released in 1985 and now widely cultivated in Central America.

Using the developments of the previous section, the following predictions were obtained.

When comparing the two varieties, note that the vertical flux ratio is essentially equal to the stalk height ratio. In contrast, the canopy work ratio and power ratio are inversely related to the flexural stiffness ratio.

Numerical simulations were then conducted to clarify the relationship between small scale plant parameters (e.g., stalk height and flexural stiffness) and larger scale behaviors e.g., canopy surface wavelength and wavecrest velocity). To illustrate the essential nature of these results, Fig. 2 shows the tall wheat canopy subjected to the action of a gust that has swept down the row from left to right. At this instant in time, the trailing edge of the gust has nearly reached the right side. The downstream stalks are still within the gust but the upstream stalks have been released and are now undergoing coordinated resonant vibration. In some regions of the canopy (e.g., between the 2 and 4 meter marks), the stalks are moving together, which serves to expel waste gas from the intracanopy space. In other regions (e.g., between the 5 and 7 meter marks) the stalks are moving apart, which serves to draw fresh air back into the canopy. Given that these motions can persist for 15 to 20 cycles (Brunet et al 1995), the process of self ventilation can involve large parts of the canopy at any instant in time. While the same mechanisms are

Table 1. Physical Behaviors of a Crop Canopy made up of one of Two Wheat Varieties (Farquhar and Meyer-Phillips 2001)

Parameter	*Kauz x rht*	*Kauz x Rht*	Ratio (*Rht/rht*)
Stalk height L	1.00 m	0.75 m	0.75
Flexural stiffness k^*	0.11 N/m	0.26 N/m	2.36
Resonant frequency f	0.9 Hz	1.4 Hz	1.56
Stalk displacement u_{max}	0.47 m	0.20 m	0.43
Surface wavelength l	5.56 m/cycle	3.57 m/cycle	0.64
Canopy work W	4.84 J/m^2	2.08 J/m^2	0.43
Canopy power P	24.2 W/m	10.4 W/m	0.43
Vertical flux Q	0.72 m^3/s/m	0.55 m^3/s/m	0.76

also at play in a semidwarf canopy, the numerical simulations indicated that the volume of air exchanged per unit time was considerably less. Similarly, the amount of energy available to drive this exchange was reduced. In both cases, this was due to the increased flexural stiffness of the shorter stalks.

IV. Discussion

Several novel findings were obtained in this model-based study. Computer simulation confirmed that the passage of a gust elicits regular canopy motions perceived as waves moving over the canopy surface. This integrated response arises from coherent phase-separated vibration of many constituent stalks. The driving force for the behavior is mechanical energy, which is captured from the non-steady flow and then stored by elastic deflection of the stalks. Once the gust has passed, the motion of the canopy facilitates an oscillatory exchange across the boundary between the exterior and intracanopy spaces. The associated air currents could exert a substantial effect on the microclimate within the canopy.

As expected, the initial deflection of the stalks increased as the gust speed increased, which increased the amount of energy stored in the canopy. This implies that the duration of the lightly damped vertical exchange is dependent on gust speed. However, the phase separation between adjacent stalks during the ensuing vibration decreased as the gust speed increased. As a result, and contrary to expectation, the maximum value of the vertical exchange is independent of the initiating gust velocity. With this insight, the volume of air exchanged per unit time can now be viewed an intrinsic canopy behavior rather than a passive response to the exterior flow.

The model indicated that the vertical exchange was substantially lower for a short stiff wheat like *Kauz x Rht* than for a more flexible taller variety like *Kauz x rht*. Similarly, the amount of energy captured from the wind was lower for the semidwarf versus tall variety. By clarifying the relationship between stalk flexural stiffness and intracanopy ventilation, the model lends credence to the idea that intracanopy gas concentration could provide the sensory feedback needed to modulate stalk properties in a useful manner. This idea could now be tested in an appropriate experimental setting.

References

Barlow PW (1995) Gravity perception in plants: a multiplicity of systems derived by evolution? Plant Cell Environment 18: 951–962

Biddington NL (1986) The effects of mechanically-induced stress in plants, a review. Plant Growth Regulation 4: 103–123

Brunet Y, Finnigan JJ, Raupach M (1995) A wind tunnel study of air flow in waving wheat: single point statistics. Bound Layer Meteo 70: 95–132

Bugbee B, Salisbury FB (1988) Exploring the limits of crop productivity. Plant Physiol 88: 869–878

Evans LT, Fischer RA (1999) Yield potential: its definition, measurement, and significance. Crop Science 39: 1544–1551

Farquhar T, Meyer H, van Beem J (2000 a) Effect of aeroelasticity on the aerodynamics of wheat. Matls Scien Engr C 7: 111–117

Farquhar T, Wood JZ, van Beem J (2000 b) The kinematics of wheat struck by a wind gust. J Applied Mechanics 67: 496–502

Farquhar T, Meyer-Phillips H (2001) Relative safety factors against global buckling, anchorage rotation, and tissue rupture in wheat. J Theoretical Biology 211: 55–65

Farquhar T, Zhou J (2002) Competing effects of buckling and anchorage strength on optimal wheat stem geometry. J Biomech Engr (in press)

Farquhar T, Zhou J, Meyer H (2002) Rht1 dwarfing gene selectively decreases the material stiffness of wheat. J Biomech (in press)

Finnigan JJ (1979) Turbulence in waving wheat. Mean statistics and Honami. Bound Layer Meteo 16: 181–236

Fischer RA, Stapper M (1987) Lodging effects on high-yielding crops of irrigated semidwarf wheat. Field Crops Research 17: 245–258

Fritig B, Legrand M (1993) Mechanisms of plant defense responses. Kluwer Acad Publ Boston MA

Gent MPN, Kiyomoto RK (1998) Physiological and agronomic consequences of Rht genes in wheat. In: Amarjit S. Basra (ed) Crop Science: Recent Advances. Hawthorne Press Binghampton NY

Inoue E (1955) Studies of the phenomenon of waving plants (Honami) caused by wind. J Agric Meteo (Japan) 11: 18–22

Jaffe MJ, Forbes S (1993) Thigmomorphogenesis: the effect of mechanical perturbation on plants. Plant Growth Regulation 12: 313–324

Mitchell C (1996) Recent advances in plant response to mechanical stress. HortScien 31: 31–35

Shaw RH (1985) Gust penetration into plant canopies. Atmos Envir 5: 827–830

Speck T, Spatz HC, Vogellehner D (1990) Capabilities of plant stems with strengthening elements of different cross-sections against weight and wind forces. Botanica Acta, 103: 111–122

Zebrowski J (1999) Dynamic behavior of inflorescence bearing triticale and triticum stems. Planta 207: 410–417

Visual Sensors and Vision

16. From Fly Vision to Robot Vision: Re-Construction as a Mode of Discovery

Nicolas Franceschini

Abstract

This chapter addresses basic issues on how vision links up with action and guides locomotion in biological and artificial creatures. The thorough knowledge gained over the past five decades on insects' sensory-motor abilities and the neuronal substrates involved has provided us with a rich source of inspiration for designing tomorrow's self-guided vehicles and micro-vehicles, which will be able to cope with unforeseen events on the ground, under water, in the air, in space, on other planets, and inside the human body. Insects can teach us some shortcuts to designing agile autonomous robots. At the same time, constructing these 'biorobots' based on specific biological principles gives us a unique opportunity of checking the soundness and robustness of these principles by bringing them face to face with the real physical world. Here we describe the visually guided terrestrial and aerial robots we have developed on the basis of our biological findings. Their architecture is akin to that of biological systems in spirit, and so is their parallel and analog mode of signal processing. As we learn more about signal processing and sensory-motor integration in nervous systems, we may eventually be able to design even better machines and micromachines than those which Nature has to offer. The millions of insect species constitute a gigantic untapped reservoir of ideas for highly sophisticated sensors, actuators and control systems.

I. Introduction

Animals and Humans are all able to move about autonomously in complex environments. These natural 'vehicles' (Braitenberg 1984) provide us with eloquent proof that physical solutions to elusive problems such as those involved in visually-guided behavior existed long before roboticists started tackling these problems in the 20th century. Over the past few years, some research scientists have been attempting to tap biology for ideas as to how to design smart visually-guided vehicles (Braitenberg 1984, Horridge 1987, Pichon et al. 1989, Maes 1991, Coombs and Robert 1992, Franceschini et al. 1992, Barlow et al. 1993, Cliff et al. 1994, Duchon and Warren 1994, Martin and Franceschini 1994, Mura and Franceschini 1994, Santos-Victor et al. 1995, Franceschini 1996, Mura and Franceschini 1996a,b, Huber and Bülthoff 1997, Srinivasan and Venkatesh 1997, Arkin 1998, Mura and Shimoyama 1998, Brooks 1999, Lewis and Arbib 1999,

Netter and Franceschini 1999, 2002, Viollet and Franceschini 1999, 2001, Chang and Gaudiano 2000, Harrison and Koch 2000, Iida and Lambrinos 2000, Möller 2000, Rind et al. 2000, Ichikawa et al. 2001, Neumann and Bülthoff 2001, Webb and Consi 2001, Ayers et al. 2002, Ruffier et al. 2003). Some authors have used both the principles and the details of biological signal processing systems to produce mobile seeing machines. Many innovations owe their existence to arthropods, particularly insects, which were largely dismissed in the past as being dumb invertebrates just able to make stereotyped manoeuvres. Insects have a rich behavioural repertoire, however, and they can teach us how to cope with complex, unpredictable environments using quite limited processing resources. Flying insects, in particular, have developed widely and account for about three quarters of all animal species. They often attain a level of skill, agility and circuit miniaturization which greatly outperforms that of both vertebrate animals and present day mobile robots. Insects' sensory-motor control systems are admirable feats of integrated optronics, neuronics and micromechatronics. Their neural circuits are highly complex, in keeping with the complex behaviour they serve, but unlike the (no less complex) vertebrate neural circuits, they can often be investigated by looking at single, *uniquely identifiable* neurons, i.e., neurons that can be reliably identified in all the individuals of the species on the basis of their location in the ganglion and their consistent electrical responses (Strausfeld 1976, Hoyle 1977, Burrows 1996).

The biologically inspired robots that we have been building since 1985 have contributed to creating the field of *Biorobotics*, in which natural principles or systems are implemented into physical hardware models (Webb and Consi 2001). In return, these models enable us to subject biological hypotheses to a rigorous test in the real world. The robots we are putting together at present are still far from the sophisticated industrial 'microsystems', which benefit from the latest collective micromanufacturing technologies. Building a complete sensory-motor control system on a chip the size of a fly-brain is still beyond our grasp, but the control systems of our robots do make use of various microsystems. In addition, their processing architecture resembles that of their biological counterparts and therefore departs considerably from the mainstream Artificial Intelligence approach to mobile robotics (software seems to be absent from animal brains: their 'intelligence' lies primarily in the layout of the *ad-hoc* adaptive analog circuits with which they are equipped). Our robots can be said to be in line with the principles of 'neuromorphic engineering' (Douglas et al. 1995) because they rely on biologically based, parallel, analog and asynchronous processing systems. They also rely heavily on *discrete* analog electronic components (Surface Mounted Devices, SMDs). Although this technology does not lead itself to the same degree of miniaturization as *analog VLSI* (Very Large Scale Integration) technology (Mead 1989, Vittoz 1994), it has several unique advantages as a means of testing biology-based principles in a physiological laboratory and achieving fast iterations between neurophysiological experiments and new robot designs and tests: low cost, low power, rapid prototyping, fast cycling between trials, no dependence on silicon brokers, opportunities for circuit tuning and component matching, etc. This approach has consistently proved to be most convenient for developing terrestrial and aerial robots endowed with insect-inspired visuomotor control systems. It is now proposed to describe some of these systems after briefly outlining some important aspects of insect vision and motion perception.

II. Fly Visual Microcircuitry

Our own laboratory pet is the fly, which belongs to the best known of all insect species. Flies are agile seeing creatures that are able to navigate swiftly through the most unpredictable environments, avoiding all obstacles without any need for sonars or laser range-finders. They process their sensory signals onboard and are not even tethered to a super-computer and an external power supply. The housefly is modestly equipped with about a million neurons (i.e., roughly 0.001% of the number of neurons present in the human brain) and views the world through its two panoramic compound eyes with only 3000 pixels each (i.e., roughly 1000 times less than a conventional digital camera and 40,000 times less than a human eye). Flies, which can be so objectionable in many ways, now also put us to shame by showing that it is definitely possible to achieve the smartest sensory-motor behavior using modest processing resources.

The front end of an insect visual system consists of a mosaic of facet lenslets (Fig. 1) and an underlying layer of photoreceptor cells which form the 'retina'. The fly seems to possess one of the best organized retinae in the animal kingdom. The fly retina has been described in exceptionally great detail, with its six different spectral types of cells, polarization sensitive cells, and sexually dimorphic cells (Franceschini 1984, Hardie 1985).

Flying insects avoid colliding with obstacles (see also chapter 17, this volume) and manage to guide themselves gracefully through their complex surroundings by processing the 'optic flow'. The optic flow field is a vector field that gives the *retinal slip speed*, based on the magnitude and the direction of each contrasting object encountered in the environment when the animal is moving and/or when something moves in the surroundings (Gibson 1958, Lee 1970, Koenderink 1986). Even when a fly is travelling through a stationary environment, the resulting optic flow field will be complex, except under special conditions such as pure translation or pure rotation. Current evidence shows that insects are able to perform the complex task of extracting the information necessary for short range navigation from the optical flow field (Collett 1978, Goulet and Campan 1981, Wehner 1981, Wagner 1982, Buchner 1984, Lehrer et al. 1988, Kirchner and Srinivasan 1989, Miles and Wallman 1993). This ability results from the fact that the visual system is equipped with smart sensors called 'motion detecting neurons', which are able to gauge the relative motion between the animal and the contrasting features of the environment.

The fly is one of the best animal models currently available for studies on motion perception (Götz 1969, Reichardt and Poggio 1976, Buchner 1984, Hausen 1984, Riehle and Franceschini 1984, Franceschini et al. 1989, Hausen and Egelhaaf 1989, Strausfeld 1989, Krapp et al. 1998). A great deal has already been learned from neuroanatomical and neurophysiological studies on the 3rd optic ganglion or *lobula plate*. This region is dedicated to (i) analysing the movement of the retinal image, i.e., the optic flow field generated when the animal is walking or flying, and (ii) transmitting the result via descending neurons to the motor neurons controlling the wing-, leg-, and head-muscles (Strausfeld 1976, 1989, Hausen 1984, Hausen and Egelhaaf 1989, Krapp et al. 1998). The 60 identifiable neurons in the *lobula plate* are collator neurons driven by numerous 'Elementary Motion Detectors' (EMDs), each of which has been assigned to a smallfield columnar neuron projecting retinotopically from the upstream ganglion, the *medulla* (Douglass and Strausfeld 1996).

Fig. 1. Head of the fly *Calliphora erythrocephala* (male) showing the two prominent panoramic compound eyes with their facetted cornea. There are as many sampling directions (pixels) in each eye as there are facets. This photograph was taken with a laboratory made Lieberkühn microscope. On the left is a photograph of the sensory-motor circuit that was developed for the robofly shown in Fig. 2a

In order to elucidate the functional principles underlying motion detection, we adopted a fairly direct approach that consisted in focusing on a single EMD in the eye of the live insect. Microelectrode recordings were performed on an individual collator neuron (H_1) in the *lobula plate* of the housefly while applying optical stimuli to single identified photoreceptor cells on the retinal mosaic (Franceschini et al. 1989). Pinpoint stimulation was applied to the minute photoreceptors (diameter $1 \, \mu m$) by means of a special instrument (a hybrid between a microscope and a telescope developed at the laboratory) in which the main objective lens was simply one of the eye's facet lenses (diameter $25 \, \mu m$, focal length $50 \, \mu m$). This optical instrument served to select a given facet lenslet, to locate the group of 7 photoreceptor distal endings in its focal plane, and to illuminate two neighboring cells *successively*. This procedure, which simulated a micro-motion ('apparent motion') in the animal's visual field, caused a conspicuous burst of nerve impulses ('spikes') in the H_1 neuron, as long as the phase relationship between the two stimuli mimicked motion in the preferred direction, and gave hardly any response when the sequence mimicked motion in the opposite, null direction (Riehle and Franceschini 1984).

Many experiments of this kind were carried out on identified cells, in which carefully planned sequences of light steps and/or pulses were applied to the two receptors. In this way, we established the functional diagram of an Elementary

Fig. 2. Three of the visually-guided robots designed and constructed at the Laboratory on the basis of our biological knowledge about sensory-motor control in flies. (a) Robofly (in French: 'Robot-mouche') with a visual system composed of a compound eye (visible at half-height) for obstacle avoidance, and a target seeker (visible on top) for detecting the light source serving as a goal. This 12 kg three-wheeled robot, which was completed in 1991 (Blanes, 1991) is fully autonomous as regards its processing and power resources (Franceschini et al. 1992, 1997). Despite its small number (116) of pixels, this artificial creature can avoid obstacles at a relatively high speed (50 cm/s) by reacting to the optic flow generated by its own locomotion. It carries a set of miniature electronic velocity sensors (visible immediately above the compound eye), each of which is inspired by the fly Elementary Motion Detector (EMD) (Franceschini et al. 1986). (b) Electrically-powered rotorcraft equipped with a frontal ventral motion-sensing visual system which enables it to follow the terrain and jump over obstacles (photo: Goetgelück). This self-sustained 0.84 kg artificial creature is tethered to a light pantographic whirling arm that allows only three degrees of freedom. It can turn around a central pole at speeds up to 6 m/s and climbs or descends depending on what it sees (from Netter and Franceschini 1999, 2002). (c) Aerial mini-robot with a sensory system that relies not only on visual motion detection but also on a microscanning process inspired by that recently found to occur in flying flies (Franceschini and Chagneux 1997). In this miniature twin-engine robot (mass 100 grams) both vision and inertial sensing are used to detect and track a target (a dark edge or a bar) moving at speeds of up to 30°/s (Viollet and Franceschini 1999, 2001). It carries all its processing and power resources onboard and can operate for as long as one hour at a time. It locks visually onto its target much more accurately than what could be achieved based on the spatial sampling abilities of the eye (from Viollet and Franceschini 2001)

Motion Detector (EMD) and characterized its dynamics and its various nonlinearities (Franceschini 1985, Franceschini et al. 1989). The overall scheme we arrived at using single neuron recording procedures combined with single photoreceptor stimulation relies on *facilitatory interactions* (and not on correlation, contra Reichardt 1987) between neighboring pixels. It is on this biological basis that we then designed a miniature electronic velocity sensor (Blanes 1986, 1991, Franceschini et al. 1986), with which the robots described below were equipped (Fig. 2). Meanwhile, a very similar principle has been proposed in the form of a smart analog VLSI chip by C. Koch's group at CALTECH, who rediscovered the principle, independently, ten years later (Indiveri et al. 1996) and patented it recently after naming it the "facilitate and sample velocity sensor" (Sarpeshkar et al. 1998)

III. Fly-Inspired Visually-Guided Terrestrial Robots

By the end of the 80's, we had designed an autonomous robot that was able to guide itself on the basis of these biologically inspired EMDs. This 12-kg robofly ('le robot-mouche'*) (Fig. 2a) steers its course, avoiding any obstacles encountered on its way to the target (an electric light) at a relatively high speed (50 cm/s) (Pichon et al. 1989, Franceschini et al. 1992). We previously interpreted the jerky, zig-zag flight paths of flies (Collett and Land 1975, Wagner 1986) as resulting from a clever strategy serving to restrict the onboard processing of optic flow to its *translational* component – which is the component depending on the range to obstacles (Gibson 1958, Lee 1970, Koenderink 1986). In the same way,

our robofly proceeds by performing a series of purely translational steps ΔL (length 10 cm), lasting 200 ms each, during which it collects the relevant visual data from the contrasting environment. The translatory steps alternate with rapid, saccade-like rotations, each of which defines a new heading direction that depends strictly on the bearings of the obstacles detected. Vision is inhibited during rotation by a process akin to 'saccadic suppression'. By the end of each translational step, the whole EMD-array has drawn up a local map of obstacles. This map is expressed in polar coordinates in the robot's eye reference frame (which is also the body reference frame because the eye turns at one with the wheels). The next course to be steered is immediately given, generating an eye + body saccade in the new direction, and a new map of obstacles is formed during the next translation, completely obliterating the previous one. No stop occurs at the end of a translation step if no steering command is issued, i.e., if no obstacle has been detected by the EMD-array. The elementary translational steps are seamlessly connected due to the high speed of the parallel, analog mode of processing used. The result is a rather jerky, 'fly-like' trajectory. The robot skirts the obstacles before reaching the target, with the advantage that it adapts automatically to unexplored or changing environments – unlike many robots based on high-level Artificial Intelligence which have to devote large computational resources to planning a safe path before daring to make a move forward. Our robofly must actually move in order to be able to see. By moving, it creates a short-lived, running representation of space in an eye- (and body-) centered frame of reference.

The robofly actually views the world through a horizontal ring of facets – corresponding approximately to a horizontal slice through the fly head (Fig. 2a). Any

---

* The Robofly (in French: 'le Robot-mouche') has been exhibited since 1994 at the Cité des Sciences et de l'Industrie, La Villette, Paris (Explora: Ilot Informatique). One of its successors, constructed by Mura and Franceschini (1996a), is now on show at the International Bionic Exhibition, Technikmuseum, Prague, Czech Republic.

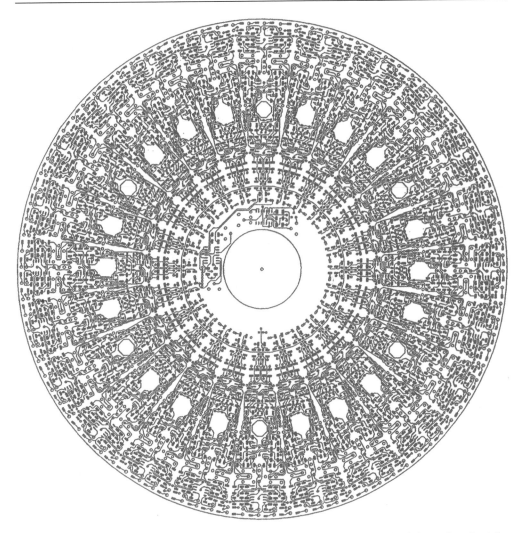

Fig. 3. Routing diagram of one face of the printed board (PCB) that integrates the information about the obstacles and the target onboard the robofly (Fig. 2a). This six-layered PCB has 210 parallel inputs (114 EMD inputs + 96 inputs from the target seeker) and a single output (near the centre). The latter gives (in Volts) the next steering angle required to reach the target while avoiding all obstacles. This side and the reverse side of the PCB are both covered with thousands of analog devices of only four kinds: resistors, capacitors, diodes, and operational amplifiers, some of which can be seen in Figure 1 (inset). The radial mosaic layout of this purely analog circuit is very different from the architecture of a von Neumann computer and is reminiscent of the neural architecture of visual areas in animal brains. This rose-window-like pattern results from the numerous repeat units and their retinotopic projections (from Franceschini et al. 1992)

two neighboring facets drive an EMD, and a total of 114 EMDs scan the optic flow field in the azimuthal plane. In addi-tion, we provided this compound eye with a *resolution gradient* such that the inter-receptor angle increases according to a sine law as a function of the eccentricity. This sine gradient compensates for the sine law inherent to the translational optic flow field (Gibson 1958, Whiteside and Samuel 1970, Nakayama and Loomis 1974, Pichon et al. 1989) and ensures that any contrasting feature will be detected with certainty if during the robot's translation by ΔL it enters the robot's 'circle of vision', whose radius is strictly depending on ΔL. Any more distant feature (which by definition will be a less dangerous obstacle) is automatically filtered out because it is most unlikely to cross the two visual axes of an EMD during the ΔL translation step. This morphological *non-uniformity* of the eye – which is reminiscent of the smooth resolution gradient across the fly's eye as observed with a telescope (Franceschini 1975) – therefore provides the robot with a kind of circular 'safety shell'. One additional advantage is that all the underlying EMDs can be built *uniformly*, each with the same time constants as its neighbors. This feature greatly simplifies both the engineering of a robot and the genetic specification of an animal nervous system.

The main reason for choosing a brain-like, parallel, analog and asynchronous mode of signal processing was the expectation that in practising nature's way of doing things, and in fighting on the same ground, we would learn more about the advantages, constraints and adaptability of this mode of 'computation', and understand how it enabled the winged insects to survive in their highly complex natural environment for 350 million years. We therefore decided to do without any von Neumann-type architecture onboard the robot, and to restrict the use of computers to the initial simulation phases in the project. Figure 3 shows the odd routing pattern connecting the thousands of analog devices used to blend together the input signals from the compound eye and those arising from the accessory target-seeking

eye, eventually delivering a single analog output: the steering angle, expressed in volts.

Simulation studies showed that a robot of this kind can not only drive around obstacles at a high speed but also automatically adjust its speed to the density of the obstacles present in the environment. This is because the robot's radius of vision turns out to be proportional to the speed – a unique and remarkable consequence of optic flow based navigation (Martin and Franceschini 1994).

IV. Fly-Inspired Visually-Guided Aerial Robots

Further simulations showed that the same motion detection principles can be used to guide a flying vehicle (Mura and Franceschini 1994), have it follow a rough terrain and land automatically (Netter and Franceschini 1999). The principle was validated onboard an experimental (tethered) miniature helicopter system having a single rotor with a variable collective pitch (Fig. 2b). This 0.85-kg rotorcraft has only 3 degrees of freedom. Mounted at the tip of a whirling arm, it lifts itself and pitches forward as the result of the aerodynamic moment created by an orientable servo-vane located in the propeller downwash. The robot reaches horizontal speeds of up to 6 m/s and climbing speeds of up to 2 m/s. It is equipped with an inertial system and a frontal ventral eye with only 20 pixels and their corresponding EMDs. Visually controlled terrain following is initiated by increasing the collective pitch in discrete steps as a function of the fused signals transmitted from the EMD array (Netter and Franceschini 2002). More recently, we showed that even with an optronic equipment reduced to its simplest form of expression (only two photoreceptors and one EMD capturing the optic flow downward) a (100-g) micro-air vehicle is able to perform some terrain following tasks robustly and reliably at 1–3 m/s (Ruffier et al. 2003).

In two other projects, we studied the advantages of a controlled retinal microscanning device such as the one we recently discovered in the fly's compound eye by performing single unit recordings on the flying animal (Franceschini and Chagneux 1997). Both projects involved a combination of microscanning and motion detection processes, both of which were based on biological findings. The first project resulted in a 0.7-kg wheeled Cyclopean robot which is able to move about in a square arena under its own visual control, avoiding the four contrasting walls despite the very low resolution of its eye (only 24 pixels) (Mura and Franceschini 1996a). The essence of this surprising ability is the microscanning process which assists the robot in *detecting* any obstacles located close to the heading direction (i.e., near the frontal 'pole' of the optic flow field). The anterograde microscanning process amounts to periodically adding a known amount of (rotational) optic flow to the very small (translational) optic flow generated by these frontal obstacles. As the amount of added optic flow is known onboard, its value can be subsequently subtracted from the overall measured optic flow. The result corresponds to the actual (translational) optic flow – that which depends on the distance to the obstacles (Mura and Franceschini 1996b).

The second process that we developed on the basis of the fly's retinal microscanner is a very small twin-engine aerial robot (Viollet and Franceschini 1999). This device is tethered to a thin, 2-meter long wire secured to the ceiling of the laboratory and is free to adjust its yaw by driving its two propellers differentially (Fig. 2c). This tiny (100-g) robot can lock onto a nearby target (a dark edge or a bar) via its micro-scanning visual system and track it at angular speeds of up to 30%/s – which are similar to the maximum tracking speed of the human eye. Target localisation occurs regardless of the distance (up to 2.5 m) and contrast (down to 10%), with an accuracy that turns out to be 40 times finer than the interreceptor angle (*hyperacuity*) (Viollet and Franceschini 2001). This robot is also equipped with a rate gyro which shows how two sensory modalities, the visual and inertial modalities, can be combined in nested control loops to enhance both the stability and the dynamic performances of a micro air vehicle.

V. Discussion: From Animal Physiology to Robot Technology and Back

Since our robots were born and raised at a laboratory where we also perform animal experiments, they have benefited from the latest biological knowledge. But in return, these physical models were called upon to improve our knowledge of biology (see also Webb 2001, 2002). The more closely a robot mimics the funtion of specific neurons or neural architectures and their signal processing modes, the more likely it will be to provide neurophysiologists and neuroethologists with useful feedback. And this is one good reason why the parallel and analog mode of signal processing characteristic of animal sensory systems is worth adopting.

One might claim that computer simulations are all we need, and that taking the trouble to *construct* a real physical robot is a disproportionately difficult means of checking the soundness of ideas gleaned from nature. There is everything to be gained, however, from completing computer simulations with real physical simulations because:

– some theories and simulations that look fine on paper (or on the computer screen) do not stand up in the real-world or lack robustness, to such an extent that they would be useless onboard an animal or robot.

- at a given level of complexity, it can actually be easier and cheaper to use a real robot in a real environment than to attempt to simulate both the robot and the vagaries of the environment (see also Brooks 1999).
- by playing with a real robot, with its dynamics, nonlinearities, noisy circuits and imperfections, one can get a better picture of the clever tricks, shortcuts, and quick remedies that the living animal has come up with. The cheap solutions devised by Nature can be fathoms apart from what the 21st century scholars were expecting (see also Pfeiffer and Scheier 2001, Nachtigall 2002).

Our 're-construction' approach has not only provided numerous insights into the subtle interactions between visual inputs and motor outputs in animals and machines, but has also inspired a number of laboratory experiments. These early efforts have spawned a new field that could be called 'Biorobot Assisted Neuroscience' (BAN), in which artificial creatures incorporating sensors, actuators and control systems that faithfully mimic physiological mechanisms serve as tools to assist neurobiologists and neuroethologists in identifying and investigating worthwhile issues. This strategy is a useful way of obtaining interesting ideas and concepts from nature, checking the soundness of biological hypotheses and raising novel questions about the whys and wherefores of the (often highly enigmatic) sensors and neural processors mounted onboard natural creatures.

VI. Conclusion

We have described the terrestrial and aerial robots we have developed on the basis of our biological findings. Their architecture is akin to that of biological systems in spirit, and so is their parallel and analog mode of signal processing. These visually guided robots make use of self-generated optic flow to carry out the humble task of detecting, locating, avoiding or tracking environmental features. This approach fits a more general framework called 'active perception' (Bajcsy 1985), in that the *ad hoc* motion of a sensor (caused here by locomotion or by retinal microscanning) is used to constrain the sensory inputs so as to reduce the computational burden involved in perceptual tasks (Ballard 1991, Aloimonos 1993).

The biorobotic approach that we initiated in 1985 is a transdisciplinary approach which is fairly demanding in terms of time, money, and human resources. This approach has turned out, however, to be most rewarding because it can kill two flies with one stone:

- It can be used to implement a basic principle borrowed from nature and check its soundness, robustness and scaling on a real physical machine in a real physical environment. This approach can lead to designing novel devices and machines, particularly in the field of cheap sensory-motor control systems for autonomous vehicles and micro-vehicles, which can thus benefit from the million-century experience of biological evolution.
- It yields valuable feedback information in the fields of neurophysiology and neuroethology, as it urges us to look at sensory-motor systems from a new angle, sheds new light here and there, challenges widely accepted facts, suggests new experiments to be carried out on the animal, and raises new biological questions which might not have been thought of otherwise (because they were too subtle) or which may simply not have been addressed (because they were seemingly too naive).

The thorough knowledge gained over the past five decades on insect sensory-motor abilities and the neuronal substrates involved has provided us with a rich source of inspiration for designing tomorrow's self-guided vehicles and micro-vehicles, which will be able to cope with unforeseen events on the ground, under water, in the air, in space, on other planets, and inside the human body. Insect neural circuits, which can be analyzed at the level of single, uniquely identifiable neurons, can teach us some shortcuts to designing the nervous system of agile autonomous robots and micro-robots. It is time for us to realize that the millions of insect species constitute a gigantic untapped reservoir of ideas for sophisticated micro-sensors, micro-actuators and smart control systems. This makes insect neurophysiology and neuroethology highly promising subsections of Information Science and Technology for the new millennium.

Acknowledgments

I am very grateful to numerous colleagues who have worked at the Laboratory over the years for many stimulating discussions: C. Blanes, J.M. Pichon, N. Martin, F. Mura, T. Netter, S. Viollet, F. Ruffier, S. Amic and M. Boyron. The English manuscript was revised by J. Blanc. This research has been supported by CNRS (Life Sciences, Engineering Sciences, Cognitive Sciences and Microsystems Program), and by EC contracts (Codest, Esprit, TMR and IST-1999-29043).

References

Aloimonos Y (1993) Active Perception. Lawrence Erlbaum, Hillsdale, USA

Arkin R (1998) Behavior-based Robotics. MIT Press, Cambridge, USA

Ayers J, Davis JL, Rudolph A (2002) Neurotechnolgy for Biomimetic Robots. MIT Press, Cambridge, USA

Bajcsy R (1985) Active perception versus passive perception. In: Proc. 3rd IEEE Workshop on Computer Vision, pp 55–59

Ballard D (1991) Animate vision. Artificial Intelligence 48: 57–86

Barlow HB, Frisby JP, Horridge AH, Jeaves M (eds) (1993) Natural and Artificial Low-level Seeing Systems. Clarendon Press, Oxford

Blanes C (1986) Appareil visuel élémentaire pour la navigation à vue d'un robot mobile autonome. DEA thesis (Neurosciences). Univ. Aix-Marseille

Blanes C (1991) Guidage visuel d'un robot mobile autonome d'inspiration bionique. Dr Thesis, National Polytechnic Institute, Grenoble

Braitenberg V (1984) Vehicles. MIT Press, Cambridge, USA

Brooks RA (1999) Cambrian Intelligence. MIT Press, Cambridge, USA

Buchner E (1984) Behavioral analysis of spatial vision in insects, In: Ali M (ed) Photoreception and Vision in Invertebrates. Plenum, New York, pp 561–621

Burrows M (1996) The Neurobiology of an Insect Brain. Oxford Univ Press, Oxford

Chang C, Gaudiano P (2000) Biomimetic Robotics (special issue on). Robotics and Autonomous Systems, 30

Cliff D, Husbands P, Meyer JA, Wilson SW (1994) From animals to animats III. In: Proc Intern Conf on Simulation of Adaptive Behavior. MIT Press, Cambridge, USA

Collett T, Land M (1975) Visual control of flight behaviour in the hoverfly Syritta Pipiens L. J Comp Physiol A 99: 1–66

Collett TS (1978) Peering: a locust behaviour pattern for obtaining motion parallax information. J Exp Biol 76: 237–241

Coombs D, Roberts K (1992) Bee-Bot: Using the peripheral optic flow to avoid obstacles. In: Intelligent Robots and Computer Vision XI, SPIE 1835, Bellingham, USA pp 714–725

Douglas R, Mahowald M, Mead C (1995) Neuromorphic engineering. Ann Rev Neurosci 18: 255–281

Douglass JK, Strausfeld NJ (1996) Visual motion-detecting circuits in flies: parallel direction- and non-direction-selective pathways between the medulla and lobula plate. J Neurosci 16: 4551–4562

Duchon AP, Warren WH (1994) Robot navigation from a Gibsonian viewpoint. IEEE Intern Conf On Syst, Man and Cybernetics, San Antonio, USA, IEEE Press, Los Alamitos, USA, pp 2272–2277

Franceschini N (1975) Sampling of the visual environment by the compound eye of the fly: fundamentals and applications. In: Snyder A,

Menzel R (eds) Photoreceptor Optics, Chap. 17, Springer, Berlin, pp 98–125

Franceschini N (1984) Chromatic organisation and sexual dimorphism of the fly retinal mosaic. In: Borsellino A, Cervetto L (eds) Photoreceptors, Plenum: New York, pp 319–350

Franceschini N (1985) Early processing of colour and motion in a mosaic visual system. Neurosci Res, Suppl 2: 17–49

Franceschini N (1996) Engineering applications of small brains. Future Electron Devices Journal, Suppl 7: 38–52

Franceschini N, Chagneux R (1997) Repetitive scanning in the fly compound eye. In: Elsner N, Wässle H (eds) Göttingen Neurobiology Rep, Georg Thieme, Stuttgart, 279

Franceschini N, Blanes C, Oufar L (1986) Passive noncontact velocity sensor (in French). Dossier Technique ANVAR/DVAR N° 51, 549, Paris

Franceschini N, Pichon JM, Blanes C (1992) From insect vision to robot vision. Phil Trans R Soc Lond B 337: 283–294

Franceschini N, Pichon JM, Blanes C (1997) Bionics of visuomotor control. In: Gomi T (ed) Evolutionary Robotics: From Intelligent Robots to Artificial Life. AAAI Books, Ottawa, Canada, pp 49–67

Franceschini N, Riehle A, Le Nestour A (1989) Directionally Selective Motion Detection by Insect Neurons. In: Stavenga DG, Hardie RC (eds) Facets of Vision, Berlin, Springer, Chap. 17, pp 360–390

Gibson JJ (1958) Visually controlled locomotion and visual orientation in animals. Brit J Psychol 49: 182–194

Götz KG (1969) Flight control in *Drosophila* by visual perception of motion. Kyb 4: 199–208

Goulet M, Campan R (1981) The visual perception of the relative distance in the wood cricket *Nemobius sylvestris*. Physiol Entomol 6: 357–387

Hardie RC (1985) Functional organization of the fly retina, In: Ottoson D (ed) Progress in Sensory Physiology 5, Springer, Berlin

Harrison R, Koch C (2000) A silicon implementation of the fly's optomotor control system. Neural Computation 12: 2291–2304

Hausen K (1984) The lobula complex of the fly: structure, function and significance in visual behaviour. In: Ali MA (ed), Photoreception and Vision in Invertebrates. Plenum, New York, pp. 523–559

Hausen K, Egelhaaf M (1989) Neural mechanisms of visual course control in insects. In:

Stavenga DG, Hardie RC (eds) Facets of Vision, Springer, Berlin, Chap. 18 pp 391–424

Horridge GA (1987) The evolution of visual processing and the construction of seeing systems. Proc R Soc Lond B 230: 279–292

Hoyle G (1977) Identified Neurons and Behavior of Arthropods. Plenum, New York

Huber SA, Bülthoff HH (1997) Modeling obstacle avoidance behavior of flies using an adaptive autonomous agent. Proc 7th Int Conf Artif Neural Networks, ICANN 97, Springer, Berlin, pp 709–714

Ichikawa M, Yamada H, Takeuchi J (2001) Flying robot with biologically inspired vision. J Robotics and Mechatronics 6: 621–624

Iida F, Lambrinos D (2000) Navigation in an autonomous flying robot by using a biologically inspired visual odometer. In: McKee GT, Schenker PS (eds) SPIE, Vol. 4196, Sensor Fusion and Decentralized Control in Robotic Systems III

Indiveri G, Kramer J, Koch C (1996) System implementations of analog VLSI velocity sensors. IEEE Micro 16: 40–49

Kirchner WH, Srinivasan MV (1989) Freely flying honeybees use image motion to estimate distance. Naturwissenschaffen 76: 281–282

Koenderink JJ (1986) Optic flow. Vis Res 26: 161–180

Krapp H, Hengstenberg B, Hengstenberg R (1998) Dendritic structure and receptive-field organisation of optic flow processing interneurons in the fly. J Neurophysiol 79: 1902–1917

Lambrinos D, Möller R, Labhart T, Pfeifer R, Wehner R (2000) A mobile robot employing insect strategies for navigation. Robotics and Autonomous Systems 30: 39–64

Lee DN (1970) The optical flow field: the foundation of vision. Phil Trans R Soc Lond B 290: 169–179

Lehrer M Srinivasan MV, Zhang SW, Horridge GA (1988) Motion cues provide the bee's visual world with a third dimension. Nature 332: 356–357

Lewis MA, Arbib M (1999) Biomorphic Robots (Special issue on). Autonomous robots 77

Maes P (1991) Designing Autonomous Agents: Theory and Practice from Biology to Engineering and Back. MIT Press, Cambridge, USA

Martin N, Franceschini N (1994) Obstacle avoidance and speed control in a mobile vehicle equipped with a compound eye. In: Masaki I (ed) Intelligent Vehicles, MIT Press, Cambridge, USA, pp 318–386

Mead CA (1989) Analog VLSI and Neural Systems. Addisson-Wesley, Reading

Miles FA, Wallman J (1993) Visual Motion and its Role in the Stabilization of Gaze. Elsevier, Amsterdam

Möller R (2000) Insect visual homing strategies in a robot with analog processing. Biol Cyb 83: 231–243

Mura F, Franceschini N (1994) Visual control of altitude and speed in a flying agent. In: Cliff D, Husbands P, Meyer JA, Wilson SW (eds) From Animals to Animats, MIT Press, Cambridge, USA, pp 91–99

Mura F, Franceschini N (1996a) Obstacle avoidance in a terrestrial mobile robot provided with a scanning retina. In: Aoki M, Masaki I (eds) Intelligent Vehicles II, pp 47–52

Mura F, Franceschini N (1996b) Biologically inspired 'retinal scanning' enhances motion perception of a mobile robot. Proc 1st Europe-Asia Congress on Mechatronics, Vol. 3, Bourjault A, Hata S (eds) ENSM, Besançon, pp 934–940

Mura F, Shimoyama I (1998) Visual guidance of a small mobile robot using active, biologically-inspired eye movements. In: Proc IEEE Intern Conf Rob Automation 3: 1859–1864

Nachtigall W (2002) Bionik, 2nd edn. Springer, Berlin

Nakayama K, Loomis JM (1974) Optical velocity patterns, velocity sensitive neurons and space perception: a hypothesis. Perception 3: 63–80

Netter T, Franceschini N (1999) Neuromorphic optical flow sensing for nap-of-the-earth flight. In: Mobile robots XIV, SPIE Vol. 3838, Bellingham, USA, pp 208–216

Netter T, Franceschini N (2002) A robotic aircraft that follows terrain using a neuromorphic eye. In: Intelligent Robots and Systems, Proc IROS-2002, EPFL, Lausanne, pp 129–134

Neumann TR, Bülthoff HH (2001) Insect inspired visual control of translatory flight. Proc Europ Conference on Artificial Life, ECAL 2001, Springer, Berlin, pp 627–636

Pfeiffer R, Scheier C (2001) Understanding Intelligence. MIT Press, Cambridge, USA

Pichon JM, Blanes C, Franceschini N (1989) Visual guidance of a mobile robot equipped with a network of self-motion sensors. In: Wolfe WJ, Chun WH (eds) Mobile Robots IV. Proc SPIE I195, Bellingham, USA, pp 44–53

Reichardt W (1987) Evaluation of optical motion information by movement detectors J Comp Physiol A 161: 533–547

Reichardt W, Poggio T (1976) Visual control of orientation behaviour in the fly, Part I: A quantitative analysis. Q Rev Biophys 9: 311–375

Riehle A, Franceschini N (1984) Motion detection in flies: parametric control over ON-OFF pathways Exp Br Res 54: 390–394

Rind FC, Blanchard M, Verschure P (2000) Collision avoidance in a robot using looming detectors from a locust. In: McKee GT, Schenker PS (eds) SPIE Vol. 4196: Sensor Fusion and Decentralized Control in Robotic Systems. Bellingham

Ruffier F, Viollet S, Amic S, Franceschini N (2003) Bio-inspired optical flow circuits for the visual guidance of micro-air vehicles. Proc Intern Symp on Circuits and Systems (ISCAS 2003), Bangkok, Thailand (in press)

Santos-Victor J, Sandini G, Curotto F, Garibaldi S (1995) Divergent stereo for robot navigation: a step forward to a robot bee. Int J Comp Vision 14: 159

Sarpeshkar R, Kramer J, Koch C (1998) Pulse Domain Neuromorphic Circuit for Computing Motion. United States Patent Nb 5,78,648

Srinivasan M, Venkatesh S (1997) From living eyes to seeing machines. Oxford Univ Press, Oxford

Srinivasan MV, Chahl JS, Weber K, Venkatesh S (1999) Robot navigation inspired by principle of insect vision. Robotics and Autonomous Systems 26: 203–216

Stavenga DG, Hardie RC (eds) (1989) Facets of Vision. Springer, Berlin

Strausfeld NJ (1976) Atlas of an Insect Brain. Springer, Berlin

Strausfeld NJ (1989) Beneath the compound eye: neuroanatomical analysis and physiological correlates in the study of insect vision. In: Stavenga DG, Hardie RC (eds) Facets of Vision. Springer, Berlin, Chap. 16: 317–359

Viollet S, Franceschini N (1999) Visual servo-system based on a biologically-inspired scanning sensor, In: Sensor Fusion and Decentralized Control II, SPIE Vol. 3839, Bellingham, USA, pp 144–155

Viollet S, Franceschini N (2001) Superaccurate visual control of an aerial minirobot. In: Rückert U, Sitte J, Witkowski U (eds) Autonomous Minirobots for Research and Edutainment. Heinz Nixdorf Institut, Paderborn, Germany, pp 215–224

Vittoz E (1994) Analog VLSI signal processing: why, where and how? J VLSI Signal Proc 8: 27–44

Wagner H (1982) Flow-field variables trigger landing in flies. Nature 297: 147–148

Wagner H (1986) Flight performance and visual control of flight of the free-flying housefly *Musca domestica*, I/II/III. Phil Trans R Soc Lond B 312: 527–600

Webb B (2001) Can robots make good models of biological behavior? Behav Brain Sci 24: 6

Webb B (2002) Robots in invertebrate neuroscience. Nature 417: 359–363

Webb B, Consi T (2001) Biorobotics. MIT Press, Cambridge, USA

Wehner R (1981) Spatial Vision in Arthropods. In: Autrum HJ (ed) Handbook Sens Physiol, Vol. VII/6C, Springer, Berlin, pp 288–616

Whiteside TC, Samuel DG (1970) Blur zone. Nature 225: 94–95

17. Locust's Looming Detectors for Robot Sensors

F. Claire Rind, Roger D. Santer, J. Mark Blanchard, and Paul F.M.J. Verschure

Abstract

Visual systems in the animal kingdom are incredibly good at extracting useful information from what can often be a very complicated world. Many of these systems can provide inspiration for the design of our own 'seeing machines' which we can then use in a variety of applications. Our own research is concerned with the detection of 'looming' or motion in depth. Our biological inspiration is the locust, *Locusta migratoria*, which possesses two uniquely identifiable neurons (the LGMD and DCMD) that respond preferentially to movements directly towards the animal. The way in which these cells are able to identify such stimuli is now becoming well understood. As such, we have been able to create a plausible computational model of the afferent inputs to these neurons that has been shown to respond in a locust-like way to looming stimuli. This model is now being used to control the movements of a mobile robot within a simplified visual environment. We aim to continue the development of this model so that it may one day function within the same visual world as the locust itself.

I. Introduction

The visual world in which we live provides a wealth of information to those organisms with an appropriate visual system to exploit the various cues. However, many robotic 'seeing machines' are unable to quickly, easily and cheaply extract this information (Indiveri and Douglas 2000), instead relying on other sensory media such as infra-red or sonar. Indiveri and Douglas (2000) suggest that instead of developing a single powerful general-purpose machine vision system, we should explore the use of multiple special-purpose vision sensors to tackle these shortcomings.

Our own research is concerned with one such 'special-purpose' system, the detection of 'looming' or movement in depth. As a model we use the locust, *Locusta migratoria*, whose own 'looming detector', the DCMD (descending contralateral movement detector) neuron, is known in some detail. This cell, and its presynaptic partner, the lobula giant movement detector (LGMD), respond to approaching objects with a rapid train of spikes (Schlotterer 1977, Rind and Simmons 1992, Simmons and Rind 1992, Hatsopoulos et al. 1995). Neither cell responds in this way to receding or translating objects. We have incorporated elements of the locust system into the control structure of a mobile robot which is now able to extract looming cues from a simplified visual environment and thus avoid potential collision with a reasonable degree of success.

A. Why Study Locusts?

The locust is a relatively simple animal possessing only a fraction of the number of neurons which we and other more complex organisms possess. Yet it is able to operate within the same visual environment, meeting the challenges set with only a million or so nerve cells. In addition, insect eyes are particularly good at seeing rapid motion rather than static detail making them ideal as inspiration for the special-purpose visual sensors previously alluded to. Most significantly though, the locust neurobiologist is equipped with a wealth of literature focusing on the animal's looming detector, the DCMD. The locust may be unique among insects in possessing such a large and accessible collision-detecting cell. One reason for this may be the animal's lifestyle; it is able to escape from aerial predators and fly long distances in dense swarms. Doubtless such demands would favour the evolution of effective collision detectors through natural selection.

B. Looming

Before discussing the neural mechanism behind the DCMD, and our attempts to model it, we should first qualify exactly what is meant by 'looming'. Looming is the motion of an object directly towards the observer and as such is one of the more valuable cues in the locust's visual environment. Such stimuli may include swooping aerial predators, other locusts on a collision course or even static objects in the insect's flight path. The benefits of being able to detect and avoid such stimuli are immediately apparent.

What then are the features of the visual environment which indicate such motion? As early as 1852 Wheatstone noted that a compelling illusion of motion in depth can be obtained by magnification of an image on the retina (Wheatstone 1852). Crucially though, as an object approaches the eye it appears to expand symmetrically and the rate at which it expands increases as the time of contact approaches. These are precisely the kinds of stimuli which excite the locust DCMD; it appears relatively uninterested in objects expanding at a linear rate. Thus if we are able to understand how the DCMD extracts these important cues we will have the necessary components for a biologically inspired collision detector.

II. A Brief Tour of the Locust Visual System

A. The Locust Eye

Insects possess compound eyes and these remarkable organs have fascinated biologists since the invention of the microscope. The surface of the eye is multi-faceted, which means that the visual environment is viewed through many thousand separate lenses, dividing the world into a regular mosaic. This structure limits the insect's spatial resolution to around 1° which is poor by vertebrate standards (Horridge 1978). However, these same eyes possess an incredible resolution in the temporal domain (Autrum 1952). To give examples, the fly retina can follow the flickering of a monitor screen up to a rate of 133 flashes/second. The bee can

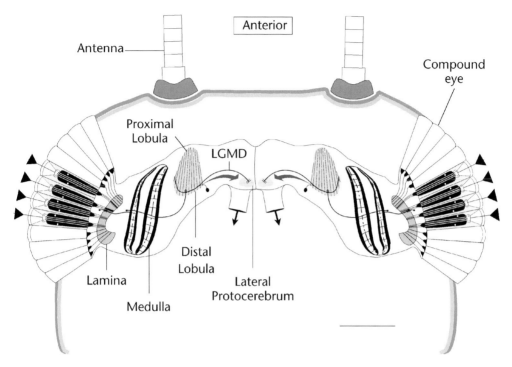

Fig. 1. A schematic diagram showing a horizontal section through the locust head to demonstrate the organisation of the visual system and the LGMD/DCMD pathway. Scale bar 1 mm

resolve more than 200 flashes/second. The advantage of such temporal resolution in collision detection should be apparent to anyone who has tried to swat a fly with a newspaper. Fortunately for those of us working on the locust system, its eye has a rather slower temporal resolution and can distinguish only around 60 flashes/second (Nilsson 1989).

Each lens of the compound eye represents the outer surface of a single ommatidial unit (see Fig. 1). A single insect eye may possess anywhere between 300 and 12,000 such units depending on the species. In the locust, the lens of each ommatidium focuses light onto 8 radially-arranged photoreceptor or retinula cells. These cells possess light sensitive pigments in finger-like microvilli which are fused down the ommatidium's central axis

into a single 'rhabdom' (Horridge 1978, Nilsson 1989). All 8 photoreceptors are concerned with the same region of the locust's visual field and send axons into the same cartridge in the underlying neuropile. This pattern of connectivity means that just as the faceted surface of the compound eye divides the world up, the retinotopic connections beneath preserve this arrangement. After all, a visual system which confused or jumbled the animal's retinotopic map of its environment would be of little functional use. As such, visual information is processed within separate, parallel channels throughout the optic lobe. The optic lobe itself is divided into three neuropiles, the lamina, medulla and lobula. Various kinds of processing occur in each of these but most relevant at this stage is the lobula. Here large

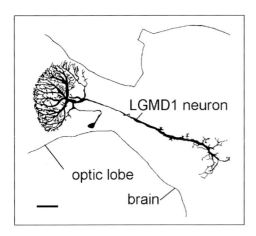

Fig. 2. The morphology of the LGMD1 neuron in the brain of the locust. The neuron has been intracellularly filled with dye and is viewed from behind. The scale bar represents 50 μm

fan-shaped neurons gather information from the entire visual field of the compound eye (see Fig. 2). One such cell is the well-studied LGMD (lobula giant movement detector) (O'Shea and Williams 1974, Rind 1996).

B. Collision Detecting Cells

The LGMD neuron projects to the brain where it makes synaptic contact with the DCMD (Rowell 1971, Rind 1984). The DCMD itself projects to the thorax where it synapses with a variety of motoneurons which are known to be involved in flying, jumping and walking (Burrows and Rowell 1973, Simmons 1980) – all behaviours elicited by a looming stimulus. Both the LGMD and DCMD characteristically respond to objects looming towards the eye with a train of very rapid spikes (Schlotterer 1977, Rind and Simmons 1992, Simmons and Rind 1992, Hatsopoulos et al. 1995). Both the frequency and number of these spikes is maximised if the looming object is approaching on a direct collision trajectory.

Deviations of just 1.8° from such a course will halve the response in terms of mean peak spike frequency in either cell (Judge and Rind 1997) (see Fig. 3).

In order to discover exactly how these cells are able to respond to looming objects they were challenged with a variety of stimuli, from clips from the STAR WARS movie to real and computer generated images. It seems selectivity for approach is achieved by responding to the increasing extent of the image, signalled by the growth of its edges, and its increasing rate of expansion on the retina (Simmons and Rind 1987). Using techniques ranging from electrophysiology to immunohistochemistry it has been possible to determine the arrangement of the LGMD's inputs which allow it to respond to such cues (Rind and Simmons 1998, Rind and Leitinger 2000). The LGMD receives excitatory inputs from a retinotopic array of small-field, spiking neurons excited transiently by changes in illumination (O'Shea and Rowell 1976). With

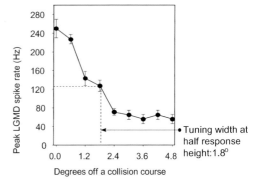

Fig. 3. Response of the locust DCMD neuron to objects approaching on collision or near-miss trajectories. Horizontal displacements away from a collision trajectory occurred in steps of 0.6° with a maximum displacement of 4.8°. Each displacement was presented six times. Simulation of object approach was via a computer monitor screen refreshed every 13.9 ms. Plot demonstrates results from a 30 × 30 mm object looming at a velocity of 2.5 m/s

repetitive stimulation synaptic transmission decrements between these afferents and the LGMD (O'Shea and Rowell 1975, O'Shea and Rowell 1976, Rowell et al. 1977). Evidence also exists for strong lateral inhibitory interactions between these afferents presynaptic to the LGMD (O'Shea and Rowell 1975, Pinter 1977, Edwards 1982, Rind and Simmons 1998) and it seems that conduction delays occur as this inhibition spreads laterally (Pinter 1977, Edwards 1982). Direct inhibitory input to the LGMD is also apparent, resulting from sudden or intense large-field stimuli such as a flash of light (Palka 1967). These inhibitory inputs have longer latencies than the excitatory PSPs evoked by the same instantaneous stimulus. Although these features have been known by neurobiologists for some time, they were not previously understood in the context of the LGMD and DCMD neuron's selectivity for approaching objects. For this reason the known circuitry was incorporated into a computational model to assess whether these alone were sufficient to explain the neuron's ability to detect looming stimuli.

III. A Computational Model

Gathering the available evidence, Rind and Bramwell (1996) were able to construct a 4-layered neural network which was thought to adequately represent the input organisation of the locust LGMD (see Fig. 4). Both the network and a mechanism for simulating looming stimuli were written in Borland Turbo C on a Research Machine PC with a 486, 33 MHz processor. Input to the model was via a series of computer generated virtual images of a moving object. These movements were controllable by the experimenter, thus allowing velocity, trajectory and even object size to be altered. The virtual image of the stimulus object was mapped onto the first layer of the LGMD network at each millisecond of simulated time.

The first layer consisted of 250 P-units, representative of the locust's photoreceptors. Each viewed a specific region of the visual field, separated from its neighbour by 3.3°. Although outside of the locust range (1.2–2.3° (Horridge 1978)), this angle was chosen to spread the field of view of all 250 'photoreceptors' in space. Each photoreceptor (P-unit) in the model responded with a brief (1 ms) excitation to a change in the level of illumination. In truth, these P-units are more accurately described as a composite between a photoreceptor and one or more postsynaptic neurons in which responses to light ON and OFF were processed to give the same signals (Autrum 1952). The excitation of these P-units was extremely transient and marked the passage of an edge with great precision. Interestingly, similar response time-courses have been observed in 'transient cells' in the locust medulla responding to light increments or decrements (James and Osorio 1996).

As in the locust optic lobe, processing is divided into channels each concerned with a specific region of the visual field and each processed separately from, but in an identical fashion to its neighbours. This is a crucial element in the wiring of any visual system and preserves the topographic map perceived by the array of photoreceptors. Account was taken of this in the computational model as each P-unit excited a single E-unit in the same retinotopic position as itself in the second layer of the network. As with other excitatory connections in this model, conduction delays were set to 0 ms. Thus excitation of any layer 2 E-unit would follow excitation of the P-unit feeding it unless it was within a 'refractory period' following excitation. Layer 2 of the network also included a second class of neuronal unit.

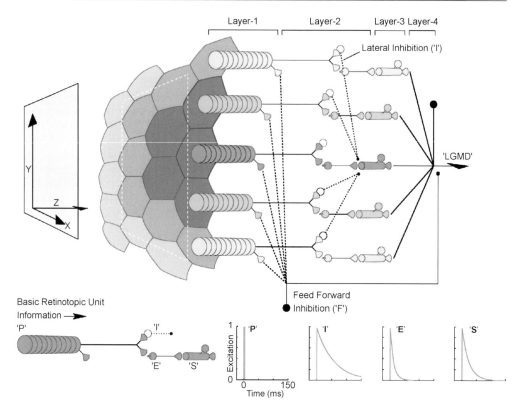

Fig. 4. A schematic representation of the Rind and Bramwell (1996) neural network model of the LGMD pathway. Layers 1–3 consist of arrays of 250 processing units each (see text for explanation). These converge onto a single 'LGMD' unit. The activation time-course of each retinotopic unit is shown in the bottom half of the figure. **Key: P-** photoreceptor unit (layer-1); **I-** laterally projecting inhibitory unit (layer-2); **E-** excitatory unit (layer-2); **F-** feed-forward inhibitory unit; **S-** summing unit (layer-3); and **LGMD-** final output unit (layer-4)

These are the inhibitory or I-units. As with the previously described E-units, each I-unit was excited by the P-unit in the same retinotopic position as itself. However, in contrast to the E-units, these 'cells' passed inhibition to the 6 neighbouring and 12 next-neighbouring processing channels with a conduction delay which could be set at between 1 and 4 ms. This inhibition was weighted so that each neighbouring channel would receive 1/6 of the total inhibition, and each next-neighbouring channel 1/12.

These excitatory and inhibitory inputs were collected and summed in layer 3 of the network by a series of retinotopically arranged S-units, one per processing channel. The E and I-unit inputs were summed linearly until a given threshold level of excitation was reached and a 'spike' was produced. Immediately following the spike voltage declined exponentially with time and was followed by a refractory period.

The final layer of the network (layer 4) consisted of a single LGMD-unit which received and linearly summed excitation from all active S-units and inhibition from a further cell, the inhibitory F-unit. This sum was expressed as a voltage.

The single F-unit constitutes a feed-forward pathway, bypassing layer 3 of the network. The F-unit was activated when a set number of P-units (normally 50) were simultaneously active. In response to activation the F-unit passed inhibition, delayed by 2–5 ms, directly to the LGMD-unit. Thus IPSPs in the LGMD could be triggered by wholefield stimuli. During each simulation a graphical display of activity in each layer of the network was available for analysis.

A. A Looming Detector?

Having modelled the locust-like network it would now be possible to test its suitability as a looming detector. The ability of the network to detect looming objects became quickly apparent (see Fig. 5), but studies of the network were also able to reveal exactly how the LGMD is able to detect collision.

The 'Critical Race' Hypothesis. The 'critical race' hypothesis was originally suggested by Simmons and Rind (1992) in their analysis of the LGMD's input connections. Studies of the computational network were able to highlight the importance of this theory. Simply put, as an object approaches the eye it appears to expand. This rate of expansion increases as the theoretical time of collision approaches. As this object expands it excites more and more P-units which in turn activate their respective processing channels. These maintain the retinotopic map of the image as viewed by the P-units and transfer it to the S-units in layer 3. However, at the same time the I-units from each excited channel are inhibiting their neighbours. As a result, just as excitation expands over the array of S-units, so too

does inhibition. Crucially inhibition from the I-units is limited by its synaptic delay (1–4 ms) whilst excitation is limited only by the rate of image expansion. Thus at the early stages of a loom the stimulus object expands rather slowly. As a result, excitation also expands slowly over the array of S-units and is effectively cancelled out by spreading lateral inhibition from the I-units. As a result, very little excitation is passed to the 'LGMD'.

Conversely, as the time of collision draws near the object, and the excitation it generates, expands more rapidly over the array of 250 S-units. As inhibition may only travel as fast as its synaptic delay permits, eventually there comes a point where it can no longer keep up with the spreading excitation. At this stage the S-units are strongly excited without the restrictive effects of I-unit mediated lateral inhibition and the 'LGMD' is strongly activated.

Feed-Forward Inhibition. It also appears that the F-unit plays a similarly crucial role in distinguishing approaching from receding objects. As this cell is only activated by a large number of transiently excited P-units being active simultaneously, the perceived image must change very rapidly in order to activate it. Such changes occur during wholefield light flashes but also at the end of a very rapid loom, or at the initiation of an object recession. As an object recedes it appears to shrink in size most rapidly at the start of movement (when it is closest to the observer). As a result, these early stages of movement activate the F-unit which in turn shuts down the LGMD-unit. Consequently the model LGMD, like its biological inspiration, gives only a brief response to a receding stimulus.

The F-unit is also able to shut down the 'LGMD' following a rapid loom. This may help to sharpen the cell's response so that it is only active whilst movement is

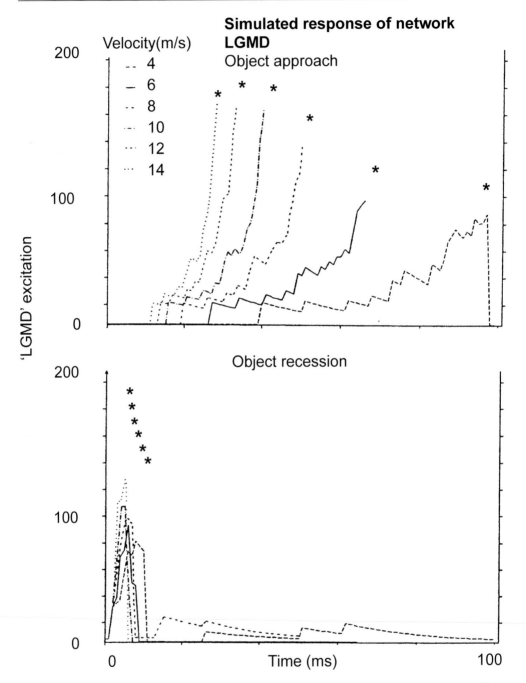

Fig. 5. The response of the model 'LGMD' to object motion. **A**: object approach on a direct collision trajectory. **B**: object recession along the same path. Velocities of 4–14 m/s were simulated and objects measured 75 × 75 mm in size. Initially the object is 500 mm removed from the 'eye' and approaches to within 100 mm. LGMD activity is given as a voltage and was plotted at 1 ms intervals throughout each simulation

occurring. This may prevent unnecessary avoidance behaviours being triggered.

IV. First Steps Towards A Locust-Inspired Collision Sensor

The previously described computational techniques have revealed the DCMD, and perhaps more relevantly in this context, our model of its input circuitry, as suitable for detecting looming objects. However, the major challenge is to produce a sensor capable of functioning in the real world. The real visual environment is dynamic in many respects. Light intensities change, shadows form, clouds move, all of which are not accurately represented by artificial looming stimuli. If indeed the locust circuitry were to be incorporated into a machine vision system, it must first be tested in this 'real world' environment.

In order to meet these aims the previously described model was realised using a more sophisticated software system, IQR$_4$21 (Verschure 1997). This software package was originally developed by Paul Verschure of the Institute of Neuroinformatics, Zürich. IQR$_4$21 is a computer program allowing the user to precisely model neural networks. Its chief advantage over earlier work is that it allows cells with realistic, neuronal properties to be modelled and these can be easily altered by the experimenter. It also permits the model network to be interfaced with a Khepera mobile robot (K-team, Lausanne, Switzerland).

Input to the model was via a miniature black and white CCD camera with a 68° field of view, mounted on the Khepera robot (see Fig. 6). Images from this camera were grabbed at each simulation timestep and relayed to the model network operating on a nearby PC via its serial port. At each timestep the camera's raw image was pre-processed to give a motion im-

Fig. 6. A Khepera mobile robot (K-team, Lausanne, Switzerland), fitted with a miniature black and white CCD camera

age. This was achieved by simply subtracting the previous timestep's image from the current one to give an 'absolute difference' image. This processed image was then mapped onto an array of 400 P-units (the number of units in each processing layer was increased in this incarnation of the model to give a better resolution of the visual environment).

In addition, rather than simply recording the model LGMD's output, it was incorporated into the control structure of the mobile robot. Thus in the absence of 'LGMD' activity the IQR$_4$21 program would instruct the robot to perform a simple exploratory behaviour (a forward movement). However, activity in the 'LGMD' above a threshold spike rate would override this exploratory behaviour and instigate a simple avoidance reaction (an anticlockwise turn). These commands were relayed to the Khepera robot via the PC's serial port. The robot was therefore able to detect and avoid potential collision using its 'LGMD' system.

In order to begin testing this model in the real world the robot was placed in a 33 cm × 40 cm arena surrounded by a white cardboard perimeter wall (see Fig. 7). Duplo blocks of various colours and contrasts were placed around the

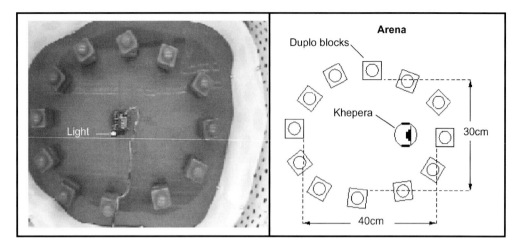

Fig. 7. The experimental environment used by Blanchard et al. (2000). **A:** The robot and its arena as seen from the overhead digital video camera used to track its behaviour during experiments. Information from this camera was fed into the TraX tracking system which automatically tracked the position of a small light mounted on the side of the robot. The robot is in the centre of its arena which is surrounded by Duplo block towers of varying colours and contrasts. **B:** Dimensions of the experimental arena. The floor of the arena is covered in clear Perspex and the perimeter wall is made of white cardboard

arena's edge for the network to detect and avoid. No specific lighting conditions were used to increase the realness of the test scenario. Although not a truly 'natural' environment, this arena represents an advance from previous computational stimuli. Within this environment the sensor must cope with shadow and contrast effects, both of which were absent in the preceding computational experiments. The robot's behaviour was tracked from above using a digital video camera in conjunction with the IQR$_4$21 TraX tracking system. This would allow both the robot's position within the arena and its 'LGMD' activity to be analysed off line.

The robot's behaviour was tested at a variety of speeds of movement ranging from 1.5–12.5 cm/s. Although very much slower than speeds encountered by the real animal, these were the highest practicable in such an arena. The robot was able to avoid collision on 91% of occasions at speeds of 2.5 cm/s, 81% at 5 cm/s and 88% at 10 cm/s (see Fig. 8).

The effectiveness of control via the 'LGMD' was significantly better than 50% at every speed tested (Blanchard et al. 2000). Although in its infancy, this initial study provides considerable encouragement for future research into the 'LGMD' as a possible collision sensor.

Current and Future Work

Real time testing of the 'LGMD' model using the IQR$_4$21 software system has also begun to reveal further interesting and locust-like properties inherent in the network. Testing of the network in a simplified arena with a single stimulus object suggests that the network shows a similar tuning to the locust for objects approaching on a direct collision course. In this series of experiments the robot travelled over a distance of 1800 mm towards a single 8 cm diameter black sphere. In this instance the robot's 'LGMD'-mediated avoidance reaction

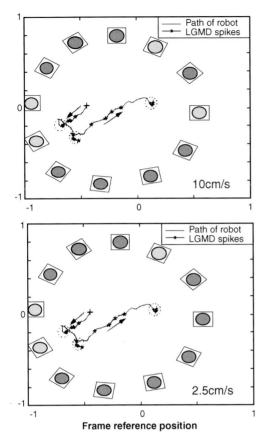

10cm/s

2.5cm/s

Frame reference position

Fig. 8. Robot collision avoidance under 'LGMD' control at 2.5 and 10 cm/s. Data were collected using the TraX system. Traces indicate the robot's path whilst performing 'LGMD'-mediated collision avoidance behaviour. 'LGMD' activity (*), the robot's starting position (+), and the robot's direction of movement (arrows), are also indicated. Dashed circles highlight 'LGMD'-mediated collision avoidance behaviours

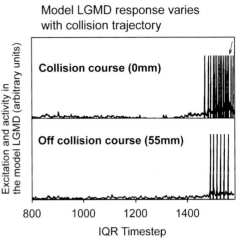

Model LGMD response varies with collision trajectory

Collision course (0mm)

Off collision course (55mm)

IQR Timestep

Fig. 9. The output of the model LGMD captured using IQR$_4$21's analysis tool. Here the simulated network was altered slightly from the previous version with the number of cells in each processing layer (P, E, I etc) increased from 400 to 1600. This produces a more locust-like resolution of the visual image. In this instance the robot travelled at 2.5 cm/s towards an 8 cm diameter dark sphere. The robot travelled 1800 mm in total although activity in the LGMD was only elicited over the final 250 mm or so. The upper plot shows the model's response on a direct collision course with the sphere. The lower trace illustrates the decrement in LGMD activity when this course deviates by just 1.75° from a collision course. Solid, vertical lines indicate spikes in the 'LGMD'. These are super-imposed on a jagged trace indicating excitation delivered to the 'LGMD' from the S units. A peak in excitation (arrow) occurs just prior to collision as a result of the looming sphere's edges leaving the robot's field of view

was disabled, meaning that it would travel towards the object regardless of visual input. The response of the model LGMD was recorded using the IQR$_4$21 analysis tool for approaches along a direct collision course and at various deviations from this approach. Results reveal that spikes in the 'LGMD' are more numerous and more frequent when the robot approaches its target on a collision trajectory compared with a non-collision trajectory (see Fig. 9). In addition, the mean number of spikes produced by the 'LGMD' declines as it deviates further from a direct collision course at speeds of 1.5 and 2.5 cm/s. Interestingly, at these speeds the model LGMD's 'tuning' is approximately 1.75° (using the same criteria as Judge and Rind 1997), which compares well with the value measured in the real LGMD (see Fig. 10).

Tuning for potential collision in the model LGMD

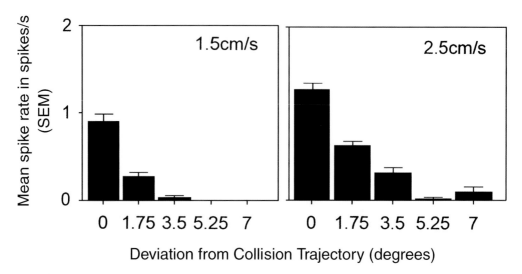

Fig. 10. Like the locust LGMD, the model also shows tuning for potential collision. In this case the robot performed no avoidance reaction and simply travelled towards an 8 cm diameter dark sphere positioned 1800 mm away. Trajectories of 0.00, 1.75, 3.50, 5.25 and 7.00° from a direct collision course were tested. The LGMD's response was recorded using the IQR_421 analysis tool. In each case $n = 6$ and bars indicate standard error. The unaltered network showed tuning for potential collision trajectories at 1.5 and 2.5 cm/s. The effect was less noticeable at higher approach speeds but could be improved by alteration of network properties, most notably synaptic delay (results not shown)

It is hoped that these results, and our continuing research into other properties inherent in the model system, will reveal a far more complete looming sensor than originally envisaged.

V. Conclusions

During the course of this brief chapter we have discussed the development of a looming sensor inspired by a well-known neuron in the locust, *Locusta migratoria*. Although still in its infancy, this project has shown considerable potential and, we believe, may eventually be used in applications as diverse as safety, robotics or vehicle design. But why should we study the locust at all?

Amazingly, the tight tuning that the DCMD shows for potentially colliding objects (3° mean half-width at half-height) actually rivals that recorded in similar cells in the pigeon (3.3° mean half-width at half-height (Wang and Frost 1992)). This, therefore, is clearly not the crude and unrefined system that those not working on invertebrate vision might expect. Why then should we not use such systems as inspiration for our own robotic designs? After all insects are a highly successful group whose nervous systems, far from being simple, are the product of millions of years of evolution. By comparison our own infrared sensors are a recent invention. In addition, by developing a battery of visual sensors detecting a variety of cues it may be possible to process a single image in a variety of ways. Thus producing a far more realistic and arguably useful vision system.

Finally, it would seem that by increasing the biological authenticity of our sensors we may be able to increase their efficiency (Harrison 2000). With this in mind we plan to continue work both on the locust itself, and on its model circuit. Hopefully this will lead ultimately to a complete, functional looming sensor for a variety of applications.

References

Autrum H (1952) Über zeitliches Auflösungsvermörgen und Primärvorgänge im Insektenauge. Naturwissenschaften 39: 290–297

Blanchard M, Rind FC, Verschure PFMJ (2000) Collision avoidance using a model of the locust LGMD neuron. Robot and Auton Syst 30: 17–38

Burrows M, Rowell CHF (1973) Connections between descending visual interneurons and metathoracic motoneurons in the locust. J Comp Physiol A 85: 221–234

Edwards DH (1982) The cockroach DCMD neurone. I. Lateral inhibition and the effects of light and dark adaptation. J Exp Biol 99: 61–90

Harrison RR (2000) An analog VLSI motion sensor based on the fly visual system. PhD Thesis, Pasadena, California Institute of Technology

Hatsopoulos N, Gabbiani F, Laurent G (1995) Elementary computation of object approach by a wide-field visual neuron. Science 270: 1000–1003

Horridge GA (1978) The separation of visual axes in apposition compound eyes. Phil Trans R Soc 285: 1–59

Indiveri G, Douglas R (2000) Neuromorphic vision sensors. Science 288: 1189–1190

James AC, Osorio D (1996) Characterisation of columnar neurons and visual signal processing in the medulla of the locust optic lobe by system identification techniques. J Comp Physiol A 178: 183–199

Judge SJ, Rind FC (1997) The locust DCMD, a movement-detecting neurone tightly tuned to collision trajectories. J Exp Biol 200: 2209–2216

Nilsson DE (1989) Optics and evolution of the compound eye. In: Stavenga DG, Hardie RC (eds) Facets of Vision. Springer-Verlag, Berlin

O'Shea M, Rowell CHF (1975) Protection from habituation by lateral inhibition. Nature 254: 53–55

O'Shea M, Rowell CHF (1976) Neuronal basis of a sensory analyzer, the acridid movement detector system. II. Response decrement, convergence, and the nature of the excitatory afferents to the fan-like dendrites of the LGMD. J Exp Biol 65: 289–308

O'Shea M, Williams JLD (1974) The anatomy and output connections of a locust visual interneurone: the lobula giant movement detector (LGMD) neurone. J Comp Physiol 91: 257–266

Palka J (1967) An inhibitory process influencing visual responses of a fibre in the ventral nerve chord of locusts. J Insect Physiol 13: 235–248

Pinter RB (1977) Visual discrimination between small objects and large textured backgrounds. Nature 270: 429–431

Rind FC (1984) A chemical synapse between two motion detecting neurons in the locust brain. J Exp Biol 110: 143–167

Rind FC (1996) Intracellular characterization of neurons in the locust brain signalling impending collision. J Neurophysiol 75: 986–995

Rind FC, Bramwell DI (1996) Neural network based on the input organisation of an identified neuron signaling impending collision. J Neurophysiol 75(3): 967–984

Rind FC, Leitinger G (2000) Immunocytochemical evidence that collision sensing neurons in the locust visual system contain acetylcholine. J Comp Neurol 423: 389–401

Rind FC, Simmons PJ (1992) Orthopteran DCMD neuron: A reevaluation of responses to moving objects. I. Selective responses to approaching objects. J Neurophysiol 68: 1654–1666

Rind FC, Simmons PJ (1998) Local circuit for the computation of object approach by an identified visual neuron in the locust. J Comp Neurol 395: 405–415

Rowell CHF (1971) The orthopteran descending movement detector (DMD) neurones: A characterisation and review. Z vergl Physiol 73: 167–194

Rowell CHF, O'Shea M, Williams JLD (1977) Neuronal basis of a sensory analyzer, the acridid movement detector system. IV. The preference for small field stimuli. J Exp Biol 68: 157–185

Schlotterer GR (1977) Response of the locust descending movement detector neuron to rapidly approaching and withdrawing visual stimuli. Canadian J Zool 55: 1372–1376

Simmons PJ (1980) Connexions between a movement-detecting visual interneurone and flight motoneurones of a locust. J Exp Biol 86: 87–97

Simmons PJ, Rind FC (1987) Responses to object approach by a wide field neurone, the LGMD2 neurone of the locust. J Comp Physiol A 180: 203–214

Simmons PJ, Rind FC (1992) Orthopteran DCMD neuron: a reevaluation of responses to moving objects. II. Critical cues for detecting approaching objects. J Neurophysiol 68: 1667–1682

Verschure PFMJ (1997) Xmorph: a software tool for the synthesis and analysis of neural systems. Technical report, Institute of Neuroinformatics, Zürich

Wang Y, Frost BJ (1992) Time to collision is signaled by neurons in the nucleus rotundus of pigeons. Nature 356: 236–238

Wheatstone C (1852) Contributions to the physiology of vision. II. Philos Trans R Soc Lond B 142: 1–18

18. Retina-Like Sensors: Motivations, Technology and Applications

Giulio Sandini and Giorgio Metta

Abstract

Retina-like visual sensors are characterized by space-variant resolution mimicking the distribution of photoreceptors of the human retina. These sensors, like our eyes, have a central part at highest possible resolution (called fovea) and a gradually decreasing resolution in the periphery. We will present a solid-state implementation of this concept. One attractive property of space-variant imaging is that it allows processing the whole image at frame rate while maintaining the same field of view of traditional rectangular sensors. The resolution is always maximal if the cameras are allowed to move and the fovea placed over the regions of interest. This is the case in robots with moving cameras. As an example of possible applications, we shall describe a robotic visual system exploiting two retina-like cameras and using vision to learn sensorimotor behaviors.

I. Introduction

This chapter describes the physical implementation in silicon of a biologically inspired (retina-like) visual sensor. It is important also to understand the context and the motivations for undertaking such a research. Over the past few years, we have studied how sensorimotor patterns are acquired in a complex system such as the human body. This has been carried out from the unique perspective of implementing the behaviors we wanted to study in artificial systems. The approach we followed is biologically motivated from at least three perspectives: i) morphology (sensors are shaped as close as possible to their biological counterparts); ii) physiology (control structures and processing are modeled after what is known about human perception and motor control); iii) development (the acquisition of those sensorimotor patterns follows the process of biological development in the first few years of life).

The goal has been that of understanding human sensorimotor coordination and cognition rather than building more efficient robots. In fact, some of the design choices might even be questionable on a purely engineering ground but they are pursued nonetheless because they improved the similarity with biological systems. Along this feat of understanding human cognition our group implemented various artifacts both in software (e.g. image processing, machine learning) and hardware (e.g. silicon implementation, robot heads).

The most part of the work has been carried out on a humanoid robot we call the *Babybot* (Metta et al. 1999, Metta 2000). It resembles the human body from the waist up although in a simplified form. It has twelve degrees of freedom overall distributed between the head, arm and torso. Its sensory system consists of cameras (eyes), gyroscopes (vestibular system), microphones (ears) and position sensors at the joints (proprioception). Babybot cannot grasp objects but it can touch and poke them around. Investigation touched aspects such as the integration of visual and inertial information (vestibulo-ocular reflex) (Panerai et al. 2000), and the interaction between vision and spatial hearing (Natale et al. 2002).

In this paper we will focus on the eyes of Babybot that mimic the distribution of photoreceptors of the human retina – we call these *retina-like cameras*. Apart from the purely computational aspects, they are best understood within the scientific framework of the study of biological systems. In our view, the retina-like camera truly represents such a thought-provoking mix of technological and biologically inspired work. Retina-like cameras have a non-uniform resolution with a high resolution central part called the *fovea* and a coarser resolution periphery.

As we hope to clarify in this chapter, the uniform resolution layout, common to commercial cameras, did not survive the evolutionary pressure. Evolution seems to be answering the question of how to optimally place a given number of photoreceptors over a finite small surface. Many different eyes evolved with the disposition of the photoreceptors adapted to the particular ecological niche. Examples of this diversity can be found in the eyes of insects (see for example Srinivasan and Venkatesh 1997 for a review) and in those of some birds which have two foveal regions to allow simultaneous flying and hunting (Galifret 1968, Blough 1979). There is clearly something to be earned by optimizing the placement of photosensitive elements. In a constrained problem with limited computational resources (i.e. the size of the brain), limited bandwidth (i.e. the diameter of the nerves delivering the visual information to the brain), and limited number of photoreceptors, nature managed to obtain a much higher acuity than what can be achieved with uniform resolution.

From the visual processing point of view, we asked on the one hand whether the morphology of the visual sensor facilitates particular sensorimotor coordination strategies, and on the other, how vision determines and shapes the acquisition of behaviors which are not necessarily purely visual in nature. Also in this case we must note that eyes and motor behaviors coevolved: it does not make sense to have a fovea if the eyes cannot be swiftly moved over possible regions of interest. Humans developed a sophisticated oculomotor apparatus which includes saccadic movements, smooth tracking, vergence, and various combinations of retinal and extra-retinal signals to maintain vision efficient in a wide variety of situations (see Carpenter 1988 for a review).

This addresses the question of why it might be worth copying from biology and which are the motivations for pursuing the realization of biologically inspired artifacts. How this has been done is presented in the following sections where we shall talk about the development of the retina-like camera. Examples of applications are also discussed in the field of image transmission and robotics. The image transmission problem is alike the limitation of bandwidth/size of the optic nerve mentioned above. The limitations in the case of autonomous robots are in terms of computational resources and power consumption.

II. Space-Variant Imaging

Babybot, shown in Fig. 1, relies on a pair of retina-like sensors for its visual processing. These sensors are characterized by a space-variant resolution mimicking the distribution of photoreceptors of the human retina. The density of photoreceptors is highest in the center (limited by the particular technology used) and decreases monotonically as the eccentricity – the distance of the photosite from the center of the sensory surface – increases. The resulting image is, consequently a compromise between resolution, amplitude of the field of view (FOV), and number of pixels. This space-variant imaging is unique because it enables high-resolution tasks

using the central region while maintaining the lower resolution periphery providing relevant information about the background. The arrangement is advantageous, for example, for target tracking: the wide peripheral part is useful for detection while the central part takes over during the tracking and performs with the highest accuracy.

Of all possible implementations of space-variant sensors what is described here is the so-called log-polar structure (Weiman and Chaikin 1979, Sandini and Tagliasco 1980, Schwartz 1980). The log-polar geometry models accurately the wiring of the photoreceptors from the retina to the geniculate body and the primary visual cortex (area V1). In this schema a constant number of photosites is arranged over concentric rings (the polar part of the representation) giving rise to a linear increase of the receptor's spacing with respect to the distance from the central point of the structure (the radius of the concentric rings). A possible implementation of this arrangement is shown in Fig. 2. Because of the polar structure

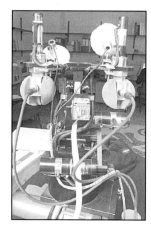

Fig. 1. The Babybot. Left: the complete setup. Middle: detail of the head, the tennis-like balls cover the eyes for esthetic purpose, the microphones and ear lobes are mounted on top of the head. Right: back view of the head showing the inertial sensors in its center

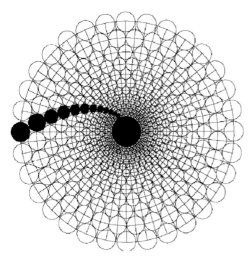

Fig. 2. Layout of receptor's placing for a log-polar structure composed of 12 rings with 32 pixels each. The pixels marked in black follow a logarithmic spiral

of the photosensitive array and the increasing size of pixels in the periphery, retina-like sensors, do not provide images with a standard topology. In formulas mapping from the retina (ρ, ϑ) into the cortical plane (η, ξ) accounts to:

$$\begin{cases} \eta = q \cdot \vartheta \\ \xi = \log_a \frac{\rho}{\rho_0} \end{cases} \quad (1)$$

where ρ_0 is the radius of the innermost circle, and $1/q$ is the minimum angular resolution. Equation (1) is easily understood by observing that the angular variable η is linearly mapped from its polar representation ϑ, and the eccentricity ρ is scaled logarithmically (with basis a). Equation (1) is related to the traditional rectangular coordinate system by:

$$\begin{cases} x = \rho \cos \vartheta \\ y = \rho \sin \vartheta \end{cases} \quad (2)$$

It is worth noting that the mapping is obtained at no computational cost, as it is a direct consequence of the arrangement of the photosites and the read-out sequence.

The topology of a log-polar image is shown in Fig. 3. Note that radial structures (the petals of the flower) correspond to horizontal structures in the log-polar image. In spite of this seemingly distorted image, the mapping is conformal and, consequently, any local operator used for standard images can be applied without changes (Weiman and Chaikin 1979).

In the Babybot, for example, images coming from the left and right channels were processed to recover the location, in retinal coordinates, of possibly interesting objects. Color is a good candidate for this purpose and it is extracted after mapping the RGB components into the Hue Saturation and Value (HSV) space. This is not normalized or scaled as for example in (Darrell et al. 2000), but proved to be enough for our experiments. The processing extracts and labels the most salient image regions. Regions are successively combined through a voting mechanism and finally the coordinates of the most "voted" one are selected as the position

Fig. 3. Left: space variant image obtained by remapping an image acquired by a log-polar sensor. Note the increase in pixels size with eccentricity. Right: Log-polar image acquired by a retina-like sensor. Horizontal lines in the log-polar image are mapped into rings to obtain the remapped image shown on the right

of the object to look at. This fusion procedure provides the attentional information to generate tracking behavior. The error – i.e. the distance of the projection of the target on the image plane from the center of the image – is used to trigger saccade-like movements or, in case of smoothly moving targets, as feedback signal for closed loop control.

The way the different visual cues are combined does not provide information about binocular disparity. In fact the procedure described above does not perform any stereo matching or correspondence search – regions are treated as 2D entities. For the purpose of measuring depth we used a different algorithm whose details can be found in (Manzotti et al. 2001). This, too, uses log-polar images and provides a measure of the global disparity, which is used to control vergence. Grossly simplifying, by employing a correlation measure, the algorithm evaluates the similarity of the left and right images for different horizontal shifts. It finally picks the shift relative to the maximum correlation as a measure of the binocular disparity. The log-polar geometry, in this case, weighs differently the pixels in the fovea with respect to those in the periphery. More importance is thus accorded to the object being tracked.

Positional information is important but for a few tasks optic flow is a better choice. One example of use of optic flow is for the dynamic control of vergence as in (Capurro et al. 1997). We implemented a log-polar version of a quite standard algorithm (Koenderink and Van Doorn 1991). According to the choice of the algorithm we defined the affine model as:

$$\begin{bmatrix} \dot{x} \\ \dot{y} \end{bmatrix} = \begin{bmatrix} u_0 \\ v_0 \end{bmatrix} + \begin{bmatrix} D + S_1 & S_2 - R \\ R + S_2 & D - S_1 \end{bmatrix} \cdot \begin{bmatrix} x \\ y \end{bmatrix}$$

(3)

where $\dot{x}$ and $\dot{y}$ are the optic flow, and x and y the image plane coordinates (with origin in the image center). Equation (3) depends on four quantities: translation, rotation, divergence and shear. The first two components u_0 and v_0 represent a rigid 2D translation, and D, R, S_1, S_2 are the four first-order vector field differential invariants: divergence, curl, and shear respectively. The details of the implementation can be found in (Tunley and Young 1994). The estimation of the optic flow requires taking into account the log-polar geometry because it involves non-local operations.

III. Technology of Solid State Log-Polar Sensors

Traditionally, the log-polar mapping has been obtained in two different ways: by means of electronic boards transforming, in real-time, standard images into log-polar ones or by building sensors with the photosites arranged according to the log-polar mapping. Electronic boards were employed first; they were used to generate log-polar images for real-time control and image compression (Rojer and Schwartz 1990, Weiman and Juday 1990, Engel et al. 1994, Wallace et al. 1994). The advantage is the use of standard off-the-shelf electronic components. The main disadvantage is the constraint introduced by the size of the original image limiting the potential advantages of the log-polar structure. This point will be clarified in the following sections. Whatever the approach used, the design of the sensor has to start from the technological limitations: the most important of which are the minimum pixel size and the maximum sensor size. Besides our realizations a few other attempts have been reported in the literature on the implementation of solid-state retina like sensors (Baron et al. 1994, 1995). So far we are not aware of any commercial device, besides those described here, that

have been realized based on log-polar retina-like sensors.

A. Parameters of Log-Polar Sensors

Starting from the technological constraints (minimum pixel size and size of
the sensor), the most important sensor's
parameter is the total number of pixels.
The total number of pixels is directly related to the amount of information acquired. For constant resolution devices
this parameter is fixed by the technological constraints in a simple way. In the
case of log-polar sensors the relationship
between these two parameters is not a
simple one (see Sandini and Tagliasco
1980, Weiman 1988, Wallace et al. 1994)
for more details). The second important
parameter, unique to log-polar sensors,
is the ratio between the largest and the
smallest pixels – we shall call it R.

For example, our first solid-state implementation was realized at the beginning
of the 90s using CCD technology (Van
der Spiegel et al. 1989). At that time, with
the technology available, the size of the
smallest possible pixel was about 30 μm
and for practical limitations the overall
sensor diameter was limited to 94 mm. A
picture of the layout is shown in Fig. 4.
This sensor is composed of 30 rings and
each ring is covered by 64 pixels. Alto-
gether the sensor had 2022 pixels, 1920
of which in the log-polar part of the sensor. In the CCD implementation R was
about 13.7 (the largest pixel was
412 μm). This parameter describes the
amount of space variance of the sensor
and is, of course, equal to 1 in standard
constant resolution sensors.

The third important parameter is the
ratio between size of the sensor and the
size of the smallest pixel. We shall call it
Q. The importance of Q can be understood by observing that its value is equal
to the size of a constant resolution image

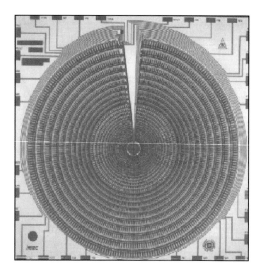

Fig. 4. Structure of the first log-polar sensor realized with CCD technology

with the same field of view and the same
maximum resolution of the corresponding
log-polar sensor. For example for the
CCD sensor shown in Fig. 4 Q is equal
to about 300 meaning that if we want to
electronically remap a constant resolution
image and obtain the same amount of information obtained from our log-polar
sensor, the original image must be at least
300 × 300 pixels.

B. CMOS Implementations

The CCD implementation described earlier, even if it was the first solid-state device of this kind in the world, had some
drawbacks mostly related to the use of
CCD technology itself. In our more recent
implementations, the CMOS technology
was used. A first version of the sensor
was realized using a 0.7 μm technology
allowing a minimum pixel size of 14 μm.
Later, improvement of technology enabled a further reduction of the minimum
pixel size to 7 μm. In fact, our most recent
CMOS implementation uses a 0.35 μm

technology. This sensor – realized at Tower in Israel – has been developed within a European Union-funded research project (SVAVISCA). The goal of the project was to build, beside the sensor, a microcamera with a special-purpose lens with a 140 degree field of view. The miniaturization of the camera is now possible because some of the electronics required to drive the sensor as well as the analog-to-digital converter is included on the chip itself. A picture of part of the layout of the sensor is shown in Fig. 5. Table 1 summarizes the main parameters of the sensor.

To compare, the first implementation had a Q equal to 300. Even if the total number of pixels of the 300×300 image was 40 times larger than the 2000 pixel retina-like image, its size was still well within the limits of standard computer hardware (e.g. bus bandwidth, memory, etc.). The latest sensor, if simulated using a software or hardware remapper would require the storage and processing of an image with a number of pixels exceeding

the current standard dimensions and, consequently, the design of special purpose hardware. The silicon solution not only requires a much smaller number of pixels but, more importantly, it also requires a lower consumption and has much faster read-out times – 33,000 pixels can be read about 36 times faster than 1,200,000 pixels. This advantage is bound to increase even more in the future when higher integration will be available.

C. The Fovea of Log-Polar Sensors

The equations of the log-polar mapping have a singularity at the origin where the size of the individual photoreceptors would theoretically go to zero. In practical terms this is not possible because – for silicon as well as biological photoreceptors – the technology limits the size of the smallest realizable photoreceptor[1]. For the sensors described here, therefore, the radius of the innermost ring of the log-polar representation is constrained by the size of the smallest pixel while the region inside this circle – the fovea – cannot follow this representation. With the CCD sensor this inner circular region was covered by a square array of pixels, all of the same size. The solution had two drawbacks. First, the square array inside the circular fovea left a portion of the visual scene unsampled; second, not only the logarithmic part of the representation was broken but also its polar components.

In the successive implementation we tried to improve on the design of the first sensor: the resulting geometry is shown in Fig. 6. The solution was to reduce the number of pixels in the foveal rings by

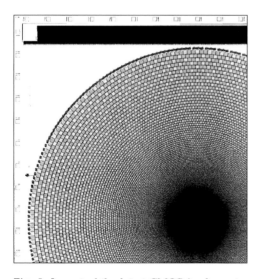

Fig. 5. Layout of the latest CMOS implementation

[1] In humans the smallest diameter of a photoreceptor in the fovea is of the order of 1.5 μm while in the sensors realized so far the size of the smallest realizable photoreceptor was 30 μm, 14 μm, 7 μm for the 2000, 8000 and 33,000 pixel sensors respectively.

Table 1. Descriptive Parameters of the Latest CMOS Sensor

Peripheral Pixels	Foveal Pixels	Total Pixels	R	Q	Size (diameter)
27720 or 252 × 110	5473	33193	17	1100	7.1 mm

halving their number when necessary. For example, as in the periphery the structure has 128 pixels per ring and the size decreases toward the center, when the size of the pixels can no longer be reduced, the successive ring only accommodates 64 pixels until the technological limit is reached again and the number of pixels per ring is halved one more time. In summary, as it can be observed in Fig. 6, the fovea contains 10 rings with 64 pixels, 5 rings with 32 pixels, 2 rings with 16 pixels and 1 ring with 8, 4, and 1 pixel respectively. It is worth noting that by employing this arrangement the polar geometry is preserved and there is no empty space between the periphery and the fovea. However, continuity in terms of spatial resolution is not preserved because, whenever the number of pixels per ring is halved, the size of the pixels almost doubles.

In our latest implementation we adopted a different solution which is optimal in the sense that it preserves the

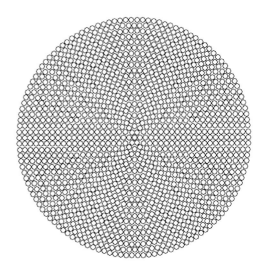

Fig. 7. Solution adopted for the arrangement of the photoreceptors in the fovea of our latest realization. With this topology the centermost region is covered uniformly with pixels of the same size

polar structure and covers the fovea with pixels of the same size. This solution is graphically shown in Fig. 7.

IV. Applications

A. *Image Transmission*

Built around a retina-like camera extensive experiments on wireless image transmission were conducted with a set-up composed of a remote PC running a web server embedded into an application that acquires images from the a retina-like camera (Giotto – see Fig. 8) and compress them following one of the recommendations for video coding over low bit rate communication line (H.263 in our case).

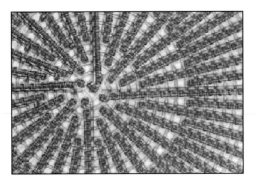

Fig. 6. Layout of the fovea of the IBIDEM CMOS sensor – 0.7 μm technology

Fig. 8. Two different prototypes of the latest version of the Giotto camera. Both cameras are shown with a wide field of view lens (140°). The rightmost model has a standard C-Mount for the lens

The remote receiving station was a palmtop PC acting as a client connected to the remote server through a dial-up GSM connection (9600 baud). Using a standard browser interface the client could connect to the web server, receive the compressed stream, decompress it and display the resulting images on the screen. Due to the low amount of data to be processed and sent on the line, frame rates of up to four images per second could be obtained. The only special-purpose hardware required is the Giotto camera; coding/decoding and image remapping is done in software on a 200-MHz PC (on the server side), and on the palmtop PC (on the client side). The aspect we wanted to stress in these experiments is the use of off-the-shelf components and the overall physical size of the receiver. This performance in terms of frame rate, image quality and cost cannot clearly be accomplished by using conventional cameras.

More recently (within a project called AMOVITE) we started realizing a portable camera that can be connected to the palmtop PC allowing bi-directional image transmission through GSM or GPRS communication lines. The sensor itself is not much different from the one previously described apart from the adoption of a companion chip allowing a much smaller camera.

B. Robotics

As far as robotics is concerned the main advantage of log-polar sensors is related to the small number of pixels and the comparatively large FOV. For some important visual tasks images can be used as if they were high resolution. Typically, for example, in target tracking the fovea allows precise positioning and the periphery allows the detection of moving targets. This property is exploited to control the robot head of the Babybot (see Sandini 1997, Metta 2000). In this implementation a log-polar sensor with less than 3000 pixels is used to control in real-time the direction of gaze of a 5 degree-of-freedom head. Aspects of the image processing required were outlined in section II.

Particularly relevant is the possibility to use vision to learn a series of diverse behaviors ranging from multisensory integration to reaching for visually identified targets. Fundamental to any other visual task is the possibility of moving the cameras so that the fovea moves and explores interesting regions of the visual field. We implemented a series of eye-head visuomotor behaviors including saccades, vergence, smooth pursuit and vestibulo-ocular reflexes (VOR). For example vergence is controlled by relying on a measure of the binocular disparity (Manzotti et al. 2001), and dynamically by using the D component of the optic flow (see equation (3)); see (Capurro et al. 1997) for more details.

In the case of the VOR the appropriate command, using a combination of visual and inertial information, is synthesized by measuring the stabilization perfor-mance. The latter measure is an example of the kind of parameters which is best estimated by means of the optic flow. In this case the translational component pro-vides information to the robot about the performance of the controller; a neural network can be trained on the basis of this information (Panerai et al. 2000, 2002). Saccades and other visual control param-eters are also learnt as shown in (Metta et al. 2000). Visual information about suitable targets is acquired in terms of eccentricity (the angular position with respect to the current gaze direction), binocular disparity, and optic flow as de-scribed in section II.

We also investigated the relationship between vision and the acquisition of an acoustic map of the environment (in neural network terms). In this case the robot, equipped with the appropriate learning rules, learnt autonomously the relationship between vision, sound, and saccade/pursuit behaviors (Natale et al. 2002). After a certain amount of training, the Babybot is able to orient towards a visual, acoustic, or visuo-acoustic target (a colored object emitting sound), thus bringing two non-homogeneous quanti-ties (vision and sound) into the same ref-erence frame.

Finally, eye-head-arm coordination was investigated. Under certain biologi-cal compatible hypotheses, we were able to show that the robot could learn how to transform visual information (coding the position of a target) into the sequence of motor commands to reach the target (Metta et al. 1999) and touch it. Also in this case, vision was the organizing factor to build autonomously and on-line an in-ternal motor representation apt to control reaching.

It is worth stressing that the whole system runs on a network of a small num-ber of PCs (four Pentium-class processors) at frame-rate and carries out on-line

learning – neural network weights are changed on the fly during the operation of the robot. While the reason this is pos-sible is related to the overall number of pixels (a limiting factor in most implemen-tations), it was not clear beforehand whether this same visual information was enough to sustain a broad range of behaviors. We believe that this implemen-tation provides a proof by existence that indeed the amount of information re-quired can be obtained visually, with a moderately computational burden, if the visual scene is sampled appropriately. Log-polar, in this sense, proved to be a good enough sampling strategy.

V. Conclusions

The main advantage of retina-like sensors is represented by the compromise be-tween resolution and field of view allow-ing high resolution as well as contextual information to be acquired and processed with limited computational power. The advantages of a silicon realization with respect to hardware and software remap-pers has been discussed showing that, as technology progresses, the retina-like ap-proach will not only maintain its current advantages, but it is bound to become even more interesting in the application areas described – and possibly in others. Different realizations of log-polar sensors have been illustrated as well as two key applications. In particular the peculiari-ties of the retina-like sensors for real-time control of gaze and, in general, for ro-botics have been presented. This is cer-tainly the most obvious use of a sensor topology that has been shaped by evolu-tion to support the control of behavior while maintaining the balance with energy consumption and computational requirements. In spite of the great variety of eyes found in nature, the conven-tional camera solution based either on increasing simultaneously the resolution and field of view or on the use of interchangeable or variable focal length lenses, has not survived. There is no doubt, in our view, that in the future, adaptable robots will have space-variant eyes or, conversely, the design will trade autonomy for batteries and computational power. The image transmission application demonstrates empirically that, in order to fully exploit a communication channel, it is better to eliminate useless information at the sensor's level than compressing information which is not used.

Acknowledgements

The work described in this paper in relation to Babybot has been supported by the EU Project COGVIS (IST-2000-29375), MIRROR (IST-2000-28159). The research described on the retina-like CMOS technology has been supported by the EU Projects SVAVISCA (ESPRIT 21951) and AMOVITE (IST-1999-11156). The research on the CMOS sensors has been carried out in collaboration with IMEC and FillFactory (Lou Hermans and Danny Scheffer), AITEK S.p.A. (Fabrizio Ferrari, Paolo Questa) and VDS S.r.l. (Andrea Mannucci e Fabrizio Ciciani). Features of the retina-like sensor topologies described are covered by patents. Giorgio Metta is Postdoctoral Associate at MIT, AI-Lab with a grant by DARPA as part of the "Natural Tasking of Robots Based on Human Interaction Cues" project under contract number DABT 63-00-C-10102.

References

Baron T, Levine MD, Hayward V, Bolduc M, Grant DA (1995) A biologically-motivated robot eye system. Paper presented at the 8th Canadian Aeronautics and Space Institute (CASI) Conference on Astronautics, Ottawa, Canada

Baron T, Levine MD, Yeshurun Y (1994) Exploring with a foveated robot eye system. Paper presented at the 12th International Conference on Pattern Recognition, Jerusalem, Israel

Blough PM (1979) Functional implications of the pigeon's peculiar retinal structure. Granda AM, Maxwell JM (eds) Neural Mechanisms of Behavior in the Pigeon (pp. 71–88). Plenum Press, New York, NY

Capurro C, Panerai F, Sandini G (1997) Dynamic vergence using log-polar images. Int J Comput Vision 24: 79–94

Carpenter RHS (1988) Movements of the Eyes (Second ed.). Pion Limited, London

Darrell T, Gordon G, Harville M, Woodfill J (2000) Integrated person tracking using stereo, color, and pattern detection. Int J Comput Vision 37: 175–185

Engel G, Greve DN, Lubin JM, Schwartz EL (1994) Space-variant active vision and visually guided robotics: design and construction of a high-performance miniature vehicle. Paper presented at the International Conference on Pattern Recognition, Jerusalem

Galifret Y (1968) Les diverses aires fonctionelles de la retine du pigeon. Z Zellforsch 86: 535–545

Koenderink J, Van Doorn J (1991) Affine structure from motion. J Optical Soc Am 8: 377–385

Manzotti R, Gasteratos A, Metta G, Sandini G (2001) Disparity estimation in log polar images and vergence control. Comput Vis Image Und 83: 97–117

Metta G (2000) Babybot: a study on sensori-motor development. Unpublished Ph.D. Thesis, University of Genova, Genova

Metta G, Sandini G, Konczak J (1999) A developmental approach to visually-guided reaching in artificial systems. Neural Networks 12: 1413–1427

Metta G, Carlevarino A, Martinotti R, Sandini G (2000) An incremental growing neural network and its application to robot control. Paper presented at the International Joint Conference on Neural Networks, Como, Italy

Natale L, Metta G, Sandini G (2002) Development of auditory-evoked reflexes: visuo-acoustic cues integration in a binocular head. Robot Auton Syst 39: 87–106

Panerai F, Metta G, Sandini G (2000) Visuo-inertial stabilization in space-variant binocular systems. Robot Auton Syst 30: 195–214

Panerai F, Metta G, Sandini G (2002) Learning stabilization reflexes in robots with moving eyes. Neurocomputing 48: 323–337

Rojer A, Schwartz EL (1990) Design considerations for a space-variant visual sensor with complex-logarithmic geometry. Paper presented at the 10th International Conference on Pattern Recognition, Atlantic City, USA

Sandini G (1997) Artificial systems and neuroscience. Paper presented at the Otto and

Martha Fischbeck Seminar on Active Vision, Berlin, Germany

Sandini G, Tagliasco V (1980) An anthropomorphic retina-like structure for scene analysis. Comp Vision Graph 14: 365–372

Schwartz EL (1980) A quantitative model of the functional architecture of human striate cortex with application to visual illusion and cortical texture analysis. Biol Cybern 37: 63–76

Srinivasan MV, Venkatesh S (eds) (1997) From Living Eyes to Seeing Machines. Oxford University Press, London

Tunley H, Young D (1994) First order optical flow from log-polar sampled images. Paper presented at the Third European Conference on Computer Vision, Stockholm

Van der Spiegel J, Kreider G, Claeys C, Debusschere I, Sandini G, Dario P, Fantini F, Bellutti P, Soncini G (1989) A foveated retina-like sensor using CCD technology. In: Mead C,
Ismail M (eds), Analog VLSI Implementation of Neural Systems (pp. 189–212). Kluwer Acad Publ, Boston

Wallace RS, Ong PW, Bederson BB, Schwartz EL (1994) Space variant image processing. Int J Comput Vision 13: 71–91

Weiman CFR (1988) 3-D Sensing with polar exponential sensor arrays. Paper presented at the SPIE – Digital and Optical Shape Representation and Pattern Recognition

Weiman CFR, Chaikin G (1979) Logarithmic spiral grids for image processing and display. Computer Graphic and Image Processing 11: 197–226

Weiman CFR, Juday RD (1990) Tracking algorithms using log-polar mapped image coordinates. Paper presented at the SPIE International Conference on Intelligent Robots and Computer Vision VIII: Algorithms and Techniques, Philadelphia (PA)

19. Computing in Cortical Columns: Information Processing in Visual Cortex

Steven W. Zucker

Abstract

The orientation hypercolumns in the primate primary visual cortex provide a rich framework for structuring early visual computations. We abstract these columns mathematically, and use this abstraction to derive the early visual computations underlying edge and line finding, stereoscopic fusion, and texture and shading analysis. Coherency within the framework further dictates interactions between these computations. Implications for understanding the neurophysiology of early vision are discussed.

I. Introduction

Visual cortex is organized largely around orientation; that is, around selective responses to local oriented bars. According to a classical observation of Hubel and Wiesel (1977), when recordings of activity in neurons are made by sampling a short distance along a tangential penetration (i.e., a penetration roughly tangent to the surface of cortex), a sequence of cells are encountered that exhibit regular shifts in orientation preference but little shift in receptive field position. Normal penetrations, or penetrations directly down from the surface of the cortex, reveal a different organization: cells with similar orientation and position preferences but different receptive field sizes. Together they define an array of orientation columns, and these columns provide a representation for visual information processing. Longer tangential penetrations reveal a jump in receptive field properties, to nearby positions and arbitrary orientations; i.e., to an adjacent column. A cartoon of one hypercolumn is shown in Fig. 1.

The question is thus raised: how can visual information processing be structured on orientation hypercolumns? Specifically, does the columnar architecture suggest which processing should take place within columns, and which between them? It is commonly held that, if receptive fields were considered as oriented filters, then intra-columnar processing amounts to scale-spaces of filters. But what information processing tasks are these filters solving, and how does the scale-space notion arise from within the task? Physiologically, in what sense does this require elaboration across different layers, rather than just additional connections? Do certain non-linearities arise

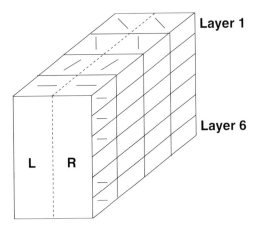

Layer 1

Layer 6

L R

Fig. 1. A cartoon view of the functional architecture of visual cortex suggested by Hubel and Wiesel and dubbed the "ice cube" model. It sketches the neuronal layout supporting the coverage of each local retinotopic area by receptive fields that span a range of orientations (tangential penetration), sizes (normal penetration), and eye-of-origin

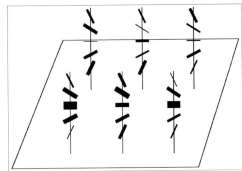

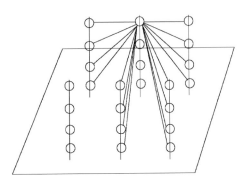

Fig. 2. Two aspects of the computational architecture of visual cortex. *Top* An interpretation of the ice-cube model into geometric terms, showing a retinotopic array (the tilted plane) and a sampling of upper-layer (layers 2–3) cells drawn as short segments along a "fibre" to show their orientation preference. The thickness of the orientated bars denotes activity of cells. (Only one eye is shown, corresponding to a tangential penetration; receptive-field size is not shown.) *Bottom* The arrangement of neighboring fibres suggests an architecture that supports interactions ("connections") between orientations. These are illustrated schematically here for only one cell whose activity is facilitated or inhibited by the activity of a number of cells in nearby columns. Physiologically, such interactions are carried by long-range horizontal connections between neurons

naturally? Is there a sense other than filtering in which receptive fields can be modeled, that is sufficiently abstract to encompass modern views of cortical physiology? These are the questions that we attempt to address from an abstract, computational perspective.

II. A Computational View of Visual Cortex

The columnar architecture suggests a particular and powerful computational view of the visual cortex. A rearrangement of the tangential penetration (Fig. 2, *top*) suggests computational elements that measure different orientations at each position, with interactions between them (Fig. 2, *bottom*). We have developed one formal model of computation in this architecture, that is consistent with the biophysics of pyramidal neurons, at least to the first-order terms in their mathematical

abstraction, and is related to relaxation labeling (in neural networks) and to polymatrix games (in operations research) (Hummel and Zucker 1983, Miller and Zucker 1999).

In this paper we concentrate on the "connections" that underlie computations in visual cortex. A clue to their understanding derives from considering the inter-columnar interactions, for which it is commonly held that, for "edge detection" and curve inference, co-aligned receptive fields should be mutually supportive. This suggests a role for long-range horizontal processes in the superficial layers of

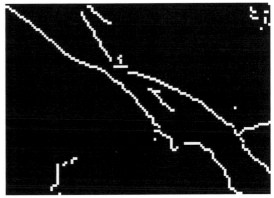

Fig. 3. An illustration of the geometric problems in interpreting standard approaches to edge detection, evaluated on a complex image (*top*). The standard edge map (*bottom left*) is obtained from the Canny (1986) operator, Matlab implementation, scale = 3. This operator resembles a receptive field convolution (or filter) followed by a hysteresis stage designed to enforce co-aligned facilitation. Thus, in some sense, it realizes the basic presumptions about physiology: oriented filtering followed by co-aligned facilitation. How should this result be evaluated? While it seems reasonable at first glance, a detailed examination of the shoulder/arm region (*bottom right*) reveals the incorrect topology. Try to determine from this "edge map" what the underlying physical object is, or even what the image intensity structure is. The full Canny edge map (*bottom left*) confirms that these observations about the incorrect topology are repeated in many different places. It can easily be shown that this is not just a scale problem, because Canny operators at different scales produce different edge maps but with the same topological problems. A sufficiency condition on edge detection, we contend, must at least be topologically correct

visual cortex in cats and primates, about which anatomical and physiological evidence is accumulating (Nelson and Frost 1985, T'So et al. 1986, Malach et al. 1993, Kapadia et al. 1995). Briefly, since the majority of connections enforce approximately the same orientation, modelers have assumed a kind of "orientation good continuation", as the Gestalt psychologists defined it (Koffka 1935, Field et al. 1993, Yen and Finkel 1997).

However, standard approaches to "edge detection" based on "orientation good continuation" are frustrating (Fig. 3), and evidence is accumulating for facilitation between cells with up to 50 deg orientation differences. How can such exceptions be explained, and should the computational structure of boundary detection be elaborated to include them?

III. The Geometry of Inter-Columnar Interactions

We attempt to answer these questions from an information processing viewpoint. We introduce a model for curve inference based on differential geometry, and argue that this captures the structure of our visual world. We will then become more specific about cortical circuitry, and will show how an interpretation of orientation selectivity as tangent estimation structures a model of columnar processing. Based on this we will then sketch a model for stereo correspondence that explicitly takes advantage of the columnar architecture. To our knowledge this is the first formal stereo model that uses monocular orientation to support 3-D differential and projective geometry. The results open up a new class of curve-based stereo algorithms, and suggest why stereo computations are distributed across multiple areas (V1 and V2) in primates. Finally, we extend the geometric framework to texture flow analysis.

A. The Geometry of Edge Detection

Local measurements of orientation are inherently ambiguous, and we adopt the mathematics of differential geometry to structure the interactions between local measurements. We first observe that, if the world consists of smooth surfaces, then the locus of positions at which those surfaces fold away from the viewer (such as the circle bounding a projected sphere) forms a smooth curve. Locally, a measurement of orientation signals the tangent to this curve, and differential geometry dictates that interactions between tangents must involve curvature (Parent and Zucker 1989) (Fig. 4, top). The question becomes how to transport a tangent at one location to a nearby location, and the analysis is not unlike driving a car: At each instant of time the axis of the car defines its (tangent) orientation, and the relationship between the orientation of the car at one instant with that at the next depends on how much the road curves. This requires that curvature must be represented jointly with orientation in the visual cortex, and we (and others) have established that another property of cortical neurons – "endstopping" – is sufficient for achieving this (Dobbins et al. 1987). In this model local circuitry between cells with large receptive fields in the deep layers and small receptive fields in the middle layers can build up a selectivity for curvature. On this basis we can derive the strength of horizontal interactions (Parent and Zucker 1989, Zucker et al. 1989), two of which are illustrated in Fig. 4 (bottom). Note how these agree with available data for straight situations (curvature = 0) (Nelson and Frost 1985, T'So et al. 1986, Malach et al. 1993), but generalize as well to explain data showing facilitation between orientations that differ by 50% (Kapadia et al. 1995). Other models (Yen and Finkel 1997) consider only the zero-curvature

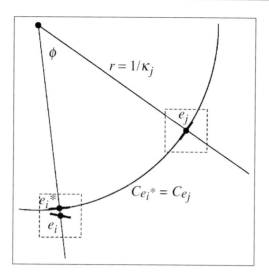

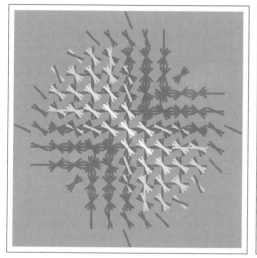 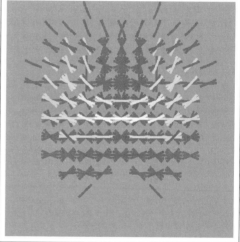

Fig. 4. The geometry of inter-columnar interactions is illustrated with nearby tangents and the osculating circle that defines the way they should interact. *Top* The relaxation labeling network selects those tangents that minimize the mismatch between e_i and e_i^*, weighted by the initial strength of match. Such networks model the neuronal interactions implemented by networks with horizontal interactions in superficial layers of visual cortex. *Bottom* Two sets of examples of the horizontal interactions between neurons. Each bar indicates a receptive field, and the "fibre" of receptive fields has been projected onto the image plane. All connections are between the central "neuron" and one of those shown in the diagram. (The full diagram indicates the complete set of horizontal interactions to the central neuron. Positive contrast signifies an excitatory connection; negative contrast signifies an inhibitory connection.) Multiple bars at the same position indicate several cells in the same orientation hypercolumn. Two cases are shown, one for an orientation about 45 deg and $\kappa_i = 0.0$, and the other for a horizontal orientation with a large amount of curvature in the negative direction. Notice in particular that many of the excitatory connections would appear to be between co-aligned cells, but in the high curvature example there are cases of excitatory connections with approx. 50 deg relative orientation. Computations by L. Iverson

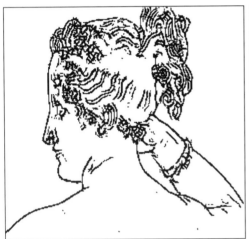 

Fig. 5. Illustration of the geometric solution to edge detection. The tangent map obtained from our biologically-motivated operators, which capture the geometry of the situation. *Bottom left* These are the initial responses from our non-linear "logical/linear" operator (Iverson and Zucker 1995) designed to capture the tangent model. We speculate that layer 5–6 (V1) (intracolumnar) interactions are suitable to implement it. *Bottom right* Tangent map that would result in layer 2/3 following geometric "intercolumnar" processing. Note differences from Canny in the edge topology, especially around the shoulder musculature and "T" junctions near the neck and chin. Computations by L. Iverson

case, and do not predict the non-co-linear data.

Transport and curvature are emergent concepts at the geometric level. They lead to information processing results that are substantially more correct geometrically than earlier attempts (Fig. 5). We refer to these results as "tangent maps," since they actually represent (an approximation to) tangents to edge curves.

B. The Geometry of Stereo Correspondence

We now generalize to interactions between two tangent maps, one for the left eye and the other for the right eye. (In effect, we now consider both of the orientation columns that make up the full hypercolumn in Fig. 1.) We have developed this product structure into an algorithm

Fig. 6. *Top* A stereo pair of Asiatic lilies. This image pair is complex to process, because the boundaries of the flowers are curves that meander through space smoothly but have many abrupt occlusions. *Bottom* Discrete tangent map computed as in previous figure, but to higher resolution in orientation

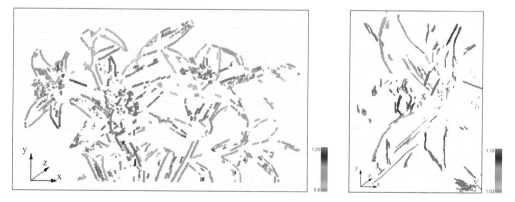

Fig. 7. The depth map associated with the stereo lily pair. The bars indicate depth scales associated with the image to their left. Each point shown is the projection of a tangent in space that is geometrically consistent with its neighbors. *Left* Full lily image. *Right* Flower detail magnified. Results computed by the algorithm in Alibhai and Zucker (2000) and reproduced here

for computing stereo correspondences (Alibhai and Zucker 2000). It required a generalization of the tangent maps for

plane curves to those for general space curves. A curve in space can be described by the relationships between its tangent,

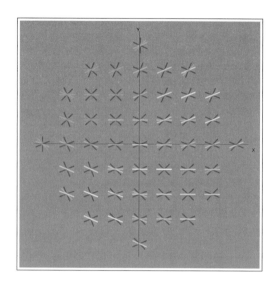

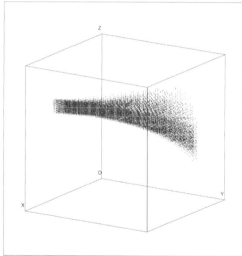
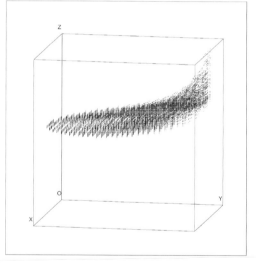

Fig. 8. The geometry of texture compatibilities. *Top* An illustration of the compatibility structure in the same format as the compatibilities in Fig. 4. Notice how the excitatory interactions are now spread out over a region (for texture flow) rather than being concentrated along a curve with inhibitory side-zones (for edge and curve detection). To provide some insight into the geometry underlying these flows, in the *bottom* figures we show the compatibility fields displayed as a three-dimensional diagram, with the x and y axes corresponding to position and the vertical axis corresponding to preferred orientation (*bottom left*) or tangential curvature (*bottom right*). Adapted from ben Shahar and Zucker (2001)

normal and binormal. The result is that the osculating circle (used for monocular analysis in the previous section) generalizes to an osculating helix, and that in addition to orientation and curvature we also require torsion (the rate at which the helix "leaves" the osculating plane, which is a higher-order "curvature" in the Frenet sense). The osculating circle was illustrated in figure 4. It is the circle that "kisses" the curve at the point of interest (technically agrees through curvature).

As the space curve moves across depth planes, there exists a positional disparity between the projection of the curve into the left image and the projection into the right image. The geometric approach also lets us take advantage of higher-order disparities, such as disparities in orientation, that occur as well, and it is this additional information that makes our geometric algorithm unique. Compatibilities are now defined by an osculating helix between nearby pairs of space tangents, and this helix projects into compatibilities in the left tangent map and the right tangent map. Thus stereo vision is based on relationships that include pairs of problems of the type solved in the previous section. Results for an example image (Fig. 6) are shown in Fig. 7, taken from Alibhai and Zucker (2000).

C. The Geometry of Texture Flows

The final example of using differential geometry to structure horizontal connections is the analysis of texture flows, or patterns derived from surface coverings with a well-defined orientation locally. Examples of texture flows include hair and fur patterns, wheat and grass patterns, and various types of surface markings from a Zebra's stripes to Dürer's etchings.

Since all of these patterns are surface coverings, there is the possibility of variation in the tangential direction (or moving

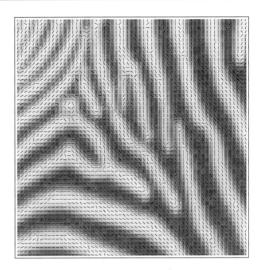

Fig. 9. An example of texture flow computation. *Top* Local measurements of orientation obtained from the image intensity gradient (rotated). Notice how such local measurements are highly variable and noisy, are missing in certain places, and sometimes do not conform to the natural flow. *Bottom* Relaxed orientation flow based on the helicoidal compatibilities. Notice how smooth the flow is almost everywhere, except near the singularities (represented as multiple orientations at the same position). Adapted from ben Shahar and Zucker (2001)

along a hair or streamline in the texture flow) and in the normal direction (or moving from one streamline to another). We model this with two curvatures, one in the tangential direction and the other in the normal direction, and derive the exact form of the curvatures using Cartan connection forms (see ben Shahar and Zucker 2001). An example of the texture compatibilities is shown in Fig. 8. It is clear from these illustrations how smooth the "compatibility surfaces" are. In this case they follow a helicoid in (position, orientation) space, which generalizes the osculating circle and osculating helix discussed previously.

As an illustration of the texture flow computation, we show an example in Fig. 9 from ben Shahar and Zucker (2001). Notice how the natural pattern formation for the zebra includes singularities, and therefore any texture flow system must be capable of capturing such discontinuities in orientation patterns.

IV. Summary and Conclusions

Starting with a view of the primary visual cortex as a computational machine organized around orientation, we have shown how tasks related to edge and curve detection, stereo, and texture analysis can be structured. The focus was on long-range horizontal connections, and we derived their forms for all three problems. The resulting forms agree convincingly with the underlying neurophysiology.

While the curve and texture tasks can easily be realized in V1 (the first visual cortex area), the monocular edge maps needed for stereo analysis had to be interpolated to higher resolution in orientation. This is necessary to realize the orientation disparity portion of the computation, and may suggest why the stereo computation is elaborated from V1 into V2 (the second visual cortex area).

Acknowledgements

I thank my collaborators S. Alibhai, A. Dobbins, P. Huggins, L. Iverson, and 0. ben-Shahar. Research supported by AFOSR.

References

Alibhai S, Zucker SW (2000) Contour-based correspondence for stereo. In Computer Vision – ECCV 2000, Lecture Notes in Computer Science 1842

ben Shahar O, Zucker SW (2001) On the perceptual organization of texture and shading flows: From a geometrical model to coherence computation. In Proc IEEE Conf on Computer Vision and Pattern Recognition

Canny J (1986) A computational approach to edge detection. IEEE T Pattern Anal 8: 679–698

Dobbins A, Zucker SW, Cynader MS (1987) End-stopped neurons in the visual cortex as a substrate for calculating curvature. Nature 329: 438–441

Field D, Hayes A, Hess R (1993) Contour integration by the human visual system: evidence for a local association field. Vision Res 33: 173–193

Hubel DH, Wiesel TN (1977) Functional architecture of macaque monkey visual cortex. Proc R Soc Lond B 198: 1–59

Hummel R, Zucker SW (1983) On the foundations of relaxation labeling processes. IEEE T Pattern Anal 6: 267–287

Iverson LA, Zucker SW (1995) Logical/linear operators for image curves. IEEE T Pattern Anal 17: 982–996

Kapadia M, Ito M, Gilbert C, Westheimer G (1995) Improvement in visual sensitivity by changes in local context: Parallel studies in human observers and in vl of alert monkeys. Neuron 15: 843–856

Koffka K (1935) Principles of Gestalt Psychology, Harcourt, Brace and World, Inc, New York

Malach R, Amir Y, Harel M, Grinvald A (1993) Relationship between intrinsic connections and functional architecture revealed by optical imaging and in vivo targeted biocytin injections in primate striate cortex. Proc Natl Acad Sci (USA) 90: 10469–10473

Miller DA, Zucker SW (1999) Computing with self-excitatory cliques: A model and an application to hyperacuity-scale computation in visual cortex. Neural Comput 11: 21–66

Nelson J, Frost B (1985) Intracortical facilitation among co-oriented, co-axially aligned simple cells in cat striate cortex. Exp Brain Res 61: 54–61

Parent P, Zucker SW (1989) Trace inference, curvature consistency and curve detection. IEEE T Pattern Anal 11: 823–839

T'So D, Gilbert C, Wiesel TN (1986) Relationships between horizontal interactions and functional architecture in cat striate cortex as revealed by cross-correlation analysis. J Neurosci 6: 1160–1170

Yen SC, Finkel L (1997) Salient contour extraction by temporal binding in a cortically-based network. Advances in Neural Information Processing Systems, 9

Zucker SW, Dobbins A, Iverson L (1989) Two stages of curve detection suggest two styles of visual computation. Neural Comput 1: 68–81

20. Vision by Graph Pyramids

Walter G. Kropatsch[*]

Abstract

To efficiently process huge amounts of structured sensory data for vision, graph pyramids are proposed. Hierarchies of graphs can be generated by dual graph contraction. The goal is to reduce the data structure by a constant reduction factor while preserving certain image properties, like connectivity. While implemented versions solve several technical vision problems like image segmentation, the framework can be used as a model for biological systems, too.

---

[*] I would like to thank Yll Haxhimusa for the improved illustrations on building an example graph pyramid. I would like to thank the reviewers and editors of this collection for their helpful comments to enhance the general readability of this chapter. This work was supported by the Austrian Science Foundation under grants S 7002-MAT, P 14662-INF and P 14445-MAT.

I. Introduction

Animals and mechanical systems are equipped with many sensors. The individual sensor elements are spatially distributed in many different arrangements. While regular *sensor arrangements* like square grids dominate the fabricated world, most of their biological counterparts show a regular distribution only on a large scale, their sensor neighborhoods are neither geometrically nor topologically regular. Figure 1 shows portions of the sensor arrangements of a typical digital camera and a monkey's retina. The sensors' positions are indicated by a small circle, the number of elements in Fig. 1 has been chosen to match approximately the 2339 sensory elements of the monkey retina. The processing of the data measured by the sensors involves a huge number of processing elements (e.g. processors, neurons) that are interconnected in very complex ways. In this paper we propose one possible interconnection network, the *graph pyramid*, that combines two data structures in an efficient way: logarithmic pyramids (see Jolion and Rosenfeld 1994 for a survey) and attributed relational graphs (for theory see Thulasiraman and Swamy 1992, for recent applications in Pattern Recognition see Jolion and Kropatsch 1998, Kropatsch and Jolion 1999, Jolion et al. 2001). Regular pyramids are built by repeatedly reducing the resolution (and the size) of the

a

b

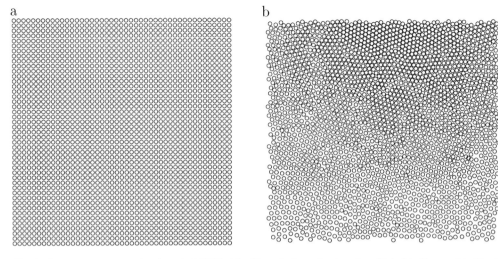

Fig. 1. Sensor arrangements, a) part of CCD chip, b) part of monkey's retina (data kindly provided by P Ahnelt)

image by a constant factor, e.g. a $1024 \times 1024 = 2^{10} \times 2^{10}$ image produces a pyramid with 10 levels above the base: 512×512, 256×256, 128×128, 64×64, 32×32, 16×16, 8×8, 4×4, 2×2, 1×1. The reduced images are computed by local (weighted) averages relating the (gray) values of the reduced pyramid level with a small number of pixels in the level below. These 'vertical' connections between the images in the pyramid stack establish short (logarithmic) connections between the apex of the pyramid and any pixel in the base image. The resulting computational efficiency has been used in many vision systems to speed up their performance. But pyramids are also used as a model to explain visual perception in biological systems (e.g. see Uhr and Schmitt 1984, Pizlo et al. 1997).

Sensors typically measure a single quantity like the light intensity but objects in the environment are connected volumetric entities, they have a 'body', a certain shape and can be *decomposed into parts*. For example, a human face is characterized by its eyes, its nose and its mouth, and, what is considered a major

structural property, the mutual relations between its parts.

Let us now study the first stages of visual information processing and focus on the data representation needed at the different levels of processing. Through the projection of scene objects of the environment into the image, object patches with same color or texture are mapped into image regions of same characteristics. By the discrete sampling of these objects, the corresponding regions are decomposed into many similar sensor signals, adjacent sensors may sense the same color, similar light intensity or even complicated texture features. 'Image Segmentation' finds homogeneous regions in the sampled image and determines the spatial relations between these regions. Region adjacency allows a system to group all region patches of one object or to find groups of objects that belong together.

A simple and efficient way to represent all three types of structures: the arrangements of sensor elements, the region adjacencies, and the part-whole relations of objects, is an *attributed relational graph* (ARG). It consists of a set of nodes or vertices V, of a set of edges E with each edge relating two vertices. Both vertices and edges may receive numerical or symbolic attribute values to specify particular properties. A pixel array can be converted to an ARG without loss of information by simply defining the pixels as the vertices of the graph with a gray value attribute. Neighboring pixels create an edge between the vertices corresponding to the pixels. But the ARG goes beyond the representational capabilities of the pixel array. Vertices can be identified with regions or also with objects. In these cases an edge can express region adjacency or mutual object relations respectively. In addition, the ARGs that are derived from images are embedded in the image plane: such graphs are called 'plane graphs'[1].

Since the sensor elements form a surface (like the eye's retina), we represent their topological arrangement by a pair of graphs, which are non-simple, plane, and dual to each other. (Large) connected regions and their topological relationships often allow the inference of spatial relations in the environment and are very robust with respect to noise and sensing inaccuracies. These properties are preserved by dually contracting the graphs, operations which can be implemented in a massive parallel and local system architecture. Repeated contraction yields a stack of successively reduced graphs: *a graph pyramid.*

This paper is organized as follows: We first examine the different processing stages of image analysis by a computer system (section II). We then study the purpose of vision both for computers and for biological systems (section III). Section IV explains the basic concepts to build a graph pyramid. This notion of

"equivalent contraction kernel" allows us in section IV.C to define all possible graph pyramids that can be built on top of a given base graph, e.g. the sensor arrangement of a retina. In the conclusion we refer to the different successful applications of pyramids and graph pyramids.

II. Processing Stages of Computer Vision

Let us consider vision as a process that is supposed to capture the essential structure of the world. The cyclic arrangement of the diagram in Fig. 2 shows the structure of the environment of a 'seeing system' ('REALITY') and the internal stages of recovery of this structure from the image. It tries to relate the structure of the WORLD on the left side with the reconstruction stages of the COMPUTER vision process on the right side. A DIGITAL IMAGE, as captured by a *sensing camera*, is an array of measurements called pixels (picture elements). Simple *image processing* consists in modifying the individual pixel values in order to stretch the image contrast, to remove noise, and to compute features used in segmentation to classify pixels as belonging to a specific image object or to the background. *Segmentation* collects all pixels of the same class into connected sets of pixels: the image REGIONS. These regions create *region adjacency* relations between the regions. Geometric region properties like "shape" allow the *identification* of SPECIFIC PARTS corresponding to the REFLECTING SURFACES of the reality. Using knowledge about the structure of objects, the specific parts can be further *assembled* into image OBJECTS. Together with their mutual relationships they constitute the SCENE DESCRIPTION describing (and interpreting) the content of the visual input.

---

[1] Plane graphs should not be confused with planar graphs for which such embedding exists. E.g. a different choice of the background face may cause a different embedding in the plane.

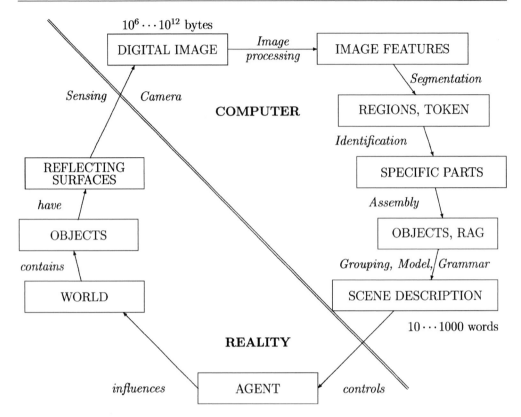

Fig. 2. The vision process as a cycle

Figure 3 illustrates some of the vision steps by means of a very simple example. 11 objects, black squares on white background, are sampled in a square grid producing 196 pixels of which 44 fall on the black object surface. Pairs of pixels are considered adjacent if they are neighbors in the same row or the same column (4-adjacency). The corresponding neighborhood graph contains 364 edges. Segmentation and connected component analysis[2] assigns each group of 4 black pixels a unique label, $1, 2, \ldots 11$, each corresponding to one object. The white background forms a single connected component having 11 holes (e.g. the 11 black squares). Since the fact that all the 11 objects are adjacent to the background is not the ultimate result (i.e. also the characters of a scanned document), spatial closeness is used to describe their arrangement in the image. The resulting object graph consists of 11 object-vertices and 18 object-relations. The later reflects the placement of the objects in the image plane[3] by relating objects which are closest in the image. Such graphs are compact and very convenient data

---

[2] A connected component is a set of pixels which have the same gray value and which are connected.

[3] Geometric object properties like coordinates or size are kept in numerical attributes of the vertices.

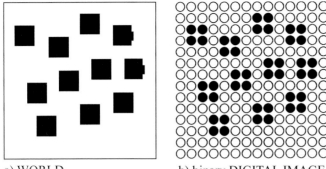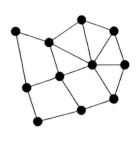

a) WORLD b) binary DIGITAL IMAGE c) SCENE DESCRIPTION

Fig. 3. Deriving the structure of 11 objects after regular sampling. *a* WORLD, *b* binary DIGITAL IMAGE, *c* SCENE DESCRIPTION

structures to describe topological relations among image entities at multiple levels in a consistent way.

III. Purpose: Smart and Efficient Decisions

Before studying the sensors and sensing processes that extract information from the environment let us ask for the purpose of these complicated and not yet sufficiently understood systems. Pizlo et al. (1997) define the goal of visual perceptions as follows: '*The goal of vision perception is to provide the observer with visual information about the 3D environment so that the observer can recognize objects, manipulate them, and navigate in the environment.*' A natural environment typically contains an enormous variety and amount of objects, even a single snap shot could not be fully interpreted in reasonable time. Hence the need to efficiently focus on those pieces of information that are *relevant* to take appropriate decisions. Less important data can be neglected, but their removal should not disturb the (probably) important spatial relations between the relevant parts. This brings us back to the formulation of Marr (1982): '*building a*

description of the shapes and positions of things from images'. We conclude that vision algorithms must be capable of *simplifying a huge amount* of sensory data without loosing the relevant information. Since sensors have different tasks the purpose varies and the respective algorithms must allow conscious control and adaptation to the actual purpose in a smart way. In this context, *adaptation* means that other weights of importance are given to the features derived from the data depending on the actual purpose. A different goal may need a different decision to be drawn from the same visual data.

IV. The Graph Pyramid

Connected objects are mapped into connected image regions if the image's projection is large enough to satisfy the discrete sampling theorem (e.g. to inscribe a circle with a radius larger than the sampling distance). Geometrical measurements derived from a digital image are sensitive to errors due to noise, discrete sampling and motion inaccuracies. However, the structural and topological relations like region-adjacency or part-whole are inherent to the objects and their arrangement in the image up to discretization. In many cases, they do not depend on the particular imaging situation. This is the background of several recent contributions describing spatial/structural representations and transformations preserving topological relations existing in the image plane. Let us enumerate a few approaches preserving structural relations:

1. The simplest most frequent representation uses coordinates as vertex attributes of an ARG. This immediate representation depends on the particular mapping geometry. For well controlled environments (e.g. geographic information systems) it is widely used due to its simplicity.

2. Rosenfeld and Nakamura (1997) consider local deformations of (digital) curves in the plane that preserve an implicitly given topology. The idea is that images showing the same topological arrangement of regions and curves can be transformed into each other continuously.

3. A pair of plane dual graphs is the base of a graph pyramid built by repeated dual graph contractions (Kropatsch 1997a). It differs from the previous approach in that the transformed data are reduced at each step by a constant reduction factor which is the origin of its computational efficiency.

4. In topological and combinatorial maps (Gareth and Singerman 1978, Lienhardt 1989) the embedding is determined by the local orientation of the structural elements. We have shown in Kropatsch and Brun (2000) how to perform dual graph contraction with combinatorial maps.

To associate a discrete image region with the corresponding surface patch of the object and to hypothesize adjacency of patches from adjacency of regions, the connectivity of the graph (or the subgraph representing the object) is an essential

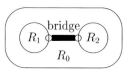

Fig. 4. The removal of the bridge would disconnect subgraphs R_1 and R_2

structural property. Since our goal is to successively remove unnecessary parts the connectivity can be lost by these operations. Before disconnecting a graph into two components these two components will be connected by a single edge which is called a *bridge* (Fig. 4). Hence bridges should not be removed, or, what is equivalent, the dual counterpart of bridges, self-loops, should not be contracted. Since our graphs are embedded in the image plane, each edge bounds two regions of the plane which are connected by the dual edge in the dual graph. The two regions bounded by a bridge are the same (R_0 in Fig. 4) and consequently the two extremities of the dual edge too: such an edge is called a self-loop.

A. Dual Graph Contraction

Let us consider an example and discuss the dual contraction of a graph by means of this example. The complete formalism will follow.

1. Example: Connected Components of an Image

Consider a simple example (Fig. 5): Pixel gray values become the attributes of the corresponding vertices of the base graph G_0 and are illustrated by the color with which the circle corresponding to the vertices are filled (Fig. 5b). Two vertices are connected by an edge (a line segment in Fig. 5b) if the two corresponding vertices share a boundary segment. The four dif-

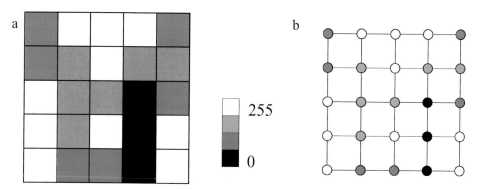

Fig. 5. Image to attributed relational graph (ARG) conversion: *a* a 5 × 5 gray level image, *b* the corresponding base graph G_0

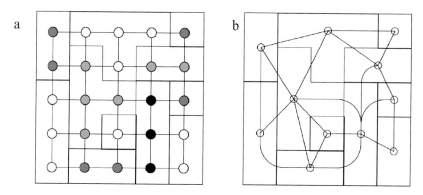

Fig. 6. Derivation of the region adjacency graph (RAG) of the connected components. *a* The 11 connected components of G_0. *b* The region adjacency graph (RAG) G_2

ferent gray values/colors of the 25 pixels form 11 groups of adjacent pixels all having the same gray value: they are called the connected components (Fig. 6). Our processing goal is to represent each such group by one vertex and to connect two vertices if the corresponding pixel groups share at least one boundary segment (see Fig. 6b).

In order to derive the smaller graph G_2 from G_0 we apply several contraction operations to the graph until the final result is reached. The primitive operation of dual graph contraction is the contraction of an edge $e = (v, w)$. It consists of the identification of the two end vertices and the removal of the edge. We can choose v

to 'survive' and substitute all appearances of w in any of the edges by v. One can visualize the process dynamically by

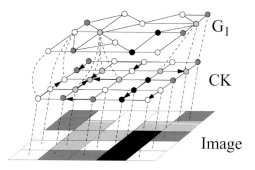

Fig. 7. Contraction kernels (CK) at the base level G_0 and the contracted graph G_1

moving w along e into v while stretching all the edges attached to w.

If we select only edges for contraction the end points of which have the same gray value, the connected components will shrink but all the connections between different gray values will be preserved. In Fig. 7 the selected edges are marked by an arrow pointing towards the surviving vertex. All the edges that contract into the same surviving vertex are called 'contraction kernel'. If these kernels do not form cycles (e.g. each is a small tree) all contractions can be executed simultaneously.

Some of the surviving vertices may become multiply connected by multi-edges. In most cases only one of the multi-edges is needed (Fig. 8). But in some cases (Fig. 9) the multi-edge represents relevant information: when the two regions meet along two or more distinct boundary segments. In all previous cases an edge between two surviving vertices was associated with one connected boundary segment between the two regions. If we would remove one of the two edges in

Fig. 9c this useful topological property would be lost. If we continue contracting one of the double edges we end up with a self-loop $e = (v, v)$ where both end points are identical. As we can see intuitively in Fig. 9d its removal would again destroy the above property. In addition we would loose the fact expressed by the self-loop that the inner region is completely surrounded by the other region. We conclude that both multiple edges and self-loops may be necessary to preserve the topological properties of the surviving components of the graph.

But how do we decide whether an edge is redundant or not? For this purpose we consider the dual graph. It can be constructed by placing a new (dual) vertex (depicted by a small square in Fig. 10) into each face of the drawn graph and connecting two new vertices across the edge of the primal graph which separates the two faces. For the pixel array this graph consists of the centers of all 2×2 blocks and the boundary segments separating two pixels (Fig. 10b). The contraction of an edge of the primal graph

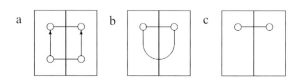

Fig. 8. Redundant double edge: a CK, b contracted, c simplified

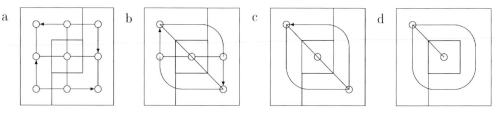

Fig. 9. Non-redundant double edge and self-loop: a $G_0 + CK_{01}$, b $G_1 + CK_{12}$, c $G_2 + CK_{23}$, d G_3 with self-loop

merges the two pixels corresponding to the edge's end points. Consequently the separating boundary segment e.g. the edge of the dual graph has to be removed. Edge contraction corresponds to edge removal in the dual graph. This allows to maintain duality between the dual graphs after 'dual graph contraction'. With the observation that the number of boundary segments of a face corresponds to the degree of the dual vertex let us reconsider the cases of multi-edge and of self-loop: the degree of the face bounded by the double edge in Fig. 8b is obviously two, the two faces between the double edges in Fig. 9c have both degree three. A similar observation holds for the self-loop: the face inside a removable self-loop has degree one, non-removable self-loops are characterized by an inside face of degree three (Fig. 9d) or higher.

A close look at the result of our first contraction (Fig. 7) shows that not all connected components have been contracted into a single vertex. We therefore repeat the selection of contraction kernels on graph G_1 and find four more edges to contract (Fig. 11). The resulting pyramid has three levels G_0, G_1, G_2 and yields at the apex the RAG of the original image. The dashed vertical lines in Figs. 7 and 11 indicate the correspondences between the vertices across the different levels. Each vertex at the top level corresponds to a connected set of vertices in the level below. Each of those corresponds to another connected set of vertices in the level below and so on. The union of all the sets of vertices in the base level that are derived from one vertex at the top level form the receptive field of this vertex.

2. Formal Definition

Figure 12 summarizes the two basic steps: dual edge contraction and dual face contraction. The base of the pyramid

consists of the pair of dual image graphs $(G_0, \overline{G_0})$. A new (reduced) level $i+1$ is computed from level i by

1. selecting the contraction kernels (S_i, $N_{i,i+1}$),
2. dually contracting the selected edges, and
3. removing redundant multi-edges and self-loops (dual face contraction).

Connectivity of surviving vertices is preserved if the contraction kernels satisfy the following definition of an irregular pyramid (see also Kropatsch 1994 [Def.5]):

Definition 1. *In a pair of dual image graphs $(G_i(V_i, E_i), \overline{G_i}(\overline{V_i}, \overline{E_i}))$, following* **decimation parameters** *(S_i, $N_{i,i+1}$) determine the contracted graphs $(G_{i+1}, \overline{G_{i+1}})$: a subset of* **surviving vertices** *$S_i = V_{i+1} \subset V_i$, and a subset of* **primary non-surviving edges**[4] *$N_{i,i+1} \subset E_i$. The decimation parameters ($S_i, N_{i,i+1}$) must be a subgraph of G_i and do not contain any circuit, e.g. ($S_i, N_{i,i+1}$) is a forest. The relation between the two pairs of dual graphs, $(G_i, \overline{G_i})$ and $(G_{i+1}, \overline{G_{i+1}})$, as established by dual graph contraction with decimation parameters ($S_i, N_{i,i+1}$) is expressed by function $C[.,.]$:*

$$(G_{i+1}, \overline{G_{i+1}}) = C[(G_i, \overline{G_i}), (S_i, N_{i,i+1})] \quad (1)$$

The connected components of the decimation parameters are called contraction kernels.

B. The Graph Pyramid and Equivalent Contraction Kernels

A contraction kernel collects all edges that can be contracted independently of each other without destroying the connectivity structure of the graph. Since the contraction operation is forbidden for self-loops the set of edges involved in

---

[4] Secondary non-surviving edges are removed during dual face contraction.

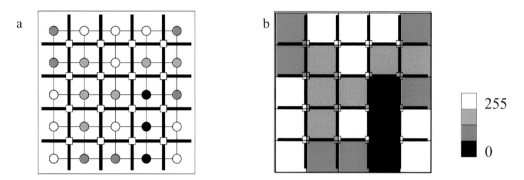

Fig. 10. The role of the dual graph $\overline{G_0}$: *a* the dual graphs $(G_0, \overline{G_0})$, *b* the dual graph $\overline{G_0}$ on the image

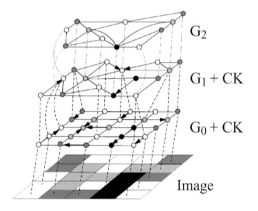

Fig. 11. The pyramid

such a sequence of contractions must not contain a circuit. Thus the set of edges involved in such a contraction may be encoded by a tree, or, a collection of non-overlapping trees spanning the given input vertices: a *spanning forest*.

Repeated dual graph contraction builds a stack of successively smaller graphs: the graph pyramid. Let us denote the contraction of a graph G_0 by a contraction kernel $N_{01} \subset E_0$ by $G_1 = G_0/N_{01}$ and the subsequent contraction by $G_2 = G_1/N_{12}$. Then there exists an *equivalent contraction kernel* $N_{02} \subset E_0$ that creates the same result in a single step: $G_2 = G_0/N_{02}$. Our

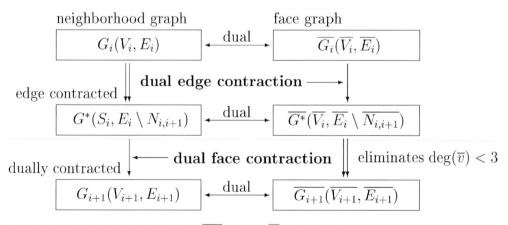

Fig. 12. Dual Graph Contraction: $(G_{i+1}, \overline{G_{i+1}}) = C[(G_i, \overline{G_i}), (S_i, N_{i,i+1})]$

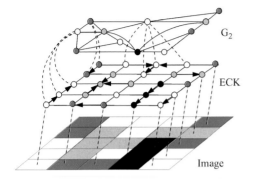

Fig. 13. Equivalent contraction kernels (ECK) contract G_0 directly into G_2

example pyramid in Fig. 11 was built using two contraction levels. The same top level graph can be derived in a single step using equivalent contraction kernels as shown in Fig. 13. Note that some kernels contain two edges to be contracted one after the other. Consequently more parallel steps are needed to compute the result. Conversely, a contraction kernel may be decomposed into two smaller ones. The successive application of the resulting contraction kernels produces the same result as the application of the initial one. The complete formalism is described in (Kropatsch 1997a).

Equivalent contraction kernels relate the data at the base level directly with any higher level in the graph pyramid

and can be seen as the receptive field of the corresponding pyramidal cell. This property has been used in Fig. 14 to show some of the 'internal' structures of the graph pyramid on the famous picture of (Bister et al. 1990) (Fig. 14a). Some of the top levels have been down-projected to the base graph by means of their equivalent contraction kernels. Since the receptive field corresponding to a lower level vertex is always included in the receptive field of its parent vertex also the boundaries of the receptive fields form such an inclusion hierarchy. Figure 14b shows the top level contours which follow the boundary between black and white in the original image but also some internal kernels which have been chosen stochastically in absense of any other control.

Since also the contents of a pyramidal cell is the result of applying a series of reduction functions there is also the equivalent way to compute the same value directly from the base. Although this is not more efficient than the iteration through the pyramid levels it offers the possibility to invert this process: given a function to compute a value for a complex decision how can this computation be decomposed into a series of local reduction functions which can be efficiently computed using the pyramid? It is clear that not all functions can be split up in such a way, but there are many that offer this

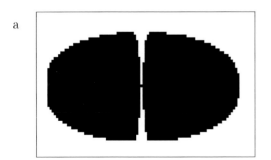

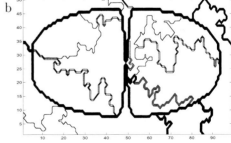

Fig. 14. Visualizing the graph pyramid. *a* binary image, *b* boundaries of some receptive fields

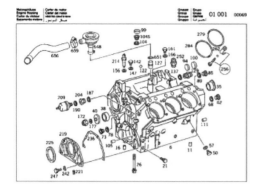

Fig. 15. Scanned technical drawing analyzed by graph pyramid

possibility and are subject of current research.

C. The Domain of All Graph Pyramids

Another question concerns the functionality of a given sensor arrangement if it is equipped with the flexibility of the interconnection network of the graph pyramid: Can I compute a given complex decision by adjusting the selection criteria and the reduction functions? What is the domain of all decision functions that can be computed?

These far reaching questions have two components: a structural component and a functional component. The structural component identifies all sensors that influence the function. The functional component involves the coding of the information and the definition of functions and methods transforming a structured set of values and codes into a more general value or code. The structural component can be studied through the concept of equivalent contraction kernel, the functional component may go beyond the scope of linear filters as in the equivalent weighting functions of (Burt and Adelson 1983) and may involve symbolic representations and learning concepts.

V. Conclusion

We summarize the conceptual framework which has been set up to perform dual graph contraction. Contraction is controlled by kernels that can be combined in many ways. Data and relations considered important for interpreting the visual input are selected to survive a repeated data reduction which, at the same time, preserves topological properties among the surviving parts. The hierarchy created in this way adapts its structure to the sensor data while integrative measures can be efficiently computed within the (sub-) hierarchy of the receptive field.

Preliminary experiments with graph pyramids have been successful in several areas, e.g., connected component labeling (Macho and Kropatsch 1995); segmentation (Kropatsch and BenYacoub 1996); '2x on a curve' (Kropatsch 1997b); line images like that shown in Fig. 15 (Burge and Kropatsch 1999); matching (Pailloncy et al. 1998); isolating moving objects from background (Kropatsch 1999); generalization preserving monotonic landscape properties (Glantz et al. 1999).

Graph pyramids have been introduced as an efficient model for visual information processing. Efficiency in graph pyramids is achieved through vertical pathways connecting the apex with the sensors in the base through the pyramidal hierarchy across the

$$\log\left(diameter(\text{base graph})\right)$$

number of (horizontal) levels (Kropatsch 1997b).

References

Bister M, Cornelis J, Rosenfeld A (1990) A critical view of pyramid segmentation algorithms. Pattern Recognition Letters, 11(No. 9), pp 605–617

Burge M, Kropatsch WG (1999) A minimal line property preserving representation of line images. Computing, Devoted Issue on Image Processing, 62, pp 355–368

Burt PJ, Adelson EH (1983) The Laplacian pyramid as a compact image code. IEEE Transactions on Communications Vol. COM-31(No.4), pp 532–540

Gareth AJ, Singerman D (1978) Theory of Maps on Orientable Surfaces. Vol. 3. London Mathematical Society

Glantz R, Englert R, Kropatsch WG (1999) Representation of image structure by a pair of dual graphs. In: Kropatsch WG, Jolion J-M (eds), 2nd IAPR-TC-15 workshop on graph-based representation. Österreichische Computer Gesellschaft, OCG-Schriftenreihe Band 126, pp 155–163

Jolion J-M, Kropatsch WG (eds) (1998) Graph Based Representations in Pattern Recognition. Computing, Suppl 12. Springer-Verlag, Wien New York

Jolion J-M, Rosenfeld A (1994) A Pyramid Framework for Early Vision. Kluwer Acad Publ Dordrecht

Jolion J-M, Kropatsch WG, Vento Mario (eds) (2001) Graph-based Representations in Pattern Recognition, GbR 2001. CUEN. ISBN 88 7146 579–582

Kropatsch WG (1994) Building irregular pyramids by dual graph contraction. Tech rep PRIP-TR-35. Institute f. Automation 183/2, Pattern Recognition and Image Processing Group, TU Wien, Austria. Also available through http://www.prip.tuwien.ac.at/ftp/pub/publications/trs/tr35.ps.gz

Kropatsch WG (1997a) Equivalent contraction kernels to build dual irregular pyramids. Advances in Computer Vision, pp 99–107, Springer series Advances in Computer Science

Kropatsch WG (1997b) Property preserving hierarchical graph transformations. In: Arcelli C, Cordella LP, Sanniti di Baja G (eds) Advances in Visual Form Analysis. World Scientific Publishing Company, pp 340–349

Kropatsch WG (1999) How useful is structure in motion? In: Chetverikov D, Szirányi T (eds) Fundamental Structural Properties in Image and Pattern Analysis 1999. Österreichische Computer Gesellschaft, OCG-Schriftenreihe Band 130, pp 35–45

Kropatsch WG, BenYacoub S (1996) Universal segmentation with pIRRamids. In: Pinz A (ed) Pattern Recognition 1996, Proc of 20th ÖAGM workshop R. Oldenburg, OCG-Schriftenreihe, Band 90, pp 171–182

Kropatsch WG, Brun L (2000) Hierarchies of combinatorial maps. In: Svoboda T (ed) CPRW2000, Proceedings of the Czech Pattern Recognition Workshop. Peršlák, CZ: Czech Pattern Recognition Society Praha. ISBN 80-238-5215-9, pp 131–137

Kropatsch WG, Jolion J-M (eds) (1999) 2nd IAPR-TC-15 Workshop on Graph-based Representation. Österreichische Computer Gesellschaft Band 126

Lienhardt P (1989) Subdivisions of n-dimensional spaces and n-dimensional generalized maps. In: Mehlhorn K (ed) Proceedings of the 5th Annual Symposium on Computational Geometry (SCG '89). Saarbrücken, FRG: ACM Press, pp 228–236

Macho H, Kropatsch WG (1995) Finding connected components with dual irregular pyramids. In: Solina F, Kropatsch WG (eds) Visual Modules, Proc of 19th ÖAGM and 1st SDVR workshop R. Oldenburg, OCG-Schriftenreihe Band 81, pp 313–321

Marr D (1982) Vision. WH Freeman, San Francisco

Pailloncy J-G, Kropatsch WG, Jolion J-M (1998) Object matching on irregular pyramid. In: Jain AK, Venkatesh S, Lovell BC (eds) 14th International Conference on Pattern Recognition, Vol. II. IEEE Comp Soc, pp 1721–1723

Pizlo Z, Salach-Golyska M, Rosenfeld A (1997) Curve detection in a noisy image. Vision Res 37(9): 1217–1241

Rosenfeld A, Nakamura A (1997) Local deformations of digital curves. Pattern Recogn Lett 18: 613–620

Thulasiraman K, Swamy MNS (1992) Graphs: Theory and Algorithms. Wiley-Interscience, New York, USA

Uhr L, Schmitt L (1984) The several steps from icon to symbol using structured cone/pyramids. In: Rosenfeld A (ed) Multiresolution Image Processing and Analysis. Springer Verlag, Berlin Heidelberg New York Tokyo, pp 86–100

Chemosensors and Chemosensing

21. Mechanisms for Gradient Following

David B. Dusenbery

Abstract

Many organisms move up or down stimulus gradients to a more favorable environment. Investigation over more than a century has revealed that a variety of mechanisms are used by different organisms. There are tradeoffs among these strategies between the number of sensors, the movements required of the searcher, and physical constraints on its orientation. Recently, theoretical analysis and computer simulation have been used to explore these tradeoffs quantitatively. This information may be useful in the design of autonomous vehicles that might also follow stimulus gradients.

I. Introduction

Biologists have long known that many single-cell organisms and small animals can guide their locomotion up or down spatial gradients of chemicals or other stimuli. With expanding development of robotic vehicles, including very small devices, engineers may be interested in the mechanisms evolution has discovered for accomplishing this task (Ferrée and Lockery 1999). Over the last century, researchers have both identified a variety of possible solutions and determined that different organisms employ different solutions, requiring different sensor capabilities (Schöne 1984: 34–44, 57–69, Dusenbery 1992: 413–431). This observation leads to the question of whether these variations in behavior correspond to differences in physical constraints on the organisms (given their size, shape, and basic mechanisms of locomotion), to differences in biological constraints on possible mechanisms of control (given molecular and cellular mechanisms), or to differences in chance events during their evolutionary history. Addressing this question has led to several quantitative analyses and exploration by computer simulation, which will be summarized here. In contrast, to the general strategies discussed here, others have focused on detailed studies of particular organisms (Tranquillo 1987, Tranquillo et al. 1988, Tranquillo 1990, Ferrée and Lockery 1999).

One basic distinction is important to make at the outset. Organisms that swim through open water face a much different situation than those moving across a surface. The former are faced with a

three-dimensional (3D) orientation problem, while the latter only have to deal
with two dimensions. Reduction of dimensionality can greatly simplify any
search problem (Adam and Delbrück
1968). In addition, the searcher moving
across a surface can attach to it and maintain its orientation indefinitely (assuming
the stimulus gradient is fixed in relation
to the surface). In contrast, the orientation
of the 3D searcher will eventually be lost
(due to Brownian motion or other distur-
bance) unless the searcher has access to
another stimulus that can be used as a
reference or a force that maintains orientation. Stimuli serving the specialized
function of providing a directional reference have been called *collimating stimuli*
(Pline and Dusenbery 1987), but their actual use has not yet been clearly demonstrated. Consequently, the following analysis will assume that collimating stimuli
are not available and 3D searchers face
limits on how much time is available to
measure intensity differences.

It should be pointed out that, in the
natural world, turbulent flows are ubiqui-
tous at size scales greater than a centi-
meter, and chemical gradients are smooth
over integration times of seconds or less
only at smaller distances (Dusenbery
1992: 76–87, Karp-Boss et al. 1996). Thus,
the organisms considered here are microscopic.

For convenience, the following basic
concepts and assumptions are useful.
For chemical stimuli, the stimulus intensity is the concentration (C) of the stimulus chemical. Position (x) is positive in the
direction of increasing C. For generality,
stimulus gradients (dC/dx) are usually
normalized to intensity, forming the *relative gradient*, $G \equiv C^{-1}(dC/dx)$, or its reciprocal, the *gradient decay length*, $L \equiv C
(dC/dx)^{-1}$. An exponential gradient is
$C(x) = e^{Gx} = e^{x/L}$. The efficiency of following a gradient is defined simply as
$E \equiv (x_{End} - x_{Begin})/\text{pathlength}$.

II. Gradient Detection

Figure 1 shows the basic situation for gradient detection. Two measurements of
stimulus intensity (concentration) are
made at two different positions, and a determination must be made as to which intensity is larger. This determination is limited by noise in measuring intensity. For
accurate determinations, the noise must
be less than the difference in intensity.
The signal-to-noise ratio (S/N) is a convenient parameter, where the signal is defined as the difference in intensity and
noise is a measure of dispersion of individual measurements around the mean value (e.g. the standard deviation). The situation is improved by increasing the signal, which is proportional to the distance

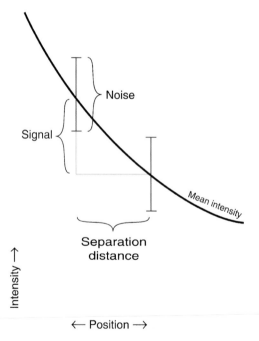

Fig. 1. The gradient detection problem. A minimum of two sensors, at two different positions
along a stimulus gradient, are required to determine the direction of the gradient. Also, the difference in intensity (signal) must exceed the measurement noise

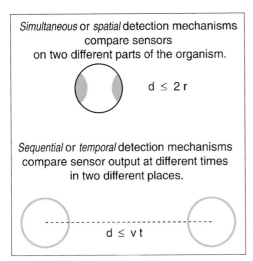

Simultaneous or spatial detection mechanisms compare sensors on two different parts of the organism.

$$d \le 2r$$

Sequential or temporal detection mechanisms compare sensor output at different times in two different places.

$$d \le vt$$

Fig. 2. The two basic mechanisms for gradient detection. Shaded (gray) regions represent sensor positions. d is the distance separating measurement positions. r is the radius of the smallest sphere surrounding the organism. v is the speed of movement of the sensor. *Simultaneous* or *spatial* detection mechanisms compare different sensors in different parts of the organism, requiring that the at least two sensor outputs be kept separate. *Sequential* or *temporal* detection mechanisms compare the output of a sensor at different times, between which it has moved from one position to another, requiring some form of memory

separating the measurement positions as well as the steepness of the gradient. Distance is increased by one of two alternative strategy types (Fig. 2).

The size of the organism may be increased so that widely-spaced sensors may make *simultaneous* measurements, and a *spatial* comparison is provided. Some animals place sensors on long, thin appendages (antennae) and obtain a large separation without making the whole animal larger. In this case, the separation distance cannot exceed the diameter of a sphere enclosing the organism.

Another way of increasing separation is to move a single sensor from one position to another. In this case, the organism makes *sequential* measurements and a

temporal comparison, requiring some form of memory. This mechanism may be accomplished either by moving part of the body or the whole body. In either case, the separation distance cannot be more than the product of the speed of movement and the time between measurements.

Alternatively, performance can be improved by decreasing sensory noise, which generally can be accomplished by averaging the measurements over longer time intervals or using a larger sensor (Dusenbery 2001b). At small size scales and low concentrations, diffusion of individual molecules to the sensor causes a fundamental lower limit on sensory noise, and this can be calculated for spherical sensors (Berg and Purcell 1977, Dusenbery 2001b) as $S/N \le (\pi rDCt)^{1/2}$, where r is the radius of the sphere, D is the diffusion coefficient of the sensed molecule in the environment, C is its numerical concentration (number of molecules per unit volume), and t is the time over which the signal is integrated. Any sensor that fits within the sphere of radius r can do no better without flow of the medium. D is relatively constant and not amenable to increase.

Size or time can be increased, but costs are associated with both. Greater size requires more material to construct, more energy for locomotion, and more visibility to enemies. Greater measurement time may cost speed in moving along the gradient. In addition, 3D searchers will find that extending integration time beyond some limit is counter-productive because they will be averaging signals from different positions as their orientation is perturbed in unknown ways. These limits are estimated in studies described below.

III. Unconstrained: Movement in Three Dimensions

When suspended in open water, without constraints, any small object is subject to

Table 1. Constraints on size for gradient following. In the constraint formulas, the parameters are grouped so that the term on the left is the signal-to-noise ratio, the first term to the right of the equal sign involves hydrodynamics, the second term involves stimulus intensity, the third term involves gradient decay length (L) and/or relative swimming speed (u), and the right-most term involves the size of the organism (radius r). See Dusenbery (1997) for definitions of other symbols. The right-hand column gives the minimum diameter an organism can have and usefully employ the indicated mechanism, assuming best estimates of parameter values in Dusenbery (1997, Table 1) and $S/N = 1$ for useful performance

Stimulus	Mechanism	Constraint formulas	$2r$, μm
Chemical	Spatial	$\frac{S}{N} = \left(\frac{4\pi\eta}{kT}\right)^{1/2}(2\pi DC)^{1/2}\left(\frac{3}{L}\right)r^3$	0.58
Chemical	Temporal	$\frac{S}{N} = \left(\frac{4\pi\eta}{kT}\right)^{3/2}(2\pi DC)^{1/2}\left(\frac{u}{L}\right)r^6$	0.65
Light	Spatial	$\frac{S}{N} = \left(\frac{4\pi\eta}{kT}\right)^{1/2}(2\pi If)^{1/2}\left(\frac{2}{L}\right)r^{7/2}$	1.77
Light	Temporal	$\frac{S}{N} = \left(\frac{4\pi\eta}{kT}\right)^{3/2}(4\pi If)^{1/2}\left(\frac{u}{L}\right)r^{13/2}$	1.05
Light	Direction	$\frac{S}{N} = \left(\frac{4\pi\eta}{kT}\right)^{1/2}(\pi If)^{1/2}\alpha r^{7/2}$	1.24
Temperature	Spatial	$\frac{S}{N} = \left(\frac{4\pi\eta}{kT}\right)^{3/4}\left(\frac{(4\pi H_T)^3}{k^2 H_c}\right)^{1/4}\left(\frac{2}{L}\right)r^{13/4}$	0.74
Temperature	Temporal	$\frac{S}{N} = \left(\frac{4\pi\eta}{kT}\right)^{7/4}\left(\frac{(4\pi H_T)^3}{k^2 H_c}\right)^{1/4}\left(\frac{u}{L}\right)r^{25/4}$	0.69

Brownian motion, which alters its orientation as well as its position, leading to rotational and translational diffusion. This random motion is probably undetectable internally and can be thought of as a kind of noise in the motor output of a swimming organism. In the cases of interest, motion is at low Reynolds number, and this motion can be rigorously calculated for simple shapes. Such calculations allow estimation of the maximum time the organism can take for integrating a sensor's signal before the organism is likely to have rotated through such a large angle that the old signal is no longer useful. These constraints are developed in the next two subsections.

A. Minimum Size

Taking advantage of the relations for rates of translational and rotational diffusion of spheres to estimate motor noise, formulas have been derived for estimating the signal-to-noise ratio for detecting gradients of light, chemicals and heat by spherical organisms (Table 1) (Dusenbery 1997).

These results lead to the surprising conclusion that there exists a very sharp size limit (about half a micrometer diameter) below which free bacteria have no use for motility. A review of published literature indicated that the prediction is supported, and about 20% of free, non-motile genera are below this size limit but the smallest of 97 motile genera was about 0.8 micrometers in length (Dusenbery 1997). This sharp limit is primarily due to the fact that the rotational diffusion rate is proportional to the third power of the radius.

Consequently, if autonomous vehicles smaller than this limit are to follow stimulus gradients, they would require physical constraints, collimitating stimuli, or higher energy density for propulsion than bacteria.

B. Optimal Shape

Perrin (1934, 1936) derived expressions for the motion of ellipsoids subject to

Brownian motion. Unfortunately, they involve definite integrals that have no solution in common functions. Approximations for high axial ratios have been much used to estimate the shapes of molecules (Tanford 1963, van Holde 1985) but do not apply to axial ratios less than 5. With computers it is now easy to evaluate the integrals, and these relations have been used to quantitatively evaluate the effects of constant-volume changes in shape (including small axial ratios) on several possible functions of altered shape (Dusenbery 1998a).

For rod-like shapes, increasing the axial ratio beyond two (as have most motile bacteria) *increases* hydrodynamic drag, because at low Reynolds number the increase in surface area increases drag more than the decrease in cross section decreases drag. Thus, drag reduction (streamlining) is not a viable hypothesis for why many bacteria are rod-shaped.

However, the results shown in Fig. 3 reveal that elongation is advantageous to gradient detection. A temporal comparison mechanism has the highest S/N for rod-like shapes moving parallel to the long axis; within the limits considered, S/N is increased 647-fold (which allows orientation to 647-fold shallower gradients). A fore/aft spatial comparison also has the highest S/N for locomotion parallel to the long axis; but S/N is increased by the lesser value of 96-fold. Lateral spatial comparison has the highest S/N for locomotion perpendicular to the long axis; and the S/N is increased only 12-fold. These results explain why bacteria 1) commonly have rod-like shapes, 2) swim parallel to their long axis, and 3) employ a temporal mechanism of comparison, although temporal comparison is not necessarily superior to spatial comparison for spherical cells (Dusenbery 1998b).

An obvious advantage of the rod-like shapes for spatial comparison is that it can increase the spatial separation of the receptors and thus the difference in stimulus intensity between them, increasing the signal. The less-obvious advantage is that elongation of a constant-volume shape strongly reduces the rate of rotational diffusion caused by Brownian motion, and this increases the time available to measure concentration gradients, decreasing noise. In all these cases, an elongated shape provides large benefits in detecting gradients, but the effect is largest for temporal comparison.

This analysis suggests that a small autonomous vehicle for following gradients should have an elongated shape and employ a temporal comparison mechanism.

C. Helical Movement

Any organism (or vehicle) propelling itself unconstrained in three dimensions without guidance is likely to follow a helical path (Purcell 1976). If the thrust is not aligned with the center of drag, the thrust will generate rotational motion, and helical motion occurs unless the direction of translation is exactly parallel or perpendicular to the axis of rotation (Crenshaw and Edelstein-Keshet 1993). In fact, many swimming microorganisms and cells are observed to swim in helical paths (Jennings 1901, Foster and Smyth 1980).

If the axis of the helix is anything but parallel to a stimulus gradient, a sensor on the organism will experience an oscillation of stimulation as the organism moves around the axis of the helix. This movement thus provides a mechanism for sampling the stimulus gradient, in a way analogous (in two dimensions) to strategy 2 of Section IV. The signal from the sensor could then be used to alter the direction of the axis of the helix, and net movement is along this axis.

Although helical swimming has been known for over a century, the difficulties

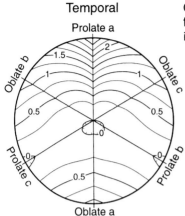

Temporal

Prolate a

Oblate b / Oblate c / Prolate c / Prolate b / Oblate a

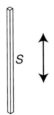

Contours are proportional to log S/N
for orientation to gradients by a temporal mechanism,
in which present simulation is compared to past stimulation.

Best sensitivity occurs with prolate or rod-like shapes
swimming parallel to their long axis.

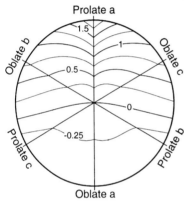

Fore / aft

Prolate a

Oblate b / Oblate c / Prolate c / Prolate b / Oblate a

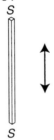

Contours are proportional to log S/N
for orientation to gradients by comparing stimulation of
leading and trailing edges.

Best sensitivity occurs with prolate or rod-like shapes
swimming parallel to their long axis.

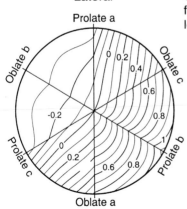

Lateral

Prolate a

Oblate b / Oblate c / Prolate c / Prolate b / Oblate a

Contours are proportional to log S/N
for orientation to gradients by comparing stimulation
left and right of swimming direction (ends of b axis).

Best sensitivity occurs with prolate or rod-like shapes
swimming perpendicular to their long axis and
comparing stimulation at the ends of the long axis.

of recording motion in 3D and analyzing motion with six degrees of freedom has discouraged study of this behavior. However, Hugh Crenshaw and associates have recently developed a rigorous mathematical analysis of helical motion (Crenshaw 1993a, b, Crenshaw and Edelstein-Keshet 1993). They find that the direction of the helical axis is largely determined by the direction of the axis of rotation of the organism; changes in the direction of the axis of rotation with respect to the organism's body cause changes in the direction of the axis of the helix. Furthermore, they demonstrate by computer simulation that the axis of helical movement is rotated into alignment with the axis of a stimulus gradient under a wide variety of conditions. It is only required that the angle between the direction of translation and the axis of rotation be modulated by stimulus intensity. For example, orientation to a stimulus gradient occurs if the rate of rotation around its body axis is proportional to the stimulus intensity encountered by a sensor. This mechanism provides a robust method of orientation to a stimulus gradient, without any constraints or other stimuli.

Consequently, helical motion is an attractive strategy for larger vehicles to employ in moving through 3D space. It would not be effective at scales so small that Brownian motion was influential.

IV. Confined to a Surface: Movement in Two Dimensions

Organisms frequently move over the surface of a much larger body to which the stimulus gradient is fixed. This situation is advantageous in that the organism can hold its position and orientation indefinitely by attachment to the surface, eliminating any constraint on integration time from loss of orientation. A variety of known and possible strategies for following gradients in this circumstance have recently been simulated and compared (Dusenbery 2001a).

Five strategies were simulated:

1. A biased random walk, with one sensor controlling the rate of change of direction (*sequential* sampling, *temporal* comparison, providing *indirect* response to the gradient, since the response is not biased in the direction of the gradient, a case called *klinokinesis*).
2. A single sensor, with constant rotation, controlling forward speed (*sequential* sampling, *temporal* comparison, providing *direct* response to the gradient, a case of *klinotaxis*).
3. A single sensor, with oscillating rotation, controlling forward speed and rotation (*sequential* sampling, *temporal* comparison, providing *direct* response to the gradient, another case of *klinotaxis*).

Fig. 3. Contour plots of performance for equal-volume ellipsoids of all shapes, with axes a, b, c. All possible equal-volume ellipsoidal shapes not too distorted from spherical are represented in these plots. The sphere (a = b = c) is at the center, and the distance from the center is proportional to the negative log of the minimum radius of curvature occurring in each ellipsoid compared to the radius of the equal-volume sphere. At the outer edge of the plots, the ellipsoides are most distorted from spherical and have a minimum radius of curvature of 1% that of the sphere. The three axes of the plots encompass shapes in which two axes of the ellipsoid are identical (ellipsoids of revolution). At one end of each axis, prolate ellipsoids, with semiaxes equal to some permutation of $(10, 10^{-1/2}, 10^{-1/2})$, resemble rods, with axial ratios of 32; and at the opposite end oblate ellipsoids, with semiaxes $(10^{-4/5}, 10^{2/5}, 10^{2/5})$ resemble disks with axial ratios of 0.063. Swimming is along the a-axis. Contours are proportional to log S/N, normalized to an equal-volume sphere. For clarity, boxes are illustrated that are just large enough to contain the optimal ellipse. Data from Dusenbery (1998a)

4. A bilateral pair of sensors, controlling rotation (*spatial* sampling, *simultaneous* comparison, providing *direct* response to the gradient, a case of *tropotaxis*).
5. Several sensors, evenly distribution around the periphery, controlling the direction of locomotion without rotation (*spatial* sampling, *simultaneous* comparison, providing *direct* response to the gradient, another case of

tropotaxis). Alternatives with 3 or 12 sensors and ability to move in any direction were simulated.

Each strategy was represented by two variants. One had a continuous and the other a discontinuous input-output relationship. Optimized values of the specific parameters in each strategy were used. Noise and biases in sensory inputs and motor outputs were simulated by choosing

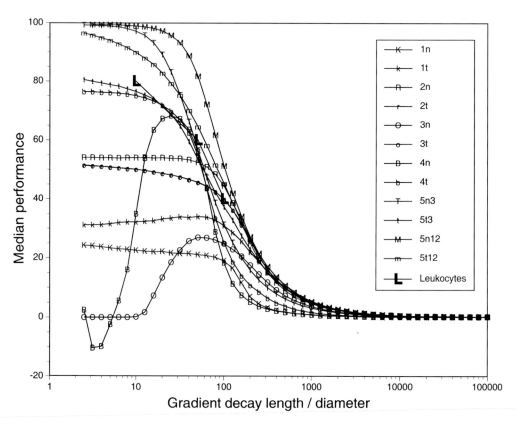

Fig. 4. Performance (efficiency) in following 2D gradients. Gradient steepness declines to the right. $S/N = 1$ at a gradient decay length of 100 searcher diameters. The gradient decay length is $C \, (dC/dx)^{-1}$, where C is concentration or other measure of stimulus intensity. Simulation data from (Dusenbery 2001a). In the graph legend, the initial digits refer to the strategies listed in the text. The following letter signifies whether the input-output relation was continuous, without a threshold (n) or discontinuous with a threshold (t). For strategy 5, the number of sensors (3 or 12) is also indicated. Data on the observed performance of leukocytes (Zigmond 1977) is included (L)

factors from a normal distribution of mean 1 and standard deviation 0.01, which placed performance in the ballpark of observed results for crawling animal cells (Zigmond 1977).

The simulation results are shown in Fig. 4. At the gradient decay length where $S/N = 1$ ($L = 100$ searcher diameters), efficiency varied over the range 18–52%. Maximum performance for the different models ranged from 24 to over 99% efficiency. The 1% efficiency threshold fell in the gradient decay range of 800 to 5,000 searcher diameters. No correlation was found between performance in steep and shallow gradients; there is no trade off for good performance in one circumstance versus the other. No clear advantage of continuous or discontinuous input-output relationships was found.

Perhaps the most surprising result of these simulations is that spatial comparison with translation in any direction was such a distinctly superior strategy, although it did employ more receptors. The two models with the best threshold response were the two models of spatial comparison with translation in any direction and the most receptors (5n12 and 5t12). The models with the best performance in steep gradients were the four models of spatial comparison with translation in any direction. In addition, these models were the most robust in that they had the fewest parameters to optimize and performed well over the widest range of gradients.

V. Magnetic Bacteria Reduce Three Dimensions to One

In 1975, a remarkable discovery was made that certain bacteria isolated from sediments contain magnets (Blakemore 1975). As a consequence, the bacteria align with the earth's magnetic field and swim along the local field lines. They have been termed *magnetotactic bacteria*. However, that term is misleading in that, in the context of locomotion, *taxis* suggests a stimulus-response mechanism, and it has clearly been demonstrated that the alignment is a passive process occurring even in dead cells (Blakemore 1982). The bacteria respond by the same mechanism as a compass needle. The magnets and earth's magnetic field provide sufficient force to overcome the forces of Brownian motion, and the bacteria are predominately aligned with the magnetic field.

One of the many interesting questions this discovery raises is: what is the adaptive value of the magnets to the bacteria that possess them? It has been argued that magnetic orientation helps the bacteria swim downward away from open water and the high oxygen levels that are toxic to them (Blakemore 1982). Supporting this view is the observation that bacteria isolated from the Southern hemisphere have the opposite magnetic polarity as those isolated in the Northern hemisphere. However, it does not explain why magnetic bacteria are found at the geomagnetic equator where the field lines are horizontal (Frankel et al. 1981). A likely explanation is provided by suggestions (Blakemore 1982, Spormann and Wolfe 1984) that magnetic orientation may allow them to follow chemical gradients more efficiently. The magnets appear to reduce the movements of the bacteria from three dimensions to one. I have used computer modeling to provide a quantitative estimate of the potential size of the benefits to a one-micrometer searcher.

Behavior was simulated using a model (Dusenbery 1989a) modified for three dimensions and using parameter values based on observed behavior of *E. coli* bacteria (Berg and Brown 1972, Block et al. 1982). The effects of Brownian motion on the orientation of an unconstrained searcher in 3D was approximated by assuming at each time step (of 0.1 s) orientation was changed by a value picked from a normal distribution of mean 0 and standard deviation of 10°. The magnetic bacteria were assumed to have insignificant deviations. Simulations included 111 runs of 1000 steps with initial orientations randomly distributed in 3D. The rate of sensory adaptation was 0.1/step.

Efficiency as a function of gradient steepness is presented in Fig. 5. Efficiency increased as the gradient got steeper, up to a point. Rotation caused by Brownian motion reduced efficiency, but confinement to one dimension (as in magnetic bacteria) increased efficiency above that of movement in 3D without Brownian motion.

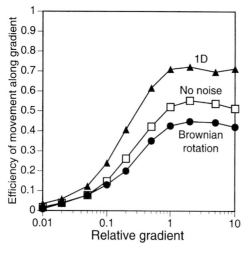

Fig. 5. Efficiency of gradient following with magnets confining movement to 1D. The relative gradient is $(1/C)(dC/dx)$, where C is concentration or other measure of stimulus intensity. Sensory noise is not included. No noise is without any deviations of the swimming path from the chosen direction. Brownian rotation is with deviations in path estimated to approximate the effects of Brownian motion. 1D is without deviations, as would occur if magnets constrained orientation to alignment with the earth's magnetic field and the gradient direction. Standard errors for all 30 points were in the range 0.006 to 0.022

This analysis supports the hypothesis that bacteria containing magnets can improve their ability to follow stimulus gradients. A more precise analysis would estimate the residual rotational diffusion from the strength of the magnets, and consider that the earth's magnetic field is not always aligned with the gradient. But this simple analysis clearly demonstrates the potential advantage of magnets, in at least some situations.

These observations may provide the engineer with novel ideas about how to improve performance in gradient following.

VI. The Importance of Sensory Adaptation

An almost universal feature of sensory systems in organisms is sensory adaptation. This is a decline in the response of the organism to steady intensities of stimulation. For chemical stimulation of small organisms, examples are well-described in bacteria (Segall et al. 1986) and the simple animals known as nematodes or round worms (Dusenbery 1980, Goode and Dusenbery 1985). An example is shown in Fig. 6. Sensory adaptation puts the emphasis on changing stimuli at the expense of maintaining information about intensity levels.

Bacteria (Armitage 1992a, b, 1999, Manson 1992) and nematodes (Ward 1973, Prot 1978, Diez and Dusenbery 1989b, Robinson 1995) both follow gradients of various kinds of stimuli and have been studied extensively. In addition to chemical stimuli, some nematodes are very sensitive to temperature gradients (Dusenbery 1988, Pline et al. 1988, Diez and Dusenbery 1989a), which they may use to navigate to particular soil depths (Dusenbery 1989b). In both these kinds of organisms, evidence has accumulated that the rate of adaptation is asymmetric,

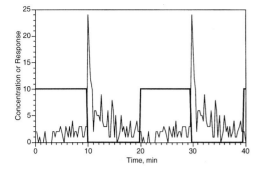

Fig. 6. Example of adaptation in a microscopic roundworm (nematode). Concentration of the stimulus (NaCl, attractant) varied cyclically in steps between 0 and 50 mM (thick line). The response is the number of reversal bouts initiated (thin line). The data are the sum for 45 cycles of stimulation. The plot repeats the sum in two successive cycles, in order to show the transitions clearly. Note an inhibition of the response after an increase in concentration. Data from Dusenbery (1980)

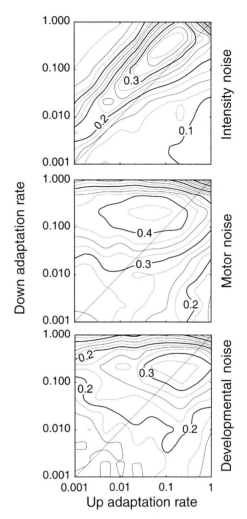

in that the rate differs between increases and decreases in stimulation (Dusenbery 1989c). Generally, adaptation is faster for stimulus changes toward less preferred conditions, which lead to increases in rates of change of direction. Such asymmetry may be seen in Fig. 6.

This observation led to the question of whether this asymmetry is beneficial in following gradients. The hypothesis was explored by computer modeling of gradient following in 2D by a sequential, temporal-comparison mechanism (like strategy 1 of Section IV, klinokinesis) (Dusenbery 1989c). The results indicate asymmetry is advantageous in some circumstances but not all (Fig. 7). The pattern of performance as the two rates varied independently depended on the type of noise assumed in the model. Three types of noise were simulated: intensity noise affected sensor outputs from one time to another, motor noise perturbed movement from one time to another, and developmental noise caused biases in the

Fig. 7. Contours are efficiency of movement along a gradient. Up and Down refer to changes in concentration of an attractant; a repellent stimulus is reversed. Each combination of rates had efficiency reproducible within about 0.01. The diagonal is the locus of equal adaptation rates. Note that the patterns are different with different kinds of noise, and they are not symmetrical around the diagonal in the first two cases. Data from Dusenbery (1989c)

tendency of each searcher to turn to one side. The results indicated that asymmetry in adaptation rates could improve performance with the first two types of noise and the optimum was in the same direction as observed in real organisms.

These results suggest that details of signal processing matter and optimal processing depends on details of movement tendencies as well as sensor performance.

VII. Conclusions

A wide variety of simple mechanisms allow an organism to move along a stimulus gradient. Evolution seems to have adopted different mechanisms in different cases. These differences probably have to do with different constraints imposed on the organisms by their environment and their basic body size, shape, and mode of locomotion. There are clearly many trade-offs between different mechanisms, but we have insufficient knowledge to understand these costs and benefits in any detail.

Nonetheless, we may extract some generalities:

- Larger size is advantageous. It allows greater separation of sensors for spatial comparison and slows rotational diffusion in 3D, allowing more time for signal averaging.
- More and larger sensors are advantageous. They acquire information faster and help reduce noise.
- Asymmetry in sensory adaptation may be advantageous.
- For 2D movement over a surface, spatial comparison with translation in any direction looks like a particularly good strategy.
- For 3D movement through open water, no mechanism is effective at sufficiently small size scales. Elongated shapes with temporal comparisons are most effective in the micrometer size range. Magnets can improve efficiency for small sizes. Helical motion is an effective strategy for larger sizes.

These studies from biology may help engineers in the design of small autonomous vehicles (Morse et al. 1998).

References

Adam G, Delbrück M (1968) Reduction of dimensionality in biological diffusion processes. In: Rich A, Davidson N (eds) Structural Chemistry and Molecular Biology. W.H. Freeman & Co., San Francisco, pp 198–215

Armitage JP (1992a) Bacterial motility and chemotaxis. Sci Prog 76: 451–477

Armitage JP (1992b) Behavioral responses in bacteria. Ann Rev Physiol 54: 683–714

Armitage J (1999) Bacterial tactic responses. Adv Microbial Physiol 41: 229–289

Berg HC, Brown DA (1972) Chemotaxis in *Escherichia coli* analyzed by three-dimensional tracking. Nature 239: 500–504

Berg HC, Purcell EM (1977) Physics of chemoreception. Biophys J 20: 193–219

Blakemore RP (1975) Magnetotactic bacteria. Science 190: 377–379

Blakemore RP (1982) Magnetotactic bacteria. Ann Rev Microbiol 36: 217–38

Block SM, Segall JE, Berg HC (1982) Impulse responses in bacterial chemotaxis. Cell 31: 215–226

Crenshaw HC (1993a) Orientation by helical motion-I. Kinematics of the helical motion of organisms with up to six degrees of freedom. Bull Math Biol 55: 197–212

Crenshaw HC (1993b) Orientation by helical motion-III. Microorganisms can orient to stimuli by changing the direction of their rotational velocity. Bull Math Biol 55: 231–255

Crenshaw HC, Edelstein-Keshet L (1993) Orientation by helical motion-II. Changing the direction of the axis of motion. Bull Math Biol 55: 213–230

Diez JA, Dusenbery DB (1989a) Preferred temperature of *Meloidogyne incognita*. J Nematol 21: 99–104

Diez JA, Dusenbery DB (1989b) Repellent of root-knot nematodes from exudate of host roots. J Chem Ecol 15: 2445–2455

Dusenbery DB (1980) Responses of the nematode *Caenorhabditis elegans* to controlled chemical stimulation. J Comp Physiol A 136: 327–331

Dusenbery DB (1988) Behavioral responses of *Meloidogyne incognita* to small temperature changes. J Nematol 20: 351–355

Dusenbery DB (1989a) Efficiency and the role of adaptation in klinokinesis. J Theoret Biol 136: 281–293

Dusenbery DB (1989b) A simple animal can use a complex stimulus pattern to find a location: nematode thermotaxis in soil. Biol Cybernetics 60: 431–438

Dusenbery DB (1989c) The value of asymmetric signal processing in klinokinesis. Biol Cybernetics 61: 401–404

Dusenbery DB (1992) Sensory Ecology. W.H. Freeman and Company, New York

Dusenbery DB (1997) Minimum size limit for useful locomotion by free-swimming microbes. Proc Natl Acad Sci USA 94: 10949–10954

Dusenbery DB (1998a) Fitness landscapes for effects of shape on chemotaxis and other behaviors of bacteria. J Bact 180: 5978–5983

Dusenbery DB (1998b) Spatial sensing of stimulus gradients can be superior to temporal sensing for free-swimming bacteria. Biophys J 74: 2272–2277

Dusenbery DB (2001a) Performance of basic strategies for following gradients in two dimensions. J Theor Biol 208: 345–360

Dusenbery DB (2001b) Physical constraints in sensory ecology. In Barth F, Schmid A (eds) Ecology of Sensing. Springer, Berlin, pp 1–17

Ferrée TC, Lockery SR (1999) Computational rules for chemotaxis in the nematode C. elegans. J Comp Neurosci 6: 263–277

Foster KW, Smyth RD (1980) Light antennas in phototactic algae. Microbiol Revs 44: 572–630

Frankel RB, Blakemore RP, de Araujo FFT, Esquivel DMS, Danon J (1981) Magnetotactic bacteria at the geomagnetic equator. Science 212: 1269–1270

Goode M, Dusenbery DB (1985) Behavior of tethered Meloidogyne incognita. J Nematol 17: 460–464

Jennings HS (1901) On the significance of the spiral swimming of organisms. Am Naturalist XXXV: 369–378

Karp-Boss L, Boss E, Jumars PA (1996) Nutrient fluxes to planktonic osmotrophs in the presence of fluid motion. Oceanographic and Marine Biology: an Annual Review 34: 71–107

Manson MD (1992) Bacterial motility and chemotaxis. Adv Microbial Physiol 33: 277–346

Morse TM, Ferrée TC, Lockery SR (1998) Robust spatial navigation in a robot inspired by chemotaxis in Caenorhabditis elegans. Adapt Behav 6: 393–410

Pline M, Diez JA, Dusenbery DB (1988) Extremely sensitive thermotaxis of the nematode Meloidogyne incognita. J Nematol 20: 605–608

Pline M, Dusenbery DB (1987) Responses of the plant-parasitic nematode Meloidogyne incognita to carbon dioxide determined by video camera-computer tracking. J Chem Ecol 13: 1617–1624

Prot JC (1978) Behaviour of juveniles of Meloidogyne javanica in salt gradients. Rev Nematol 1: 135–142

Purcell EM (1976) Life at low Reynolds number. In: Huang K (ed) Physics in Our World: A Symposium in Honor of Victor F. Weisskopf. American Institute of Physics, New York, pp 49–64

Robinson AF (1995) Optimal release rates for attracting Meloidogyne incognita, Rotylenchulus reniformis, and other nematodes to carbon dioxide in soil. J Nematol 27: 42–52

Schöne H (1984) Spatial Orientation. Princeton University Press, Princeton, NJ

Segall JE, Block SM, Berg HC (1986) Temporal comparisons in bacterial chemotaxis. Proc Natl Acad Sci USA 83: 8987–8991

Spormann AM, Wolfe RS (1984) Chemotactic, magnetotactic and tactile behavior in a magnetic spirillum. FEMS Microbiol Letts 22: 171–177

Tanford C (1963) Physical Chemistry of Macromolecules. Wiley, New York

Tranquillo RT, Lauffenburger DA (1987) Stochastic model of leukocyte chemosensory movement. J math Biol 25: 229–262

Tranquillo RT (1990) Theories and models of gradient perception. In Armitage JP, Lackie JM (eds) Biology of the Chemotactic Response. Cambridge University Press, Cambridge, pp 35–75

Tranquillo RT, Lauffenburger DA, Zigmond SH (1988) A stochastic model of leukocyte random motility and chemotaxis based on receptor binding fluctuations. J Cell Biol 106: 303–309

van Holde KE (1985) Physical Biochemistry. Prentice-Hall, Englewood Cliffs, NJ

Ward S (1973) Chemotaxis by the nematode Caenorhabditis elegans: Identification of attractants and analysis of the response by use of mutants. Proc Natl Acad Sci USA 70: 817–821

Zigmond SH (1977) Ability of polymorphonuclear leukocytes to orient in gradients of chemotactic factors. J Cell Biol 75: 606–616

22. Representation of Odor Information in the Olfactory System: From Biology to an Artificial Nose

John S. Kauer and Joel White

Abstract

The olfactory systems of animals as diverse as insects and primates are well-known for having extraordinary sensitivity while, at the same time, exhibiting broad discriminative abilities. These properties, often mutually exclusive in other chemical recognition systems, appear to arise from the parallel, distributed nature of the processes that underlie how odors are encoded at each level in the olfactory pathway in the brain. In this paper we describe how we have tried to characterize the physiological aspects of these processes in biological experiments, capture these processes in a computational model, and then to use these observations to design and build a biologically inspired artificial device. The Tufts Medical School Nose has achieved a degree of sensitivity and discriminability that, for certain compounds under defined conditions, approaches that of its biological parent.

I. Introduction

Many molecular recognition systems, such as those involved in neurotransmission and gene regulation, trade response diversity for specificity and sensitivity. The more sensitive a receptor system (for example, pheromone receptors in insects or high affinity receptors in the brain), generally the less able it is to recognize a diversity of molecular species. High sensitivity usually implies high selectivity at the cost of broad-band response. If the structure of a compound (or compounds) to be detected is defined, an effective way to build a detector (or detectors) is to make receptors highly specific for that compound (for example, to recognize pheromones). If, however, one requires that a system can recognize a wide diversity of molecular structure and be flexible enough to discriminate among compounds without prior knowledge of their identity, then reliance on highly specific receptors is not advantageous. To have broad discrimination, even with the use of extensive arrays, multitudes of specific receptors must be produced, one for each compound of interest, with the risk that receptors for compounds outside the range of the target set (or that might have future significance), would not be generated. In the vertebrate olfactory pathway the dichotomous requirements of high sensitivity *and* broad-band response seem to have been met by pressures that have evolved an unusual, but highly effective and flexible, chemical detector system.

In this paper we briefly outline a number of features of the air-breathing, vertebrate olfactory system. We then describe a computational model we have devised to represent some of these properties in mathematical terms, and, finally, we show how we have used these properties to build a device designed to carry out a real-world, low-concentration odor detection task – finding landmines. The first artificial nose developed by Persaud and Dodd (1982) was also inspired by two basic attributes of olfactory function: cross-reactive sensors and pattern recognition. We have extended this approach such that we have incorporated more than 20 characteristics of olfactory function into our device. The underlying theme of this work is that the requirements for a flexible, highly sensitive, broad-band olfactory system have been satisfied by complementary interactions between primary odorant transduction processes involving relatively non-specific (with regard to overall molecular structure) receptors linked to biochemical cascades and the brain circuits that extract the information necessary for odor-guided behavior of the organism. Some of these ideas that relate to biological olfactory function have also been discussed in other reviews (see Kauer 1980, 1987, 1991, Kauer and Cinelli 1993, Shepherd 1994, Buck 1996, Hildebrand and Shepherd 1997, Mori et al. 1999, Kauer and White 2001).

II. From Biology to an Artificial Nose

A. Odor Coding Mechanisms in the Biological Olfactory System

The prevailing hypothesis about how odorant molecules are encoded by the nervous system (see reviews above), is that chemical structure is represented by a combinatorial process involving distributed recognition of multiple molecular subcomponents, not by a binding to the global molecular structure of the odorant. The subcomponents of even monomolecular odors (variously called 'odotopes', Shepherd 1987, in analogy with epitopes in the immune system; 'olfactophores' (Ham and Jurs 1985); or 'recognition elements') bind to arrays of olfactory receptors, converting chemical information into electrical neuronal activity. This encoded information is sequentially propagated to olfactory circuits first in the olfactory bulb and then to higher order olfactory areas in a form that generates spatially and temporally distributed activity patterns at each level of the pathway. This encoding process does not imply that the activity is diffuse, non-specific, chaotic, nor lacking in component structure, rather that, even for monomolecular stimulus compounds, many cells at each level in the brain pathway participates in these odor encoding, re-encoding, and integration processes.

Distribution of activity in this way is reminiscent of 'neural networks' (as this term is used in computer theory; see Kauer 1991) and provides for many of the well-documented aspects of olfactory function that include high sensitivity accompanied by broad-band response, adaptability to new environments, and fault tolerance upon exposure to injury. This encoding scheme is substantively different from the way chemical recognition is thought to occur in the specific interactions between individual ligands and their high affinity, cognate receptors. We have termed this process 'distributed specificity', in which the fidelity of recognition for a particular odor at the perceptual level emerges from complex events widely distributed in space and time at more peripheral levels. We believe this hypothesis is not only important for studying olfactory function, but also may serve as a paradigm for understanding how other systems in the brain represent information in a distributed fashion.

1. Structure and Function of the Peripheral Olfactory Pathway

This section briefly describes a number of the general features of the peripheral olfactory system as they are seen in the animal model we have worked on, the tiger salamander *Ambystoma tigrinum* (Kauer 1973). An interesting feature of olfactory systems is the conservation of structure and function across many phyla, such that properties described for the salamander are relevant for a diversity of other species. A large literature on the olfactory system of the tiger salamander (some 160 papers) has been generated over the last 30 years providing a wealth of data for making direct comparisons among anatomical, physiological, biochemical, molecular biological, and behavioral information from a single species. This animal has numerous experimental advantages including easily obtained physiological signals from a variety of brain regions, robust tissue that tolerates *in vitro* study, large, experimentally accessible neurons, nasal cavity structure that permits controlled odorant stimulation, easily measured odor-guided behavior, relatively well-defined primary receptor and neurotransmitter systems, and molecular receptors similar to those found in the mammal.

2. The Olfactory Epithelium

Figure 1 shows a simplified diagram of the first stages in the olfactory pathway, the olfactory epithelium (OE) and olfactory bulb (OB). Although in the salamander the nasal chamber is a flattened sac that contrasts with the convoluted and aerodynamically complex nasal cavity of the mammal, the anatomy and physiology of the olfactory sensory neurons (OSNs) and the distribution of odorant receptors (ORs) are remarkably similar. OSNs send dendrites to the epithelial surface where they end in sensory cilia embedded in a layer of mucus. The biochemical machinery for primary odorant transduction, including the ORs and second messenger cascades, is in the cilia. The initial recognition event occurs via binding of an odorant to members of a large family of 7 transmembrane, G-protein linked receptors, that ultimately open cyclic nucleotide gated channels via an adenylate cyclase mediated cascade (see reviews Anholt 1993, Ache 1994, Buck 1996, Schild and Restrepo 1998, Mombaerts 1999).

In the fish, amphibians, and mammals studied thus far, OSNs expressing one OR phenotype are found scattered within broad, but defined, OE regions or zones (Ressler et al. 1993, Vassar et al. 1994), not in contiguous clusters. Distribution of odorant responses across the OE has been observed using several different recording methods and have consistently shown that, as for the distribution of the ORs, they are broadly, but non-homogeneously distributed across the OE. Precisely how the scattered distributions of ORs relate to the broadly distributed odorant response patterns is still not clearly defined.

In general, OSNs distributed in zones, sequentially depolarize as the bolus of odor moves through the nasal cavity during a sniff. This generates barrages of temporally patterned action potentials in multitudes of OSN axons that project to the first synapse in the glomeruli of the OB (see boxes A–E, Fig. 1). After binding to ORs in the cilia, the chemical information in the odorant stimulus is now represented in these distributed firing patterns. The general finding is that individual OSNs respond to multiple odorants and that OSNs expressing different ORs can respond to the same odorant. This is the fundamental observation that underlies the combinatorial complexity of the first stage encoding process.

It has been shown that OSNs turn over in all vertebrates studied (Graziadei 1973) including the salamander (Simmons and Getchell 1981), yet it is thought that olfactory function generally remains more or less stable throughout life. Any hypothesis of how odorants are encoded by the nervous system must accommodate this turnover process that continues for the lifetime of the organism.

3. Connections to the Olfactory Bulb

Unlike the regular topographic mapping long recognized as a canonical attribute of other sensory modalities, there has been a question of how the projections of OSN axons in the OE map onto their glomerular targets in the OB. Recent molecular marking experiments (Mombaerts 1999) have confirmed and extended earlier anatomical and physiological data (Kauer and Moulton 1974, Kauer 1980, 1987, Cinelli and Kauer 1995, Cinelli et al. 1995) and show that the spatially distributed populations of OSNs expressing one OR converge onto one or two glomeruli. These convergence patterns are an essential element for understanding how odorants are encoded and we have included them in our models and engineered devices.

4. The Olfactory Bulb

Much is known about the structure of the OB circuit, the distribution of neurotransmitters, and odorant and electrical response properties of single OB cells in a number of different animals (for reviews see Hildebrand and Shepherd 1997, Christensen and White 2000), including the salamander. These studies show that the OB circuit and responses of the output mitral/tufted (M/T) cells are, as for OSN properties, remarkably conserved through phylogeny. Upon stimulation with an odorant, subsets of M/T cells across the extent of the OB are activated with a limited number of temporal response patterns which relate to odorant concentration and recording site (see examples of intracellular records in Fig. 2 IIAb, Bb, Cb; IIIAb, Bb). Each of these temporal response patterns is seen with all odors tested and thus firing pattern *per se* does not encode odor quality. Like OSNs, each M/T cell generally responds to a number of different compounds in ways that suggest they are encoding some molecular attribute of the odorant. For example, recently it has been observed that rabbit M/T cells (and certain OB locations observed using imaging methods) respond with higher firing rates to molecules (especially fatty acids) having a particular carbon chain length (Slotnick et al. 1997, Friedrich and Korsching 1998, Mori et al. 1999, Rubin and Katz 1999, Uchida et al. 2000). These kinds of data support the hypothesis that the characteristic to which these cells respond is likely to be a molecular sub-component of the compound (i.e. carbon chain length), not the entire molecule.

Overall distributions of OB responses have been observed using electroence-phalographic, 2-deoxyglucose (2DG), c-fos, and voltage-sensitive dye (VSD) re-cording (see reviews Hildebrand and Shepherd 1997, Christensen and White 2000, Kauer and White 2001). Results from these studies consistently show that activity appears in widely distributed loci within and across the layers of the bulb after stimulation with odor. Patterns of activity to defined odorants are relatively consistent across individuals of the same species and generally become more widely distributed with the application of high concentrations.

5. Higher Olfactory Centers: The Piriform Cortex, Anterior Olfactory Nucleus, Amygdala, Olfactory Neocortex

There is still relatively little known about how odor information is processed in olfactory areas beyond the OB. Although many details about pyriform cortex cir-cuits have been elucidated (Haberly 2001, Zou et al. 2001), there are few data on odor responses at either single cell or ensemble levels in this structure. Recent experiments in the mammal have shown that adaptation to odorants in the OB and pyriform cortex is complex and depends on the frequency of odor delivery and on the state of feedback circuits from one level to another (see Wilson 1998, 2000, Young and Wilson 1999). These experi-ments emphasize how sniffing (odorant access) is likely to be an important variable in the coding process. We have included control over stimulus access using brief, defined 'sniffs' in our artificial device (see also Chapter IV 3, this volume).

6. Odor-Guided Behavior

There have been many studies on odor-related behavior in rodents, salamanders, and other animals. Among the most sig-nificant for understanding how the system encodes molecular odor information have been those in which odor detection or re-cognition has been tested after pathway lesion (Hudson and Distel 1987, Slotnick et al. 1997, Hudson 1999). In general, these studies show that animals do well on odor detection tasks even with up to 85% of their OB's removed, indicating a remarkable redundancy in OB circuits in which widely distributed information un-derlies a robust fault tolerance. We have attempted to capture this property of fault tolerance and redundancy in our compu-tational model and artificial device.

7. An Odorant Coding Hypothesis

A schematic diagram of a number of the attributes of olfactory function and how they relate to our odor coding hypothesis is shown in Fig. 1.

The essential elements of this hypoth-esis are as follows:

- Individual OSNs respond to several different odorant compounds. This is because they express ORs (a,b,c... in Fig. 1) not responsive to the overall shape, but rather to a substructure (i.e. a recognition element) of the odor-ant. Recognition of a number of such subcomponents, taken together, are necessary to characterize even single, monomolecular compounds.
- OSNs express one, or a small number of, ORs such that one OSN population expressing one OR phenotype responds to one recognition element; other OSN populations expressing other ORs re-spond to other recognition elements. A number of different OSN populations thus respond to any one odorant and,

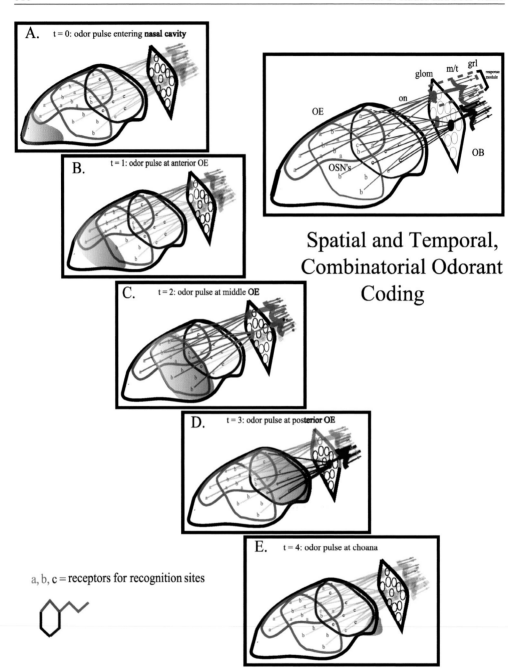

Fig. 1. Simplified schematic diagram of spatial and temporal properties of odorant encoding mechanisms in the peripheral olfactory system of the salamander. The large box to the upper right shows an outline of the olfactory epithelium (OE) at the left, with olfactory receptor neurons (OSNs) expressing

conversely, one OSN population responds to as many different odorants as share the recognition element to which its ORs are sensitive, even compounds which may perceptually smell different. The definitive number of ORs expressed and functional at any one time to encode one odorant species is unknown.

- The cells comprising an OSN population expressing one OR are distributed within broad, but defined regions of the OE ('zones'; see encircled regions in the OE in Fig. 1), mixed in among other OSN populations expressing other ORs. Sensory cells expressing one OR are not found as clusters of contiguous cells in uniform patches of epithelium, but are intermixed within a zone (see overlap of encircled regions in the OE in Fig. 1). The relationships of each OSN population expressing one OR to the temporally and spatially distributed responses elicited by a particular odorant are not well understood.

- Since any one odorant interacts with multiple OSN populations, how much these populations overlap spatially, determines the patterns of distributed activity in the OE that have been observed using 2-deoxyglucose, electroolfactogram, and voltage-sensitive dye recording.

- The OE does not map onto the OB in the point-to-point fashion that other senses map onto the central nervous system. Rather, axons from distributed OSNs that express the same OR converge onto one or two glomeruli. See lines coming from a's, b's, and c's converging onto glomeruli (glom) in Fig. 1.

- Groups of M/T cells, with associated interneuronal circuitry, connect with functionally related groups of glomeruli (Fig. 1) and serve as building block modules making up the overall response to an odorant. The ensemble of activated modules (with temporally complex responses, as shown in the descending panels, A–E of Fig. 1 depicting activation of different OB neurons over time) encodes the ensemble of recognition elements expressed by the odorant stimulus in a way similar to how computer 'neural networks' encode distributed information. Distributed activation of the modules is the basis for the widespread patterns of activity observed in 2-deoxyglucose and voltage-sensitive dye recordings from the OB. The relationships of these modules to specific OSN populations and to particular molecular features is beginning to be determined.

- A module responding to a recognition element consists of M/T (and other OB) cells that are both excited and inhibited in limited numbers of temporal patterns that depend on their role in the circuit. Concentration is coded by temporal patterning of the responses and by the number of cells responding.

---

single olfactory molecular receptors (a,b,c . . .). These are distributed within defined zones (encircled regions in the OE) and relate to subcomponents of a stylized molecule as shown below left. OSNs connect to the olfactory bulb (OB) with olfactory nerve (on) axon projections that converge from distributed OSNs of one type onto single glomeruli (glom). Other bulbar elements shown include mitral/tufted (m/t) output cells and inhibitory interneuron granule (grl) cells. Inhibitory interneuron periglomerular cells are not shown. The mitral/tufted and granule cells associated with a glomerulus that receives input from a group of OSNs defined by the molecular receptor they express, form a response module. Boxes A–E depict a stylized time series of how the system responds to a bolus of odorant vapor progressing through the nasal cavity from left to right, sequentially interacting various OSN populations that in turn respond with different spatial and temporal patterns response in different OB modules

I.

A

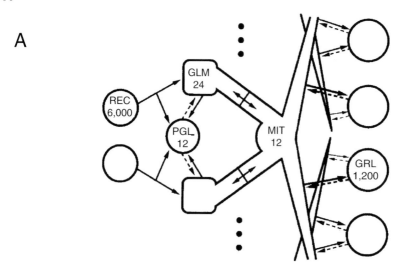

B   Olfactory Epithelium   Olfactory Bulb

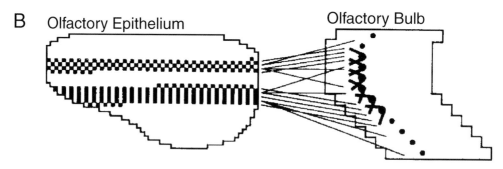

Fig. 2. Simulated olfactory epithelium and olfactory bulb circuit (I) and computed and measured responses from salamander mitral/tufted cells (II, III). IA shows the cellular elements used to compute the responses to electrical and odor stimulation shown in II and III. The circuit consists of 6000 receptor cells (REC), 24 glomerular elements (GLM), 12 inhibitory periglomerular cells (PGL), 12 mitral/tufted output cells (MIT), and 1200 inhibitory granule cells (GRL). PGL cells are connected to mitral primary dendrites in the glomeruli via reciprocal dendrodendritic synapses and GRL cells are connected to mitral secondary dendrites via reciprocal dendrodendritic synapses. IB shows the convergent/divergent connectivity patterns from defined epithelial zones onto olfactory bulb glomeruli. IIA illustrates computed (a) and real (b) intracellular recordings after stimulation of the olfactory nerve with a low intensity electrical shock. Note the similar depolarization with single spike, following hyperpolarization sequence in both computed and measured records. IIB and C show similar comparisons for higher intensity olfactory nerve shock and with antidromic activation of the output axons of mitral/tufted cells. IIIA and B show similar comparisons for 'odor' stimulations. The important observation here is that the dramatic and complex change in temporal patterning occurring with increased odor concentration in the real intracellular recording (compare IIIAb with IIIBb), is replicated in the computed output (compare IIIAa – MIT 7 with IIIBa – MIT 7)

II.

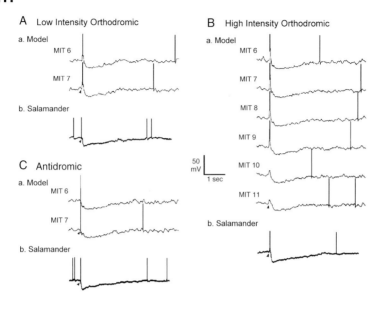

III.

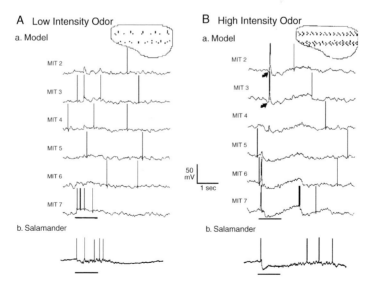

Fig. 2 (continued)

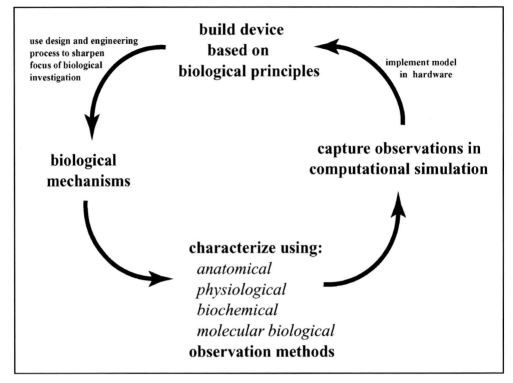

Fig. 3. Outline of the procedural approach we have used to implement biological principles in an engineered, artificial olfactory system. This process extends from studying biological mechanisms using standard investigation procedures (anatomy, physiology, etc) to capturing these events in computational models, to implementing biological mechanisms in hardware, and importantly, learning from the engineering process to focus investigation back to the biology

This simplified characterization of OE and OB function describes the initial components of odor encoding as being a kind of parallel, distributed processor. We have tried to capture a number of these elements in the formal, mathematical model described in the next section.

B. Representation of the Coding Hypothesis in a Computer Simulation

Figure 2 shows a brief overview of the OE/OB circuit that we have represented in a computer simulation (I) and illustrates some of the comparisons of the computed results with actual intracellular recordings (II and III). A more complete description of this model can be found in White et al. (1992). IA shows the essential components of the circuit including 6000 receptor cells (REC), 24 glomerular (GLM) projection targets of the receptor cell axons, 12 periglomerular (PGL) inhibitory interneurons, 12 mitral/tufted (MIT) cells, and 1200 granule (GRL) inhibitory interneurons. The number of cellular elements represented in the model are in similar ratios to one another as are the neurons in the salamander olfactory system, but the abso-

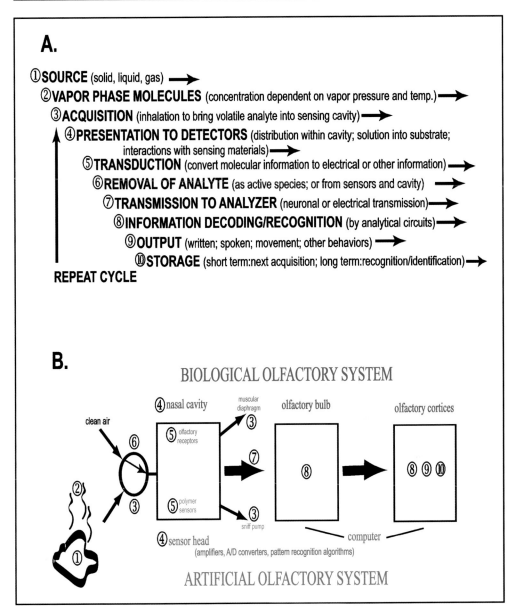

Fig. 4. A. Sequence of steps that occurs in a vapor phase chemical recognition process. **B.** Block diagram of where the steps in A occur in biological and artificial olfactory systems

lute numbers are reduced by about one thousand fold (for example, there are estimated to be about 12,000 mitral/tufted cells in the salamander). The connections

and spatial relationships among the different cells is a simplified, but accurate, representation of the connections as they are known in this and other animals. The goal of these experiments was to evaluate the degree to which a simplified representation of the peripheral olfactory circuit could generate computed cellular activity that was commensurate with responses observed in real biological neurons. Such comparisons are shown in Fig. 2 II. and III for computed and electrophysiologically measured mitral/tufted cell behavior.

Figure 2 II Aa is an example of a comparison between computed intracellular recordings from two mitral/tufted cells (MIT 6 and MIT 7) and an actual intracellular recording from an identified salamander M/T cell (Fig. 2 II Ab) in response to a single, low intensity, orthodromic electric shock to the afferent olfactory nerve (input to the circuit). One can see that the computed record is remarkably similar to the intracellular record in showing a brief depolarization with an action potential, followed by a period of hyperpolarization. Favorable similarities are also shown for an antidromic electrical stimulation (Fig. 2 IICa,b) and for application of a higher intensity orthodromic shock in Fig. 2 IIBa,b. These records indicate that the simulated circuit captured sufficient detail of the spatially and temporally distributed excitatory and inhibitory interactions that the computed and measured outputs look similar for simple electrical activation of the circuit.

Another test of the ability of the simulation to capture more complex physiological details is shown in Fig. 2 III. Here the computed output represented the results of applying a simulated 'odor' to an array of simulated OSNs in spatial patterns depicted by the distributions of the dots on the OE outlines seen in Fig. 2 III Aa and Ba top. Activating the simulated OSNs with these patterns once again generated simulated membrane potential changes in one of the mitral/tufted cells (Fig. 2 III. Aa – MIT 7) that were remarkably similar to that seen in a real intracellular recording after odor stimulation (Fig. 2

III. Ab). In addition to the similarity seen at one concentration, when the simulated odor stimulation was applied at a 'higher concentration' the temporal pattern of the computed output changed in a complex way similar to the temporal pattern in the intracellular recording (Fig. 2 IIIBa,b).

Although these simulations are clearly oversimplifications of real biological circuits we were gratified that a number of the essential elements generating spatially and temporally distributed patterns in M/T cells appeared to have been captured. These simulations made us realize that we could compute spatial/temporal patterns of activity in biologically inspired circuits that would represent both odor quality and concentration. We thus had the basis for a simple 'artificial olfactory system'. What remained was to add an input stage consisting of sensors that could respond to vapor phase chemicals in the environment, a way to deliver the vapor, and a mechanism to interpret the response patterns.

Among several requirements we had for such sensors was that they be broadly responsive, that is, intentionally non-specific, and that we could generate many different kinds of them. These requirements were met by a class of optically interrogated sensing materials that change their fluorescence when exposed to odorants (Barnard and Walt 1991, White et al. 1996). These materials consist of intrinsically fluorescent polymers or polymers impregnated with fluorescent dyes that can be made into arrays of discrete sensor sites as described in the next section.

C. Design of an Artificial Olfactory System Based on Biological Principles

The design and development of our artificial olfactory system has been broadly based on the scheme outlined in Fig. 3. After many years of research, we started (Fig. 3, left) with observations on the biological attributes of the olfactory system in a defined experimental model animal, progressed to generating computational simulations of these observations, used computer simulation as a basis for building working hardware prototypes, and finally used (and continue to use) the entire process to focus attention on details of the biological process that might have escaped study had we not been confronted by them in building the engineered device. In this endeavor, we have extended the approach of Persaud and Dodd (1982), who were the first to build a biologically inspired artificial nose.

To identify the components required for building an artificial olfactory system we tried to characterize the essential steps in an olfactory recognition event as shown in Fig. 4A. Figure 4B illustrates approximately where each of these steps occurs in biological and artificial systems. We have designed our device to also carry out these steps, albeit more crudely than occurs in the olfactory pathway. Explicit characterization of these events has been an important guide for making the artificial device function effectively and accurate definition of these steps continues to focus future development on critical mechanistic issues.

As shown in Fig. 5, the essential elements of the Tufts Medical School Nose consist of an air sampling chamber in which resides an array of optically interrogated vapor phase detectors, a means for delivering vapor phase samples to the sensing array as short 'sniffs' (sniff pump), electronic circuits (amplifiers, analogue/digital converters) to detect changes in the optical signals from each sensor (Barnard and Walt 1991), and a computer to evaluate the matrix of signals from each sensor over time. In this figure vapors from the odor source, gated into

discrete pulses by valves (not shown), are drawn from below by a sniff pump (right). The vapor stream passes over the sensing materials (polymers and dyes) which are caused to fluoresce by illumination at the appropriate wavelengths from the array of light emitting diodes (LEDs, at the bottom of the sensing chamber). Changes in fluorescence over the time course of the sniff as a result of the vapor interacting with the sensing sites are detected by the array of photodiodes (PDs, at the top of the sensing chamber). There is one LED and one photo-diode for each sensing site. This configuration is essentially an array of micro-spectrofluorimeters, each of which is tuned to the wavelengths appropriate for its sensing material. The output of this sensor/detector array consists of a 2 dimensional matrix of numbers in which each column is the signal from one sensor over time. Figure 6 shows such response matrices plotted as topographical surfaces for six different vapor exposure situations. The magnitude of fluorescence change in either the increasing or decreasing direction is plotted on the y-axis. The number of the sensor is plotted on the x-axis (in this case there were 32 sensors) and the time over which data were taken (about 1 s) is plotted on the z-axis (going into the page). Note that that responses are obtained with short sniffs (about 0.5 s), that the sensing materials respond rapidly, and that the response surfaces are characteristic and different for each compound tested. The response matrix for each odorant or mixture of odorants is then stored in memory.

As used for a real-world odor detection/identification task, the device is 'trained' by storing a library of response profiles obtained after sniffing known sources – either pure compounds or complex mixtures such as beer, coffee, or a landmine. When using the device for detecting and identifying unknown odors,

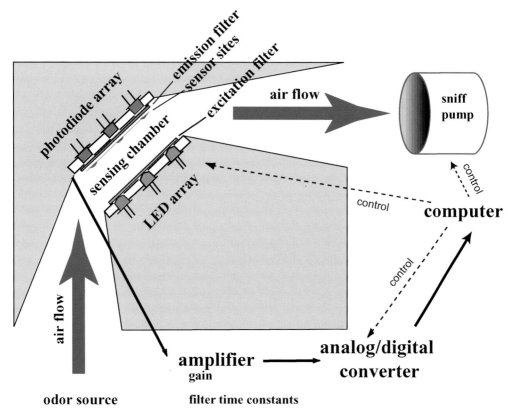

air flow

sniff pump

control

computer

control

control

analog/digital converter

amplifier
gain

air flow

odor source   **filter time constants**

Fig. 5. Schematic diagram of the Tufts Medical School artificial olfactory system. The sensing chamber consists of an airway into which are drawn (by the sniff pump, left) the vapors (odor source) to be detected and identified. The sensors consist of an array of light emitting diodes (LEDs), excitation filters, the polymer/dye sensing materials, emission filters, and fluorescence detector photodiodes. In the situation shown here there was a 3 × 3 array of detectors. The changes in fluorescence detected by the photodiodes are amplified, digitized and stored in computer memory. All aspects of the process including sniff generation, modulation of the light sources, modulation of the signals, digitization, and storage of the data are under feedback computer control

sniffs of the unknown source are taken and the response profile is compared to those stored in memory. The matrix in memory to which the profile of the unknown matches best sets the criterion for identification. Statistical methods can be applied to assess the accuracy of the match and a probability that exact identification has been achieved. This is important for knowing the level of certainty when identifying dangerous targets such as landmines having low concentra-

tion, ephemeral odor signatures. We have used a number of algorithms for this matching process and we use a variation of the OE/OB model described above in order to separate concentration from quality information.

We have applied the general approach described above to the problem of finding buried landmines by their odor signature (George et al. 1999). In this task we included fluorescent sensing materials specifically designed by Tim Swager of MIT

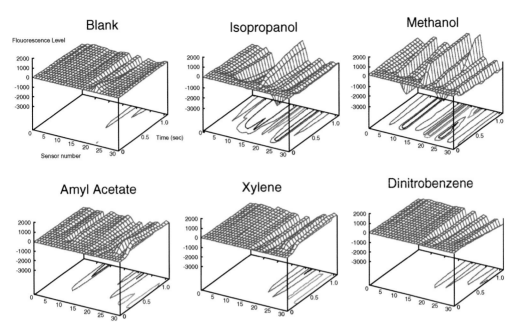

Fig. 6. Surface plots of response matrices for six different vapor exposure (stimulus) conditions: blank room air, isopropanol, methanol, amyl acetate, xylene, and dinitrobenzene. This detector array had 32 sensors. Data are plotted with sensor number on the x-axis, change in fluorescence on the y-axis, and time of odor exposure (sniff time) on the z-axis going into the page. Dinitrobenzene was included in this test series because it is a compound found associated with the explosive TNT in landmines. This sensor array included some materials specifically designed for detecting nitroaromatic compounds (see text) as well as other sensors with broader response spectra

(Yang and Swager 1998) for detecting nitroaromatic compounds such as TNT (trinitrotoluene) and DNT (dinitrotoluene), along with other, more broadly responding, materials that allow us to discriminate genuine DNT/TNT from other compounds such as methanol that also give responses in the nitroaromatic sensors.

Under highly controlled and calibrated laboratory conditions at Auburn University (Hartell et al. 1998), we have tested how well our device can detect controlled sources of DNT, TNT and landmine vapors, and discriminate them from background air or methanol. In the tests performed under these conditions the device can detect and discriminate DNT at about 300 parts per trillion. This is below the threshold of most dogs (about 1 part per

billion) that have been tested in the same apparatus.

In preliminary trials at a test facility set up by the Defense Advanced Research Projects Agency at Fort Leonard Wood, MO, we have been able to identify the presence of buried landmines (without fuses) when conditions are optimal. Optimal conditions include appropriately warm temperatures, relatively high humidity, and no ambient air movement (wind). Although the device is not yet ready for detecting live mines in real mine fields such as in Afghanistan, these gratifyingly encouraging field results have focused our attention on how to improve the sensitivity and noise immunity of the device. We continue to be guided in this effort by studying and implementing detailed mechanisms of olfactory function revealed in our biological studies.

III. Conclusion

We have developed an artificial olfactory system based on design principles that have been observed in investigations into the biological sense of smell. At the present time we have incorporated more than 20 such principles into our device (see Table I). These have allowed us to build an apparatus with sufficient sensitivity and discriminability to permit, under certain defined conditions, the detection and identification of nitro-aromatic and other compounds associated with landmines in both laboratory and field settings. Our experience shows that the use of anatomical and physiological observations to guide formulation of mathematical models

Table 1.

1. Sensors have broad response spectra distributed within the space of the sample chamber.
2. Sensing device exerts control over environmental attributes of the vapor phase stimulus – humidity, temperature, position in the ambient environment (restricted zone to which sniffing is applied).
3. The vapor phase stimulus is delivered to the detectors in a temporally controlled manner (onset, rise time, duration, fall time, individual sniff frequency, sniff bout frequency).
4. Temporal response profiles are used in pattern analysis algorithms. Temporal patterns are governed by frequency and depth of sniffing, by the response characteristics of the detection circuitry, by duration, intensity, wavelength, rise and fall times of light exposure, and by response properties of sensing materials.
5. Feedback control is exerted over the odorant delivery process in real time when using multiple sniffs per test trial; later sniffs are modulated based on information arriving in early sniffs.
6. Control over the adaptation characteristics of the detectors is accomplished by using short pulse stimuli and by brief and controlled exposure of the detectors to the light source (normalized and controlled bleaching).
7. Long term changes in sensors/detectors are monitored to compensate for changes or deterioration.
8. Consideration of which sensing materials to use is based on their shapes and sizes in 'odor space'. Design of detection system and analytical algorithms is based on the nature of the odorants to be detected.
9. Maximize gain and distance from other sensors in 'odor space' by using measurements of detector response that change non-linearly and are orthogonal to one another along 'odor space' dimensions.
10. Control over sensor material sensitivity is accomplished by a) increasing the surface area of the sensor material; b) increasing surface area and properties of the photodetectors; c) improving intrinsic sensitivity of sensor materials.
11. Sensor responses are stabilized by continuous oscillatory application of ambient odors and of target analyte. Continuous sniffing throughout test session.
12. Multiple sensor mechanisms are used for detection (intrinsic fluorescence of designed materials; fluorescence provided by addition of extrinsic dye).
13. Data from sensors are selectively weighted by feedback from analytical algorithms to optimize the information content that each sensor contributes to defining the odor signature.
14. Gain compression.
15. Ambient odorant interference is controlled by subtraction of background (adaptation).
16. Information transfer from the detectors to analytical circuits is clocked by the sniff   cycle.
17. Convergence of input from multiple identical sensors is used to improve signal to noise ratio.
18. Analytical algorithms are based on biological neuronal circuits
19. Recognition and identification emerge from repeated (and continuous) training with feedback ('reward' contingency).
20. Recognition templates are stored in a library in random access memory.
21. Output is by spoken word.

and the incorporation of these formalized biological principles into a hardware device, can be powerful biomimetic approaches for developing engineered solutions to real-world problems. These studies further indicate that the engineering process required to develop hardware can be an important heuristic process for reciprocally informing the investigation of biological machines (Fig. 3).

Acknowledgements

We gratefully acknowledge support for this work from the National Institutes of Health (NIDCD), the Office of Naval Research, and the Defense Advanced Research Projects Agency.

References

Ache B (1994) Towards a common strategy for transducing olfactory information. Semin Cell Biol 5: 55–63

Anholt RR (1993) Molecular neurobiology of olfaction. Critical Reviews. Neurobiology 7: 1–22

Barnard SM, Walt DR (1991) A fibre-optic chemical sensor with discrete sensing sites. Nature 353: 338–340

Buck LB (1996) Information coding in the mammalian olfactory system. Cold Spring Harbor Symp Quant Biol 61: 147–155

Christensen TA, White J (2000) Representation of olfactory information in the brain. In: Finger TE, Silver WL, Restrepo D (eds) The Neurobiology of Taste and Smell. Wiley, New York, pp 197–228

Cinelli AR, Hamilton KA, Kauer JS (1995) Salamander olfactory bulb neuronal activity observed by video-rate voltage-sensitive dye imaging. III. Spatio-temporal properties of responses evoked by odorant stimulation. J Neurophysiol 73: 2053–2071

Cinelli AR, Kauer JS (1995) Salamander olfactory bulb neuronal activity observed by video-rate voltage-sensitive dye imaging. II. Spatiotemporal properties of responses evoked by electrical stimulation. J Neurophysiol 73: 2033–2052

Friedrich RW, Korsching SI (1998) Chemotopic, combinatorial, and noncombinatorial odorant representations in the olfactory bulb revealed using a voltage-sensitive transducer. J Neurosci 18: 9977–9988

George V, Jenkins T, Leggett D, Cragin J, Phelan J, Oxley J, Pennington J (1999) Progress on determining the vapor signature of a buried landmine. Proc 13th Ann Int Symp Aerospace/Defense Sensing, Sim Controls 258–269

Graziadei PPC (1973) Cell dynamics in the olfactory mucosa. Tissue & Cell 5: 113–131

Haberly LB (2001) Parallel-distributed processing in olfactory cortex: new insights from morphological and physiological analysis of neuronal circuitry. Chem Senses 26: 551–576

Ham CL, Jurs PC (1985) Structure-activity studies of musk odorants using pattern recognition: monocyclic nitrobenzenes. Chem Senses 10: 491–506

Hartell M, Myers L, Waggoner L, Hallowell S, Petrousky J (1998) Design and testing of a quantitative vapor delivery system. Proc 5th Int Symp on Anal Det Explosives

Hildebrand JG, Shepherd GM (1997) Mechanisms of olfactory discrimination: converging evidence for common principles across phyla. Ann Rev Neurosci 20: 595–631

Hudson R (1999) From molecule to mind: the role of experience in shaping olfactory function. J Comp Physiol A 185: 297–304

Hudson R, Distel H (1987) Regional autonomy in the peripheral processing of odor signals in newborn rabbits. Brain Res 421: 85–94

Kauer JS (1973) Response properties of single olfactory bulb neurons using odor stimulation of small nasal areas in the salamander. University of Pennsylvania 142p

Kauer JS (1980) Some spatial characteristics of central information processing in the vertebrate olfactory pathway. In: van der Starre H (ed) Olfaction and Taste VII. pp 227–236

Kauer JS (1987) Coding in the olfactory system. In: Finger TE, Silver WL (eds) Neurobiology of Taste and Smell. Wiley, New York, pp 205–231

Kauer JS (1991) Contributions of topography and parallel processing to odor coding in the vertebrate olfactory pathway. TINS 14: 79–85

Kauer JS, Cinelli AR (1993) Are there structural and functional modules in the vertebrate olfactory bulb? Micr Res and Tech 24: 157–167

Kauer JS, Moulton DG (1974) Responses of olfactory bulb neurones to odour stimulation of small nasal areas in the salamander. J Physiol (Lond.) 243: 717–737

Kauer JS, White J (2001) Imaging and coding in the olfactory system. Ann Rev Neurosci 24: 963–979

Mombaerts P (1999) Molecular biology of odorant receptors in vertebrates. Ann Rev Neurosci 22: 487–509

Mori K, Nagao H, Yoshihara Y (1999) The olfactory bulb: coding and processing of odor molecule information. Science 286: 711–715

Persaud K, Dodd G (1982) Analysis of discrimination mechanisms in the mammalian olfactory system using a model nose. Nature 299: 352–355

Ressler KJ, Sullivan SL, Buck LB (1993) A zonal organization of odorant receptor gene expression in the olfactory epithelium. Cell 73: 597–609

Rubin BD, Katz LC (1999) Optical imaging of odorant representations in the mammalian olfactory bulb. Neuron 23: 499–511

Schild D, Restrepo D (1998) Transduction mechanisms in vertebrate olfactory receptor cells. Physiol Rev 78: 429–466

Shepherd GM (1987) A molecular vocabulary for olfaction. Ann NY Acad Sci 510: 98–103

Shepherd GM (1994) Discrimination of molecular signals by the olfactory receptor neuron. Neuron 13: 771–790

Simmons PA, Getchell TV (1981) Neurogenesis in olfactory epithelium: loss and recovery of transepithelial voltage transients following olfactory nerve section. J Neurophysiol 45: 516–528

Slotnick BM, Bell GA, Panhuber H, Laing DG (1997) Detection and discrimination of propionic acid after removal of its 2-DG identified major focus in the olfactory bulb: a psychophysical analysis. Brain Res 762: 89–96

Uchida N, Takahashi YK, Tanifuji M, Mori K (2000) Odor maps in the mammalian olfactory bulb: domain organization and odorant structural features. Nature Neurosci 3: 1035–1043

Vassar R, Chao KC, Sitcheran R, Nunez JM, Vosshall LB, Axel R (1994) Topographic organization of sensory projections to the olfactory bulb. Cell 79: 981–991

White J, Hamilton KA, Neff SR, Kauer JS (1992) Emergent properties of odor information coding in a representational model of the salamander olfactory bulb. J Neurosci 12: 1772–1780

White J, Kauer JS, Dickinson TA, Walt DR (1996) Rapid analyte recognition in a device based on optical sensors and the olfactory system. Anal Chem 68: 2191–2202

Wilson DA (1998) Habituation of odor responses in the rat anterior piriform cortex. J Neurophysiol 79: 1425–1440

Wilson DA (2000) Odor specificity of habituation in the rat anterior piriform cortex. J Neurophysiol 83: 139–145

Yang J-S, Swager TM (1998) Fluorescent porous polymer films as TNT chemosensors: electronic and structural effects. J Am Chem Soc 120: 11864–11873

Young TA, Wilson DA (1999) Frequency-dependent modulation of inhibition in the rat olfactory bulb. Neurosci Lett 276: 65–67

Zou Z, Horowitz LF, Montmayeur J-P, Snapper S, Buck LB (2001) Genetic tracing reveals a stereotyped sensory map in the olfactory cortex. Nature 414: 173–179

23. The External Aerodynamics of Canine Olfaction

Gary S. Settles, Douglas A. Kester, and Lori J. Dodson-Dreibelbis

Abstract

Following a review of precedent literature, flow visualization techniques are used to observe external canine olfactory airflows. This reveals the canine nostril as a variable-geometry aerodynamic sampler, being alternately a potential-flow inlet during inspiration and an outlet flow diverter during expiration. Close nostril proximity to a scent source is important. Separate flow pathways are provided for the inspired and expired air by way of nostril flexure. During sniffing, the nostril midlateral slits open to direct the expired air rearward and to the sides, away from the object being scented. If particulates are present on a surface being scented, they are readily disturbed by these expired jets and can be subsequently inspired. These and other results are brought to bear upon aerodynamic sampling for purposes of chemosensing, in which a sampler or sniffer acquires the airborne trace signal and presents it to an appropriate detector. Preliminary results from a laboratory-prototype sniffer are given.

I. Introduction

A. Literature on the External Aerodynamics of Olfaction

There is abundant literature on olfaction, but little on the external aerodynamics thereof. Some initial insight was gained from rabbits (Glebovskii and Marevskaya 1968), where nostril flare was seen accompanying inspiration, lowering the nasal passage resistance to airflow. However, resistance rose again sharply with expiration, when the nostril flare relaxed. The first known data on the external aerodynamics of olfaction were obtained by Bojsen-Moeller and Fahrenkrug (1971), who observed expired airflows through moisture condensation on a cold "Zwaardemaker mirror." This showed that the expired streams are directed laterally and downward from the nostrils in the rat and rabbit. Studies of canine nostrils reveal that they also flare during inspiration (Syrotuck 1972), and that the sniffing pattern of dogs depends upon the scent concentration: "short sniffs" being the norm and "long sniffs" occurring in the case of weak or inaccessible scents (Zuschneid 1973).

However, despite this background, there exists little or no literature on the external aerodynamics of olfaction for canines or, for that matter, any species except humans (Haselton and Sperandio 1988). Thus aerodynamic sampling – a key factor in the extraordinary olfactory acuity of canines – remains virtually unexplored.

We have attempted to fill this gap with a series of experiments with live dogs described in Section II of this chapter.

B. Literature on Aerodynamic Sampling Technology

The state of understanding of aerodynamic sampling technology, e.g. for traces of explosives, drugs, and contraband, is likewise very rudimentary. Airborne samplers for particles and air pollutants (impactors, cyclones, etc., e.g. Liu and Pui 1986) are highly developed, following decades of environmental funding, but that is a stationary approach that is inappropriate for present purposes. A search of the technical and patent literature for "sniffers" yields mostly leak-detection equipment with long hoses and hand-held, pointed tips (e.g. Jackson et al. 1998). This approach is intended for the detection of leaking gases such as helium. It does not address standoff distance, required airflow rate, deposition loss in the transport line, or the need to cover large areas in limited time. In short, any sort of mechanical sniffer that mimics a dog's nose appears entirely missing from the literature.

There is, however, important precedent material in the industrial ventilation literature (Baturin 1971, Heinsohn 1991, Goodfellow and Tähti 2001, Heinsohn and Cimbala 2002). Aerodynamic inlets are used in local exhaust hoods to capture welding fumes, cooking effluents, and the like. The "reach" of an inlet is defined as the upwind region from which all the air ultimately enters the inlet, while "capture velocity" refers to the airspeed in front of a hood inlet required to overcome a crosswind and thus capture airborne particulates (Heinsohn and Cimbala 2002).

While such industrial inlets are large and have very high airflow rates compared to what is considered here as a "sniffer," nonetheless there are several common characteristics. For example, a strong distinction is made between the limited reach of an inlet and the much-stronger "throw" of a blowing airjet, even when the volume flux of air is the same for both (Heinsohn and Cimbala 2002). Thus, close proximity of an exhaust hood to an airborne contamination source is of recognized importance. Further, the idealization of potential flow theory is known in the precedent literature as an essential means of calculating the expected performance of a specific inlet configuration. Finally, the effect of crosswind interference on inlet performance has at least been considered (e.g. Baturin 1971), if somewhat crudely.

What is desired here, however, is something much smaller and more mobile, for field use as the aerodynamic sampling "front-end" of a chemosensory apparatus. After covering what we learned from dogs, our goal is thus to define the requirements and parameters of sniffer function for such purposes, and to suggest some simple solutions that can meet these demands.

II. Canine Olfaction Experiments

Lacking much original technology for mobile indoor/outdoor airborne chemical trace sampling, we turned to the outstanding evolutionary example set by

canines. In studying the external aerodynamics of canine scenting, we hope to
learn principles that underlie the appropriate design of any mimicking device.

A series of live canine experiments was
conducted following procedures ap-
proved by the Penn State Institutional
Animal Care and Use Committee, and
using non-intrusive airflow visualization
(Merzkirch 1984). High-speed schlieren
videography of thermal air currents
(Settles 2001), light scattering by airborne
particles, and direct imaging of nostril
motion have been applied in order to
achieve a better understanding of the external aerodynamics of canine olfaction.

We studied both pet animals and
trained detection dogs ranging from 1–5
years old, including a female Golden Retriever, a male Yellow Labrador Retriever,
a female German Shepherd, a male Airedale, a female Malinois (trained for explosive detection), and a male Doberman
Pinscher (trained for firearms and drug
detection). Canine nares airflows were
observed in both open-air scenting and
scenting near a ground plane, with emphasis on the latter. A variety of scent
sources of a few mm in size was used,
ranging from food to neutral objects and
including TNT and marijuana scents for
the trained animals. In the majority of experiments, the dogs were encouraged to
investigate scent sources placed at the
center of the camera field-of-view upon
a ground plane. Edible treats were used
first to train the dogs in this routine. Later, inedible and unusual scent sources
were used to encourage investigative behavior. In the special case of the trained
dogs (whose handlers were present), the
drug and explosive scents were randomly
used on occasion, along with other scent
sources, to determine whether they elicited any distinctive olfactory behavior
(no such distinction was observed). In
some cases the animal was trained to
place its head in a headrest in order to

fix the nostril location for the camera,
whereupon an airborne scent was presented to it.

Initial observations were made with a
sensitive schlieren optical system (Settles
2001; see also chapter 23, this volume)
that revealed thermal variations in the olfactory airflows without distracting the
animals. Warm or cold scent sources ther-
mally tagged the scent-laden air in these
experiments. Schlieren video records at
up to 1000 frames/second revealed the
following: During panting respiration a
large turbulent jet is expired from the
mouth, obscuring any scent-bearing air
currents in the vicinity. The dog must
therefore normally stop panting in order
to sniff. Ordinary or short sniffs occurred
at a regular frequency of 3–5 Hz, with
each sniff cycle being composed of an inspiratory and an expiratory phase.

Inspired air enters the canine nares
from a distance of up to 10 cm or more
from the scent source, though the animal
narrows this distance to essentially zero if
allowed. The nature of the nostril inlet
airflow is inherently omni-directional: a
potential sink flow in which the velocity
varies inversely with some power of the
radial distance. Thus the detailed spatial
distribution of a scent source can only be
discerned when the nostril is brought into
very close proximity with it. We believe
such behavior is an evolutionary adaptation of canines, who depend upon dis-
cerning detailed olfactory "messages" to
a much greater extent than humans
(Thomas 1993).

To further illustrate this, sample potential-flow calculations were done of the airflow into a symmetric bulbous nares-like
inlet at various standoff distances from a
ground plane (as simulated by opposed
source-sink doublets of opposite sign).
In Fig. 1a, streamlines and lines of constant velocity potential are shown for
one such case, with an auxiliary plot on
the left showing the behavior of the

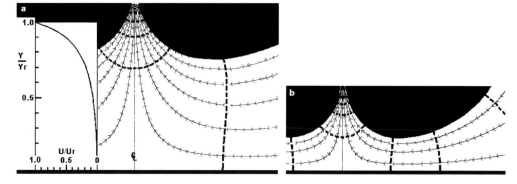

Fig. 1. Potential-flow air streamlines for simulated canine olfaction with nostril in close proximity to a perpendicular ground plane. The distance from the doublet singularity to the ground plane is halved between Figs. 1a and 1b. Inspired streamlines are arrowed, while lines of constant velocity potential are dashed

centerline airspeed. Compared to a reference location (Yr) deep within the inlet, the relative airspeed at half this distance to the ground plane falls to about 10%. This illustrates that the "reach" of such an inlet is extremely limited. Moreover, our flow visualization experiments show that olfactory inspiration at large standoff distances is readily disrupted by crossflows due to ambient wind motion. However, in Fig. 1b the standoff distance between inlet and ground plane is halved. Now more of the inspired airflow sweeps the local surface area directly beneath the inlet, and surface scents are diluted with a smaller volume flux of extraneous air from the surroundings.

Note that Fig. 1 is not meant to represent canine olfaction exactly, but only in principle. The 2-D computation and the lack of a wall viscous constraint in potential flow, among other factors, fail to model reality perfectly. Nonetheless, as noted earlier, relatively-simple potential flow calculations like this are quite valuable in studying inlet performance. Heinsohn and Cimbala (2002) cover the topic in detail.

Expiration, on the other hand, is capable of being vectored by the geometry of

the nostril, and can have many times the "reach" of a potential-flow inlet. In all cases we observed turbulent expired air jets directed to the sides of the dog's nose and downward, as in the case of rats and rabbits observed previously. A schlieren image of our $1\frac{1}{2}$-year-old Golden Retriever, Bailey, in profile illustrates the downward component of the expired airflow in Fig. 2.

The volume flux of expired air, estimated from the flow visualization results, is about 30 ml/s per nostril. This yields an expired-jet Reynolds number of roughly 260: high enough to expect turbulent flow. Indeed a very short region of laminar flow, if any, is followed by rapid transition to turbulence in the expired jets, seen clearly in Fig. 2.

The capacity for variable geometry is inherent in the anatomy of the canine external nares (Evans 1993), shown in Fig. 3. The canine nostrils are more than just simple orifices leading to the inner nose. The bulbous *alar fold* obstructs the nasal vestibule, so air must flow around it. Our simultaneous direct nostril videography and schlieren imaging revealed that, during inspiration, nostril dilation allows a pathway to open above the alar fold,

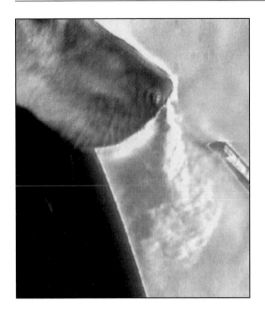

Fig. 2. Schlieren image showing vectored orientation of expired turbulent canine nostril air jets. Black object is headrest. Scent source is held in forceps

nares: It avoids discharging expired air back against a scent source, which would disrupt the inspiratory aerodynamics of olfaction. Instead, it diverts the expired streams away from the scent source. The same slit-nostril anatomy is found in other species (e.g. bears) that depend on olfaction, and is clearly very ancient: a successful adaptation that was naturally selected long ago.

In cases where they were presented with an inaccessible scent source, our canines displayed the markedly different sniff frequency and airflow pattern referred to earlier as a "long sniff" (Zuschneid 1973). Here the sniff frequency was only $1/3–1/2$ Hz and the expired jets were directed less ventralaterally than in the case of short sniffs. However, this behavior was never observed during free investigation of a scent source on a ground plane. We thus focus the remaining discussion on the olfactory aerodynamics of normal or "short" sniffing.

While sniffing a scent source on a ground plane, the orientation of the dog's nose is such that the expired air jets are directed to the rear and sides along the ground plane (Fig. 4). The turbulent mixing of these jets entrains the surrounding air, drawing an air current toward the nostrils from perhaps several cm ahead along the ground plane. This extends the aerodynamic "reach" of the inspiratory olfaction phase and helps, in some scenarios, to draw scents forth from concealed locations. It is a form of aerodynamic ejector or inducer (see also Fig. 5a).

Upon approaching a scent source on a ground plane, several of our test animals displayed a behavior that we call "scanning" (Fig. 5b): Instead of pointing the nose directly at the source, the nose was initially lowered to close nares proximity with the ground plane before reaching the source. The dog then moved its nose horizontally toward the scent source, pausing when the nostrils were directly

which we refer to as the "upper orifice." However, during the expiratory phase of olfaction, this upper pathway closes and the nostril "wings" (*nasal ala*) flare outward and upward, thus opening the mid-lateral slits (*nasal sulcae*) which lie directly beneath them. The location and geometry of the mid-lateral slit with respect to the alar fold is therefore responsible for diverting the expired airflow laterally and downward.

The dorsal nature of the observed inspiratory airflow through the "upper orifice" suggests that it may be channeled upward toward the olfactory epithelium. The epithelium is well inside the dog's nose (Evans 1993), lies above the normal respiratory flow region, and is in direct contact with the brain.

This observed variable-geometry, alternating aerodynamic inlet and outlet is a key evolutionary adaptation of the canine

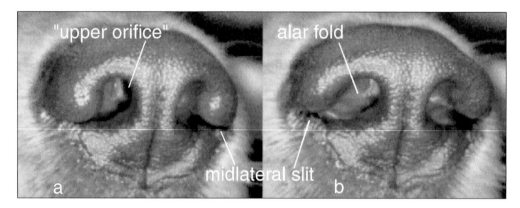

Fig. 3. The canine external nares during inspiration (**a**) and expiration (**b**). The bulbous alar fold is seen as an obstruction just inside the nostril

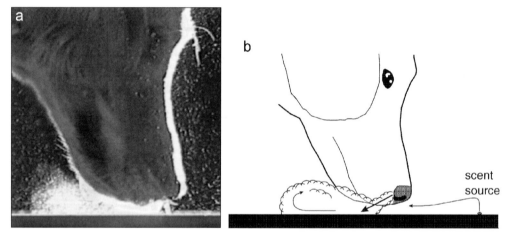

Fig. 4. Olfaction at a ground plane, side view, visualized using particle light scattering (**a**). Aerodynamic "inducer" effect due to entrainment by expired air jets (**b**)

above it, sniffing all the while. Often the nose was scanned past the scent source, allowing the expired air jets to impinge directly upon it. Finally the nose was returned to a position directly above the scent source for few more sniff cycles. This behavior promotes visual as well as olfactory inspection of an object or surface, and allows a local "survey" of spatial scent distribution. It also has the effect of disturbing any fine particles in the vicin-

ity of a scent source, through the impingement of the expired air jets.

In order to explore the interaction of canine olfactory airflows with surface particles, a light-scattering flow visualization technique was employed (Merzkirch 1984). The ground plane near the scent source was dusted with talcum powder having a median particle size of $2\,\mu m$. A spotlight was then directed obliquely across the ground plane toward, but not

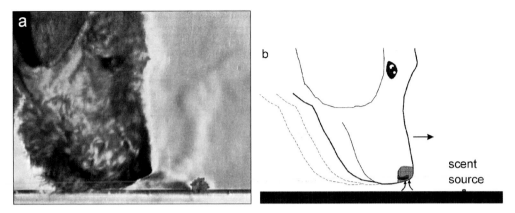

Fig. 5. (a) Schlieren image showing aerodynamic induction of scent into Airedale's nostrils. Scent source (a small flower) was warmed to make airflow visible. (b) Diagram of olfactory "scanning" of the ground plane

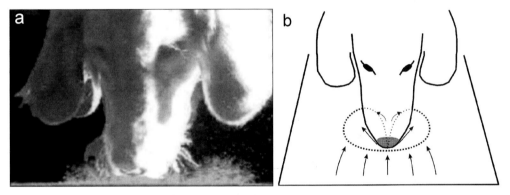

Fig. 6. Olfaction of a scent source on a ground plane, front view, visualized using particle light scattering (a). Diagram of the pattern of expired-air jet impingement upon the ground plane, showing the directions of entrained air streamlines (b)

directly into, the camera lens. With the camera aimed at a dark background, airborne particles became visible by scattered light.

Using this approach, we observed a strong interaction between the expired olfactory airjets and surface particles aft and to the sides of the nostrils (Figs. 4a and 6a). Impinging upon the ground plane, these airjets produce the classical aerodynamic phenomenon of wall jets initiated by starting vortices (Glauert 1956).

While most disturbed particles are blown away, our test animals often inspired some particles that were rendered airborne in this way. Particle inhalation is most prevalent when the dog "scans" past a scent source and returns, but some particles are inspired purely through the interaction of the inspiratory airstream and the particles on the ground plane.

Experiments with various particle sizes revealed that silica particles smaller than about 100 µm can be made airborne by the impingement of the expired airjets. Particle streak velocimetry at the canine nares further yielded an approximate inspired airspeed of 1 m/s. The characteristic time for the expired airjets to travel the typical 0.5 cm distance from nares to ground plane was estimated at 5 milliseconds. Since the expiratory phase of short sniffing lasts some 30 times longer than this, one may assume that the airjet impingement phenomenon shown in Figs. 4 and 6 reaches a quasi-steady state.

These results lead to a new view of the canine nostril as a variable-geometry aerodynamic sampler that functions alternately as a potential-flow sink inlet and an outlet flow diverter. Given the current level of interest in biomimicry (Benyus 1997), electronic noses, land mine detection (Strada 1996), and the extraordinary canine olfactory acuity in scenting explosives and drugs, these observations need to be considered in the design of aerodynamic samplers for chemosensing applications.

III. The Design of an Aerodynamic Sniffer

A. Background

Applied to a sniffer design, these canine results provide a basic framework of rules under which such a sniffer should function. First of all, it is clear that the sniffer inlet must get as close as possible to the scent source (remote sensing of airborne chemical plumes is a different topic not covered here). This is not only because of the limited "reach" of a potential-flow inlet, but also due to the rapid dissipation of an airborne trace signal by ambient air motion. Thus a proper aerodynamic sniffer needs to be able to approach and examine surfaces and terrain, and to provide a certain level of isolation from disruptive air currents (to be discussed later).

The design of the sniffer inlet plays a key role in optimizing its performance. A simple open-ended tube is not optimum, since it draws in air from the rear and sides as well as from the forward direction. From the field of industrial ventilation (Baturin 1972, Heinsohn 1991), it is well known that fitting such an inlet tube with a "collar" or flange limits the inspired airstream to a hemispherical capture zone in front of the inlet, and improves the inlet's "reach". Even better, evolution provides the bulbous canine nares with a faired "bellmouth" entrance, which conforms to the natural shape of potential-flow streamlines (as in Fig. 1), avoids flow separation at sharp edges, and thus has an entry loss coefficient only a few percent of that of a sharp-edged, open-ended tube (see Goodfellow and Tähti 2001).

Next comes the issue of exhaling the inspired airstream after the chemosensing step is finished. Dogs have evolved a complex variable-geometry nostril for this, but here biomimicry breaks down: No animal larger than a microbe has managed to evolve turbomachinery (Vogel 1994), but mankind has built small, light, quiet fans and blowers to move airstreams. Lacking these, the dog depends upon a bellows action that is complicated and unnecessary to mimic for the purpose of chemosensing. Instead, the inspired airstream should be exhausted elsewhere, after the detection step, via a once-through system.

Further comes the issue of particulates. In sniffing the ground for landmines, for example (Jenkins et al. 2000, Settles and Kester 2001), it is very likely that surface particulates carry adsorbed explosive-related traces that may be a thousand-fold more potent than traces in the surrounding air. Whether or not a dog collects and desorbs these particles during sniffing is an issue of current debate, but our observations of canine sniffing, given earlier, definitely show particles disturbed and rendered airborne by the exhaled nostril airjets. This can certainly be mimicked, and particle collection/desorption is not to be ignored in some realistic chemosensing applications.

The means to disturb surface particles artificially by way of auxiliary airjets attached to a sniffer inlet are relatively straightforward. Additional information on this topic can be found, for example, in Smedley et al. (1999).

Nevertheless particle collection and desorption may come at a high cost in terms of the time required to disturb surface particles, inhale them, collect them, desorb them, and direct the desorbed vapors to the detector.

The separation of solid-phase particulates from an airstream is a known technology, mentioned earlier in Sec. I.B of this chapter. It nonetheless requires adaptation to the present problem of sniff-sampling for chemosensing. Depending upon airflow rate and other circumstances, a cyclone separator uses centrifugal force to separate the heavy particles from the air, or else an impactor uses the inertia of the particles to collect them at a sharp turn while allowing the air to turn the corner (Liu and Pui 1986). Particles in the μm range and above can be removed from the sampled air using such devices. Thus the design challenge of integrating particle removal into an aerodynamic sniffer is within the current state-of-the-art. Even if such particles are not to be desorbed and sensed, their removal may still be necessary to avoid clogging the inlet orifices of some chemosensing detectors.

Inside the canine nose and that of other animals (Cheng et al. 1990), the mucous lining serves to trap particulates. This may be the natural way of sampling and chemosensing aerosol-borne trace substances.

B. A Basic Experiment on Aerodynamic Sniffing

Despite what was learned in our canine olfaction experiments, some basic questions still linger about the aerodynamics of sniffing, e.g. what flow rate is required, as a function of distance from a chemical trace source on a perpendicular surface, in order to acquire a detectable signal? Commercial sampler and industrial ventilation technology, reviewed earlier, does not address such questions. In fact, basic data on sniffer performance, as a function of such variables as sniffer-tube diameter, scent source diameter, standoff distance from a ground plane, and lateral displacement, have apparently never been obtained previously.

A basic experiment was thus designed to investigate the aerodynamic phenomena and performance of sniffing (Fig. 7). A stable thermal layer on a horizontal plane was used as a "scent" source according to the principle of Reynolds Analogy between heat and mass transfer (see, e.g. Heinsohn 1991). The detector was a thermocouple inside a simple sniffer tube.

The quantitative results of this experiment confirm the importance of sniffer proximity to localize an airborne trace source. For example, in steady-flow operation our experimental sniffer achieved a maximum sampled signal level at a 5 cm standoff distance h from the surface being sampled when the flow rate Q through the sampler was about 1.5 liters/second (Fig. 8). However, when h was reduced to 2.5 cm the required flow rate Q dropped by a factor of 5 and a higher maximum signal level was obtained.

Flow patterns produced by the sniffer of Fig. 7 were once again observed by the schlieren optical method. An example schlieren image is shown in Fig. 9a. Here, air in contact with the ground plane is symmetrically drawn into the flanged

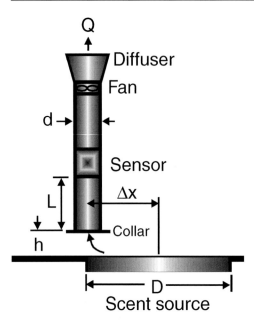

Fig. 7. Conceptual design of an aerodynamic sniffer (Settles and Kester 2001). Given a surface contamination zone ("scent source") of diameter D, the sniffer inlet of diameter d is positioned at standoff distance h above the ground plain. A fan draws airflow rate Q through the inlet, part of which is sampled by the sensor. After sampling, the airflow is discarded through a diffuser for pressure recovery. Lateral separation between the sniffer and the scent source is indicated by distance Δx

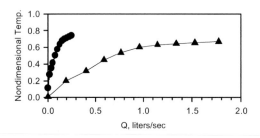

Fig. 8. Data from the basic sniffer aerodynamics experiment. $D = 25\,cm$, $d = 2.5\,cm$, $\Delta x = 0$, ● $h = 2.5\,cm$, ▲ $h = 5\,cm$

inlet of the sniffer. (Fig. 9a is inverted for clarity; due to the buoyancy of the warm thermal boundary layer on the ground

plane, the experiment was done upside-down.)

In transient sniffer operation a surprising behavior was observed: the sampled signal rose quickly to a "spike" at the sniff onset (Fig. 9b), followed by signal decline due (apparently) to depletion of the available trace-saturated boundary layer on the surface. It was also observed that steady-flow sniffing shows extreme sensitivity to transient disruptive air currents, which destroy the symmetric flow pattern of Fig. 9a and literally "blow the signal away."

These results suggest aerodynamic sampler design criteria for chemosensing and electronic-nose devices. Unfortunately the experiments had to be terminated before measurements could be made at larger standoff distances or with unstably-stratified trace-bearing layers and their resulting thermal plumes, and before the effect of ambient wind could be quantified.

C. Other Considerations in Aerodynamic Sniffer Design

Nonetheless, more can be said regarding design issues of the aerodynamic sampling "front-end" required to mate with a suitable detector for the purpose of chemosensing trace detection. A discussion of two specific topics follows.

1. Sniffer/Detector Impedance Matching

Detector characteristics play an important role in determining a sniffer design. For example, some chemosensory detectors may allow or even take advantage of a relatively-high air volume flux Q. In this case no pre-concentration step is called for, and high Q also means an extended inlet "reach." Most detectors, however,

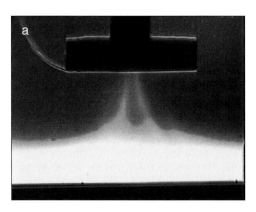
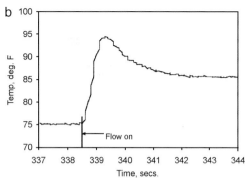

Fig. 9. a Schlieren image of stable thermal boundary layer being drawn into sampler inlet during basic experiments on sniffer aerodynamics. **b** Detected signal strength as a function of time from beginning of inhale airflow, demonstrating initial spike and signal depletion effect

can accept only a very small airflow input. Mass spectrometers and ion-mobility spectrometers, for example, input only milliliters/minute or less from a total sampled airflow rate that needs to be at least 10^4 to 10^5 times that value for effective aerodynamic sampling, based on Fig. 8. This leads to what we have dubbed, by electrical analogy, the "impedance matching" problem between the high airflows that must be sampled in practice and the miniscule airflows that actually feed the chemosensory trace detection step. If impedance matching is not done at all or is done poorly, most of the available signal is discarded. Since the signal level is a mere trace to begin with in applications like land mine detection (Jenkins et al. 2000) and anti-terrorism (Gowadia and Settles 2001), detection is likely to fail in such cases.

One solution to the impedance-matching problem is to pre-concentrate the trace material by passing the sampled airflow through an appropriate pre-concentrator, then discarding the main flow and desorbing the pre-concentrated trace material with a separate clean-gas flow matching the input requirements of the detector. This works well for aviation

security portals (Settles and McGann 2001, Settles 2000) but may prove too slow for other applications like land mine detection. Fast pre-concentration is thus a fruitful topic for further research.

In some cases, re-sampling from a relatively-high inspired airflow rate Q is needed in order to provide a very-low flow rate to the chemosensing detector. If the originally-inspired airstream is not homogeneous, trace detection can be defeated at this step. Under such circumstances flow mixing vanes or "turbulators" are called for between the sampler inlet and the chemosensor, not unlike the *ethmo-turbinates* inside a dog's nose.

2. Ambient Wind Isolation

Finally, it is especially important to provide some aerodynamic isolation for a sniffer that must operate, either outdoors or indoors, under conditions of a crossbreeze. As already noted, both in our experiments with dogs and in our basic sniffer experiments, an open, unshrouded inlet becomes extremely sensitive to crosswind disruption as the standoff distance is increased. A practical compromise thus becomes necessary among standoff distance to accommodate terrain and vegetation (in outdoor applications), suction airflow rate through the sniffer, and effectiveness in a crosswind situation.

Our preliminary experiments have shown some success in shrouding the sniffer inlet using soft brush bristles or a fine-mesh screen in order to reduce the effects of ambient crosswind on sampling efficiency. Such aerodynamic isolators work by producing a lateral pressure drop and thus a resistance to the crossflow, and they have the added advantage of providing a soft rather than a hard contact when the sniffer contacts surfaces under examination.

More work is needed on this topic, however. While no specific studies of the aerodynamic isolation problem are known, there is nonetheless relevant information in the fluid dynamics literature (e.g. Cant et al. 2002).

IV. Conclusion

Little precedent technology was found for mobile indoor/outdoor airborne chemical trace sampling. By studying the external aerodynamics of canine scenting, we have learned some principles that underlie the appropriate design of a mimicking device. Unable to improve upon the inherently-short range of a potential-flow inlet, evolution has instead given the canine an agile platform with which to bring its aerodynamic sampler into close proximity with a scent source.

Based further on experiments with a laboratory-prototype sniffer, we have attempted to state some guiding principles for aerodynamic sniffer design and practice. This work is preliminary, however, and more needs to be done on such issues as the effects of large sniffer standoff distance, lateral separation between sniffer

and trace source, unstable thermal stratification, and aerodynamic crosswind isolation.

Acknowledgement

This work was funded by the DARPA Unexploded Ordnance/Dog's Nose program, directed by Drs. R. E. Dugan and T. Altshuler. The assistance of J. D. Miller is gratefully acknowledged. Sec. IV is adapted from material originally presented in Settles and Kester (2001). We appreciate the assistance of C. J. Fahey and family, Dr. and Mrs. L. R. Bason, and Makor K-9 Inc., Napa CA, in the canine olfaction experiments.

References

Baturin VV (1971) Fundamentals of Industrial Ventilation. Pergamon, NY

Benyus JM (1997) Biomimicry. Morrow, NY

Bojsen-Moeller F, Fahrenkrug J (1971) Nasal swell bodies and cyclic changes in the air passages of the rat and rabbit nose. J Anatomy 110: 25–37

Cant R, Castro I, Walklate P (2002) Plane jets impinging on porous walls. Expts Fluids 32: 16–26

Cheng YS, Hansen GK, Su YF, Yeh HC, Morgan KT (1990) Deposition of ultrafine aerosols in rat nasal molds. Toxicology & App Pharm 106: 222–233

Evans HE (1993) Miller's Anatomy of the Dog, 3rd ed., Saunders, Philadelphia

Glauert MB (1956) The wall jet. J Fluid Mech 1(5): 625–643

Glebovskii V, Marevskaya A (1968) Participation of muscles of the nostrils in olfactory analysis and respiration in rabbits. Fiz Zhur SSSR 54: 1278–1286

Goodfellow H, Tähti E (eds) (2001) Local ventilation. Ch. 10 of Industrial Ventilation Design Guidebook. Academic Press, NY

Gowadia HA, Settles GS (2001) The natural sampling of airborne trace signals from explosives concealed upon the human body. J Forensic Science 46(6): 1324–1331

Haselton FR, Sperandio PGN (1988) Convective exchange between the nose and the atmosphere. J Appl Physiol 64(6): 2575–2581

Heinsohn RJ (1991) Industrial Ventilation. Wiley, NY

Heinsohn RJ, Cimbala JM (2002) Indoor Air Quality Engineering. Marcel Dekker, NY

Jackson CN, Sherlock CN, Moore PO (eds) (1998) Leak testing. In: Nondestructive Testing Handbook. Vol. 1, 2nd ed. Amer Soc. For Nondestructive Testing

Jenkins TF, Walsh ME, Miyares PH, Kopczynski JA, Ranney TA, George V, Pennington JC, Berry TE (2000) Analysis of explosives-related chemical signatures in soil samples collected near buried land mines. Technical Report ERDC TR-00-5, US Army Corps of Engrs CRREL

Liu BYH, Pui DYH (1986) Aerosol sampling and sampling inlets. In: Lee SD (ed) Aerosols. Lewis Publishers Inc, Chelsea, Michigan, USA

Merzkirch W (1984) Flow Visualization, 2nd ed. Academic Press, Orlando, Florida, USA

Settles GS (2000) Chemical trace detection portal based on the natural airflow and heat transfer of the human body. US Patent 6,073,499

Settles GS (2001) Schlieren and Shadowgraph Techniques. Springer-Verlag, NY

Settles GS, Kester DA (2001) Aerodynamic sampling for landmine trace detection. SPIE vol. 4394, paper 108

Settles GS, McGann WJ (2001) Potential for portal detection of human chemical and biological contamination. SPIE vol. 4378, paper 1

Smedley GT, Phares DJ, Flagan RC (1999) Entrainment of fine particles from surfaces by gas jets impinging at normal incidence. Expts Fluids 26: 324–334

Strada G (1996) The horror of land mines. Sci Am 278(5): 40–45

Syrotuck WG (1972) Scent and the Scenting Dog. Arner Pubs, Rome NY

Thomas EM (1993) The Hidden Life of Dogs. Houghton Mifflin, Boston

Vogel S (1994) Nature's pumps. Amer Scientist 82(5): 464–471

Zuschneid K (1973) Die Riechleistung des Hundes. Doctoral Diss, Vet Med, Free Univ Berlin

24. Microcantilevers for Physical, Chemical, and Biological Sensing

Thomas Thundat and Arun Majumdar

Abstract

Recent advances in designing and fabricating microcantilever beams capable of detecting extremely small forces, mechanical stresses, and mass additions offer the promising prospects of chemical, physical, and biological sensing with unprecedented sensitivity and dynamic range. Molecular adsorption on a cantilever surface can be detected by measuring the shift in the cantilever resonance frequency. If the adsorption is confined to one surface of a cantilever, the resulting differential surface stress leads to cantilever bending, thus providing an additional method of detecting molecular adsorption. Differential stresses can also be created by mismatches in thermal expansion of two cantilever materials, thus resulting in highly sensitive temperature and radiation sensors. By functionalizing cantilever beams with receptors, and by using the selectivity of chemical and biochemical receptor-ligand interactions, chemical and biological sensing can be achieved with high specificity and sensitivity. Here, we review the development of microcantilever sensors and sensor arrays, and present illustrative applications of chemical, physical, and biological sensing based on microcantilevers.

I. Introduction

Microcantilevers are the subject of considerable interest because of their potential as physical, chemical, and biological sensors. Figure 1 shows an electron micrograph of a typical single cantilever that is approximately 200 μm long, 1 μm thick and 40 μm wide. The deflection of such cantilevers can be measured by various techniques (optical, piezoresistive, capacitive etc.) which have resolution in

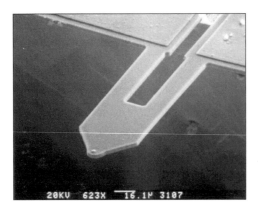

Fig. 1. Electron micrograph of a silicon micro-cantilever. The cantilever is about 200 μm long, 1 μm thick and 40 μm wide

the sub-nanometer range (Sarid 1994). Microcantilever spring constants are generally on the order of 10^{-3}–10^0 N/m, such that extremely small forces ($\approx 10^{-12}$–10^{-9} N) can be detected (Albrecht et al. 1990). To appreciate the level of sensitivity, it is worth noting that the force to break a single hydrogen bond is on the order of 10^{-12} N. The ability to detect such small forces makes the microcantilever an ideal device for detecting surface chemical phenomena such as molecular adsorption or binding, as well as physical phenomena such as thermal expansion mismatches (Thundat et al. 1994, 1995a,b; Chen et al. 1995). These are described in detail in this chapter.

In addition to static cantilever deflections, shifts in cantilever resonance frequency can be used to detect changes in cantilever mass or spring stiffness, both of which can change due to molecular adsorption. Because the thermal mass of microcantilevers is extremely small, they can be heated and cooled with a thermal time constant of less than a millisecond. This is advantageous for rapid reversal of molecular absorption processes and regeneration purposes. Finally, changes in the quality factor (Q-factor) of cantilever resonance can also be a sensitive method

of detecting variations in cantilever damping, which arise from processes such as changes in viscosity or some form of thermodynamic dissipation. While other sensors can generally be used for either physical (temperature, pressure, viscosity), chemical (gases, pH) and biological (DNA, proteins) sensing, a major advantage of using a microcantilever is that it provides a common platform for detecting various types of external stimuli as long as it can be "programmed" or functionalized to detect the stimuli. In addition, microcantilevers are readily adapted for fabricating multi-element sensor arrays, thus allowing high-throughput multi-analyte detection. It has recently become clear that the microcantilever platform offers an unparalleled opportunity for the development and mass production of extremely sensitive, low-cost sensors for real-time sensing of many chemical and biochemical species. In this review, we will discuss principles of operation, relevant theory, and the emergence of functional microcantilever sensors.

II. Theory

Microcantilever sensors can be operated in two different modes: resonance response variation and cantilever bending. The following provides some theoretical background for these modes of operation, and discusses the dominant noise sources that limit detection resolution.

A. Resonance Frequency

For a rectangular diving-board shaped cantilever, the spring constant for vertical deflection as derived for a load at the end is given by Eq. (1) (Sarid 1994):

$$K = \frac{F}{h} = \frac{Ewd^3}{4L^3} \tag{1}$$

where F is the load applied at the end of the cantilever and h is the resulting end deflection, E is the modulus of elasticity for the cantilever material and w, d, L are the width, thickness, and length of the beam, respectively. The fundamental resonance frequency can be written as

$$f = \frac{1}{2\pi}\sqrt{\frac{K}{m^*}} = \frac{d}{2\pi(0.98)L^2}\sqrt{\frac{E}{\rho}} \quad (2)$$

where ρ is the density of the cantilever material and m^* is the effective mass of the cantilever. The effective mass can be related to the mass of the beam, m_b, through the relation $m^* = nm_b$, where n is a geometric parameter. For commercially available silicon nitride cantilevers with spring constants of 0.06 and 0.03 N/m, the values of n are 0.14 and 0.18, respectively, and for the case of a diving-board shaped rectangular bar, n is 0.24. Assuming that surface adsorbed molecules have no influence on the spring constant, K, the mass of the adsorbed material can be determined from the initial and final resonance frequency and the initial mass of the cantilever as

$$\frac{(f_1^2 - f_2^2)}{f_1^2} = \frac{\Delta m}{m} \quad (3)$$

where f_1 and f_2 are the initial frequency and the final frequency, respectively, and Δm and m are adsorbed mass and initial mass of the cantilever, respectively. If the adsorption is confined to the free end of the cantilever, Eq. (3) needs to be modified to take the effective mass of the cantilever into account.

In certain cases when adsorption is uniform over the entire cantilever, the spring constant of the cantilever can change. From Eq. (2) it is clear that the resonance frequency can change with changes in mass as well as changes in spring constant in a competitive fashion:

$$df(m^*, K) = \left(\frac{\partial f}{\partial m^*}\right)dm^* + \left(\frac{\partial f}{\partial K}\right)dK$$

$$= \frac{f}{2}\left[\frac{dK}{K} - \frac{dm^*}{m^*}\right] \quad (4)$$

By designing cantilevers with localized adsorption areas at the terminal end of the cantilever (end loading), the contribution from differential surface stress [the dK/K term in Eq. (4)] can be minimized. In such cases, changes in resonance frequency can be attributed entirely to mass loading. The change in spring constant can be due to a number of causes, such as changes in the elastic constant of the surface film and dimensional changes of the cantilever or coatings.

B. Mass Sensitivity

If a discrete mass, m_d, is added at the end of a rectangular cantilever, the effective mass of the cantilever-adsorbate system is given by $m_{eff} = m^* + m_d$, where m^* is the effective mass of the cantilever. This results in an expression for the resonance frequency of the cantilever given by

$$f = \frac{1}{2\pi}\sqrt{\frac{Ewd^3}{4L^3(m_d + 0.24\rho wdL)}} \quad (5)$$

To compare the microcantilever to other gravimetric sensing devices, it is necessary to determine the sensitivity of the sensor. The mass sensitivity of a sensor is given by (Ward and Buttry 1990, Thundat et al. 1997).

$$S_m = \lim_{\Delta m \to 0} \frac{A}{f}\frac{\Delta f}{\Delta m} = \frac{A}{f}\frac{df}{dm} \quad (6)$$

where Δm and dm are normalized to the active sensor area of the device and A is the area of the cantilever. As can be seen from this expression, the sensitivity is the fractional change of the resonance frequency of the structure with addition of mass to the sensor. When applying this definition to the case of the microcantilever sensor, the sensitivities are

Table 1. Gravimetric Sensitivity Comparison of Acoustic Wave Devices

Device	Frequency f_o (MHz)	Sensitivity S_m (cm^2/g)	Minimum detectable mass density (ng/cm^2)
Microcantilever (end loading)	0.01–5	10,000	0.02
Microcantilever (distributed load)	0.01–10	5,000	0.04
Surface acoustic wave (SAW)	112	151	1.2
Quartz crystal microbalance (QCM)	6	14	10
Shear wave	104	65	1.0
Flexural wave (Lamb)	2.6	951	0.4

defined as

$$S_m = \begin{cases} \frac{1}{\rho d} & \text{distributed load} \\ \frac{-\xi}{2\rho(\xi d_d + 0.24d)} & \text{end load} \end{cases} \quad (7)$$

where ξ and d_d are the fractional area coverage and thickness of the deposited mass for the end load case. As one might expect, adding mass to the end of a cantilever results in a decrease of the resonance frequency of the device; and as mass is added in a distributed loading situation (corresponding to a cross-sectional thickness increase), the resonance frequency increases as indicated by the sign of the sensitivity expressions in Eq. (7). Table 1 shows the sensitivity comparison for a number of acoustic devices.

C. Adsorption-Induced Cantilever Deflection Approach

A fascinating and often overlooked aspect of molecular adsorption on a surface is the fact that these reactions are driven by free energy reduction of the surface. A free energy reduction leads to a change in surface tension or surface stress. While this produces no observable macroscopic change on the surface of a bulk solid, the adsorption-induced surface stresses are sufficient to bend a cantilever beam if the adsorption is confined to one surface of the beam. Adsorption-induced forces are applicable only for monolayer films and should not be confused with bending

due to dimensional changes such as swelling of thicker polymer films on cantilevers. Adsorption-induced stress sensors have sensitivities three orders of magnitude higher than frequency variation (for resonance frequencies in the range of tens of kHz) based on adsorbed mass. In addition, adsorption-induced cantilever bending is ideal for liquid-based applications where viscous damping reduces the sensitivity to detect shifts in resonance frequency.

Using Stoney's formula (Stoney 1909), we can express the radius of curvature of cantilever bending due to adsorption as (Butt 1996)

$$\frac{1}{R} = \frac{6(1 - \nu)}{Ed^2}\sigma \quad (8)$$

where R is the cantilever's radius of curvature; ν and E are Poisson's ratio and Young's modulus for the substrate, respectively; d is the thickness of the cantilever; and σ is the differential surface stress. The differential surface stress is the difference between the surface stresses on the top and bottom surfaces of the cantilever beam in units of N/m or J/m^2. A relationship between the cantilever displacement and the differential surface stress can be expressed as (Butt 1996)

$$h = \frac{3L^2(1 - \nu)}{Ed^2}\sigma \quad (9)$$

where L is the length and h is the deflection. Equation (9) shows a linear relation between cantilever bending and differential surface stress.

D. Thermal Motions of a Cantilever

The resolution of all measurements is ultimately limited by noise. When optical techniques are used to detect cantilever deflections (Sarid 1994), then the dominant noise source is thermal vibrational motion of the cantilever. It can be shown from statistical physics (Salapaka et al. 1997) that for off-resonance frequencies, thermal vibrations produce a white noise spectrum such that the root-mean-square vibrational noise, h_n, can be expressed as

$$h_n = \sqrt{\frac{2k_B TB}{K\pi f_o Q}} \qquad (10)$$

Here, k_B is the Boltzmann constant $(1.38 \times 10^{-23} \, \text{J/K})$, T is the absolute temperature (300 K at room temperature), B is the bandwidth of measurement (typically about 1000 Hz for dc measurement), f_o is the resonance frequency of the cantilever, K is the cantilever stiffness, and Q is the quality factor of the resonance which is related to damping. For resonance frequencies, the root-mean-square amplitude of vibration is obtained by multiplying the right hand side of Eq. (10) by the quality factor, Q. While many of the sensors described earlier become more sensitive with increased cantilever length, L, it is clear from Eq. (10) that lower spring stiffness, K, and resonance frequency, f_o, resulting from increased length also increase the thermal vibrational noise. Hence, there is an optimum geometry that maximizes the signal to noise ratio. Table 2 shows the properties of materials commonly used in chip fabrication. Using these values, the vibrational noise density, $h_n/\sqrt{B}$ of most widely used cantilevers (see Fig. 1) fall in the range of 10^{-13}–$10^{-12} \, m/\sqrt{Hz}$.

Table 2. Properties of Materials Commonly Used in Chip Fabrication

Material	Density, ρ [kg/m^3]	Elastic modulus, E [10^{11} N/m^2]
Silicon nitride	2,400	3.20
Silicon dioxide	2,300	0.72
Silicon	2,328	1.00
Gold	19,300	0.73
Aluminum	2,702	0.80

E. Q-Factor

The quality factor, or Q-factor, of a resonator is a measure of the spread of the resonance peak, Δf, and is thus related to energy loss due to damping $(Q = f_o/\Delta f)$. The lower the Q-factor, the more damped the oscillator. The Q-factor depends on parameters such as cantilever material, geometrical shape, and the viscosity of the medium. Typically, the Q-factor of a rectangular silicon cantilever in air is approximately 30. However, in liquids the Q-factor decreases by a factor of 10 making the resonance peak very broad. Therefore, measurements of adsorbed mass based on resonance frequency variation suffer from low resolution in liquid environment.

III. Experimental Setup

A. Cantilevers

Silicon or silicon nitride cantilevers, such as those used in atomic force microscopy, are typically 100–200 μm long, 20–40 μm wide, and 0.6 μm thick (available from Digital Instruments, CA, and Park Scientific, CA). For achieving chemical selectivity, the cantilevers are modified with appropriate chemically selective coatings. When cantilevers are used for detected adsorption using bending, the receptors are attached to only one cantilever surface. This is generally achieved by depositing a thin film of metal on the silicon or silicon nitride cantilevers and then using the chemical selectivity of attaching receptors to either the metal or the silicon/nitride surface.

B. Excitation Techniques

For dynamic measurements, the cantilever must be excited to its resonance frequency. The frequency of the excitation wave is scanned in a given frequency range, and the frequency of maximum cantilever amplitude is taken as the resonance frequency. The frequency spectrum from a position sensitive detector (PSD) shows the fundamentals as well as the harmonics of cantilever vibration. Cantilevers can also be excited into resonance by photothermal means — for example, by exposing the cantilevers to rectangular pulses of light or a heat source. In many cases the ambient temperature causes thermal motion in the cantilever and can thus be used as a passive excitation source.

C. Readout Techniques

The most common readout technique for cantilever motion is the optical beam deflection technique, which can detect cantilever motion with sub-Angstrom resolution limited only by the thermal noise that is determined by Eq. (10). A light beam from a laser is focused at the end of the cantilever and reflected onto a PSD (Sarid 1994). The bending of the cantilever changes the radius of curvature of the cantilever, resulting in a large change in the direction of the reflected beam. The dc signal provides the cantilever bending, while the ac signal yields the resonance frequency and the Q-factor.

Another attractive readout technique is based on piezoresistivity, whereby the bulk electrical resistivity varies with applied stress. Doped silicon exhibits a strong piezoresistive effect (Tufte and Stelzer 1993). The resistance of a doped region on a cantilever can change reliably when the cantilever is stressed with deflection. This deflection can be caused by changes in adsorption-induced stress or by externally applied forces. The variation in cantilever resistance can be measured using an external, dc-biased Wheatstone bridge (Boisen et al. 2000).

IV. Physical Sensing Using Single Microcantilever Sensors

Microcantilevers have been used to detect parameters such as force, pressure, acceleration, acoustic radiation, flow rate, viscosity, and temperature. Here, a few examples of detection are described.

A. Viscosity

The viscosity of gases and liquids can be determined by using the resonance response of microcantilevers. We have carried out viscosity measurements in varying concentrations of glycerol in water with viscosity varying over five orders of magnitude (absolute viscosities ranging from $\eta = 1.81 \times 10^{-4}$ to $14.9\,\mathrm{g\,cm^{-1}\,s^{-1}}$). Figure 2 shows the change in resonance response with the ratio of glycerol in water (Oden et al. 1996a). The sharp resonance peak in the figure corresponds to the ambient excitation in air. The resonance frequency, the amplitude, and the Q-factor decrease for the cantilever immersed in the glycerol–water mixture and the resonance response varies as a function of glycerol concentration. Both the viscosity and the density of the liquid can be determined from the resonance response.

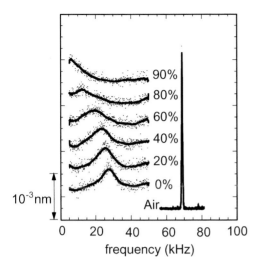

Fig. 2. Amplitude response versus frequency of a bar-lever for various concentrations of glycerol in water (expressed as % glycerol) as well as in air

Table 3. Comparison of Theoretical and Experimental Magnetic Susceptibilities

Material	Adsorbed mass (ng)	χ_s/χ_{ref} (expt.)	χ_s/χ_{ref} (theory)
ErO_3	1.43–1.57	1	1
$FeSO_4$	1.5–2.1	0.21–0.23	0.17
MnO_2	4–12	0.18–0.21	0.19
$CoCl_2$	1.53–4.79	0.17–0.19	0.17
$NiCl_2$	3.8–28	0.082–0.094	0.083

B. Magnetic Susceptibility

Microcantilevers can be used to detect extremely small magnetic forces which can be used to characterize the magnetic properties of materials (Cowburn et al. 1997, Finot et al. 2001). The magnetic force acting on a sample attached to a cantilever can be measured as a change in resonance frequency or bending of the cantilever. The shift in resonance frequency of the cantilever is proportional to the field gradient, whereas the deflection of the cantilever is proportional to the magnetic force. This concept can be used to measure magnetic susceptibilities (χ) of nanogram quantities of materials (Finot et al. 2001). In this case the sample is attached to the free end of a cantilever, and the resonance response and cantilever bending are measured with and without the magnetic field. Similar experiments were also conducted with a standard sample whose magnetic susceptibility is known. The ratio of the force on the sample relative to the standard

sample can be used to obtain the magnetic susceptibility of the sample by using Eq. (11):

$$\frac{\chi_s}{\chi_{ref.}} = \frac{m_{ref}M_s\Delta h_{ref}}{m_s M_{ref}\Delta h_s} \qquad (11)$$

where M and Δh are the molecular weight and the cantilever deflection in a magnetic field, respectively. The subscripts s and ref refer to the sample and the standard (reference) sample, respectively. The mass of the sample can be calculated from the resonance frequency variation with and without sample. Table 3 shows a comparison of experimental and theoretical values of the ratio of susceptibilities. The experimental magnetic susceptibilities were calculated from the ratio of $\frac{\chi_s}{\chi_{ref}}$ using ErO_3 as a standard ($\chi_{ErO_3} = 73.9\times10^{-6}$ c.g.s).

C. Temperature and Infrared Radiation Sensing

A highly sensitive micromechanical temperature sensor can be created by coating one side of a cantilever with a material of vastly different thermal expansion coefficient (Barnes et al. 1994, Gimzewski et al. 1994, Thundat et al. 1994, Datskos et al. 1996, Wachter et al. 1996, Lai et al. 1997, Varesi et al. 1997). A temperature change produces a differential thermal expansion of the two cantilever materials, which leads to cantilever bending. Cantilever deflection, h, as a function of temperature change ΔT can be expressed as (Lai et al. 1997)

$$h = 3(\alpha_2 - \alpha_1)\left(\frac{n+1}{K}\right)\left(\frac{L^2}{d_1}\right)\Delta T$$

where $K = 4 + 6n + 4n^2 + \phi n^3 + \dfrac{1}{\phi n}$;

$$n = \frac{d_2}{d_1}; \phi = \frac{E_2}{E_1} \qquad (12)$$

where subscripts 1 and 2 refer to the two cantilever materials, d and E are the thickness and elastic modulus of each layer, respectively, α is the thermal expansion coefficient, and L is the cantilever length.

Temperature changes on such a sensor can be induced by absorption of radiation power or by an exothermic or endothermic chemical reaction. For an incident IR flux modulated at frequency ω, $q(t) = q_o e^{-i\omega t}$, the modulated temperature change of the cantilever can be expressed as $\Delta T = q_o A\psi/G(1+\omega^2\tau^2)^{1/2}$, where ψ is the absorptance, A is the IR absorbing area of the pixel, G is the thermal conductance between the cantilever and the surroundings, and $\tau = C/G$ is the thermal response time constant. Here, C is the heat capacity of the pixel: $C = \sum(\rho w d L c)_i$ where ρ is the density and c is the heat capacity. At low frequency ($\omega\tau \ll 1$), $\Delta T \approx q\psi A/G$, whereas, at high frequency ($\omega\tau \gg 1$), $\Delta T \approx q\psi A/(\omega C)$. The conductance G arises from three main contributions, $G = G_{leg} + G_f + G_{rad}$, where G_{leg} is the conductance due to heat conduction through the cantilever leg and to the chip substrate, G_f is the conductance by the surrounding fluid, and G_{rad} is the thermal radiative conductance between the pixel and its surroundings. Conductive losses through the cantilever leg can be expressed as $G_{leg} = \sum(kwd/L)_i$ where k is the thermal conductivity of the leg material, and w, d, and L are the width, thickness, and length, respectively, of the

cantilever leg. The fluid conductance, G_f, is generally due to heat diffusion in the fluid and can be estimated as $G_f = k_f L_c$, where k_f is the thermal conductivity of the fluid and L_c is a characteristic dimension, such as the cantilever length. The radiative conductance is given as $G_{rad} = 4\sigma(\varepsilon_1 + \varepsilon_2)AT^3$ where σ is the Stefan-Boltzmann constant, T is the cantilever temperature and ε_1 and ε_2 are the emissivities of the two cantilever surfaces. Barnes et al. (1994) demonstrated that the cantilever detector had an observed sensitivity of 100 pW, corresponding to an energy resolution of 150 fJ, and proposed using the sensor as a femtojoule calorimeter. Later, Varesi et al. (1997) demonstrated that resolution can be improved to 10 pW for radiation power, 20 fJ for energy, and 10 µK for temperature.

The idea of using bimaterial cantilevers for radiation sensing was further developed by designing and fabricating cantilever arrays (Manalis et al. 1997, Perazzo et al. 1999) for infrared (IR) imaging (Oden et al. 1996b). Figure 3 shows an electron micrograph of a cantilever array that was designed to detect IR radiation in the 8–14 µm wavelength range, relevant for night vision (Perazzo et al. 1999). Thermal design of each pixel minimized the cantilever conductance in order to maximize the cantilever temperature rise for a given incident flux. Thermomechanical design of the cantilever maximized the cantilever deflection given by Eq. (12) by optimizing the thickness ratio and cantilever length. Finally, the optical readout for measuring deflections of an array of 300 × 300 bimaterial cantilevers relied on Fourier optics (Zhao et al. 1999). Also shown in Fig. 3 is an IR image of a human hand using the cantilever array (Perazzo et al. 1999, Zhao et al. 1999). At present, the best noise-equivalent temperature difference (NETD) that such an imaging system can detect is about 200 mK, although calculations

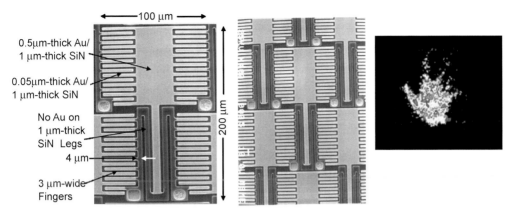

Fig. 3. Electron micrograph of a small region of a focal plane array (FPA) for infrared imaging, showing the structure of each pixel and the pixel tiling pattern. Also shown is an infrared image of a human hand obtained by the cantilever-based IR imaging system

suggest that the NETD could be improved to 10 mK.

D. Charged Particle Detection

Microcantilevers have been successfully demonstrated as micromechanical radiation detectors for alpha particles (Stephan et al. 2002). In this experiment alpha particles were allowed to impinge on an electrically insulated metallic wire of 1 mm diameter. A microcantilever kept at a fixed distance of a few nanometers was deflected as a function of residual charge accumulation on the surface. In addition to bending, the resonance frequency of the microcantilever shifted due to electrostatic force (non-uniform fields produce a change in resonance frequency because the force constant of the cantilever is modified as the result of field gradients). The damping rate of the resonance also changed because of the presence of a non-uniform field gradient. Figure 4 shows the variation in the damping constant c as a function of alpha particle exposure time. Here, the damping constant c represents the degree of damping such that, in response to a step input, the can-

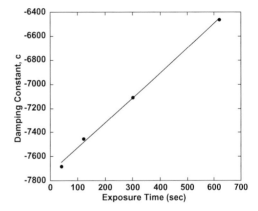

Fig. 4. Cantilever damping parameter as a function of the exposure time of the detector system to the alpha particle beam. The damping constant c represents the degree of damping such that in response to a step input, the cantilever oscillates as $h = h_o \sin(2\pi f_o t) \exp(ct)$, where f_o is the resonance frequency

tilever oscillates with an amplitude, $h = h_o \sin(2\pi f_o t) \exp(ct)$, where f_o is the resonance frequency. The minimum detectable particle fluence using this technique was calculated to be around 1000 particles in air, or about 3.2×10^{-4} pC. The device can be optimized by using large area collectors and by increasing the resolution of frequency measurement.

V. Chemical Sensing Using Single Microcantilever Sensors

Despite its high sensitivity, the cantilever platform does not offer any intrinsic chemical selectivity. Therefore, the use of a cantilever as a chemical sensor requires modification of one of its surfaces by application of a chemically selective layer. Many polymeric materials developed to provide selectivity in chemical sensors based on surface acoustic wave (SAW) devices and quartz crystal microbalances (QCMs) can be used as selective coatings on microcantilevers (Grate et al. 1993, McGill et al. 1994, Ballantine et al. 1997). Although many of these polymeric coatings provide high sensitivity and selectivity, obtaining reproducible sensor performance remains a challenge. This is mainly due to the lack of techniques and methods by which these polymers can be reliably and uniformly applied on cantilever surfaces. It is clear from the sensitivity equations that thinner cantilevers provide higher sensitivity because they have lower force constants. As microcantilevers continue to be made smaller, efficient ways of coating them with thin layers of selective materials will continue to be a challenge. Many of the challenges faced with selective coatings, however, can be overcome by modifying the cantilevers with chemically selective self-assembled monolayers (SAMs).

We have demonstrated many extremely sensitive microcantilever chemical sensors based on selective coatings (metals, polymers, and SAMs) on the cantilever for detection analytes in the vapor as well as the liquid phase. Vapor phase analyte detection has been demonstrated by many groups (Baller et al. 2000, Battiston et al. 2001, Betts et al. 2000, Boisen et al.

2000, Jensenius et al. 2000). We provide below a brief synopsis of chemical sensing using microcantilevers.

A. Detection of pH

Bimaterial microcantilevers can be excellent pH sensors (Fritz et al. 2000 b, Ji et al. 2001). Micromechanical pH detection is based on the deflection of cantilevers due to charging of the surface species when in contact with a liquid solution. As the surfaces accumulate charges proportional to the pH of the surrounding liquid, the cantilever undergoes bending due to differential surface stress created by the electric double layer. The cantilevers can be modified with selective layers that will respond sensitively to certain ranges of pH. A sensitivity of 10^{-3} pH units was achieved using silicon nitride cantilevers for the pH ranges of 1 to 4 and 10 to 14. The sensitivity for other ranges was significantly lower. Therefore, an array-based approach is essential for covering a wide pH range. Experiments are presently under way to develop different coatings to enhance sensitivity and pH range. The influence of other ions was found to be negligible below a concentration of 10^{-2} M.

B. Hydrogen

It is well-established that palladium selectively absorbs hydrogen, resulting in structural changes in the palladium film. Microcantilevers with a thin layer of palladium show excellent sensitivity to hydrogen gas (Hu et al. 2001). Since cantilever bending is due to diffusion of hydrogen into palladium film, thinner films reach equilibrium faster than thicker films. Repeated exposure to 1% hydrogen gives responses repeatable to better than 95%. Hydrogen concentration can be obtained by measuring either the maximum deflection signal of a cantilever sensor, which can take a long time, or the rate of bending, which is much faster, both of which follow Sievert's law.

C. Mercury Vapor Adsorption

Unlike hydrogen absorption in palladium film, which is a bulk absorption system, mercury adsorption on gold film is an excellent example of surface adsorption. It has been reported in the literature that, at room temperature, mercury atoms reside only on the top layer or near the surface of a gold film. At higher temperatures, adsorbed mercury atoms diffuse into the deeper layers of a gold film, up to several nanometers from the surface. Experiments have shown that the cantilever bends away from the gold surface when mercury vapor is adsorbed, indicating a relaxation of tensile stress on the gold surface. Also, the rate of cantilever bending is directly related to mercury concentration. Once the flow of mercury vapor is stopped (nitrogen gas flows continuously at the same flow rate), the cantilever stops bending (Thundat et al. 1995b). Although the mercury concentration is not a linear function of the cantilever bending amplitude, the bending rate shows a linear response to mercury concentration.

Adsorption-induced surface relaxation implies that negative stress increases with the coverage of the monolayer. The observed surface relaxation in the gold film, induced by mercury adsorption, reflects a decrease in the free energy on the gold surface. By preparing gold films of different thicknesses and exposing them to the same mercury concentration, it has been shown that the cantilever bending is independent of film thickness (Thundat et al. 1995b). Therefore, mercury sensors based on gold films are controlled primarily by surface area available for adsorption, not the gold film thickness. Various

experimental methods (including ours) have confirmed that the adsorption of mercury atoms on a gold surface is a continuous process until about one monolayer of coverage is reached. As a consequence, the adsorption rate decreases with coverage.

Mercury adsorption on cantilevers with a thin layer of gold on one side shows that the resonance frequency increases as a function of exposure to mercury vapor (Thundat et al. 1995b). However, when the gold film is confined to the free end of the cantilever, a linear decrease in resonance frequency is achieved. The increase in resonance frequency of a cantilever with a uniform layer of gold on one side is most likely due to changes in the spring constant. Figure 5 shows the resonance frequency variation of a cantilever coated with gold along its entire length (cantilever mass $= 34.6$ ng) when exposed to mercury vapor (adsorption rate for Hg $= 553$ pg/min). A calculation carried out using Eq. (3) and the data in Fig. 5 shows that the spring constant K increases at a rate of about 0.001 Nm^{-1} min^{-1}. The frequency response observed in Fig. 5 appears to be influenced by variations both

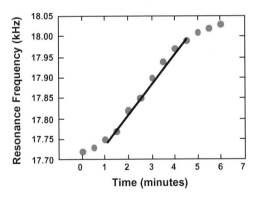

Fig. 5. Resonance frequency variation of a microcantilever with uniform gold film on one cantilever surface, as a function of mercury vapor exposure. The resonance frequency increases due to variation in the spring constant of the cantilever

in the spring constant and the mass loading. The resonance frequency response can be explained by the competing effects between spring constant variation and mass loading.

D. Benzene

Gas sensing can be achieved by adsorbing gases on chemically selective polymer films coated on a cantilever surface. The response of such functionalized microcantilevers depends on the volatility of the vapor and the strength of the coating-vapor solubility interactions (e.g., hydrogen bonding, dispersion, and dipole–dipole interactions). Since the microcantilever response to a given analyte depends on the functional end-groups of modifying polymers, judicious selection of polymers can lead to significant differences in the cantilever response. The selection of coatings for an array is based mainly on the sensitivity and selectivity of analyte-coating solubility interactions. The response of a polymer-coated microcantilever to a particular analyte depends on the partition coefficient, Γ, which represents the equilibrium vapor-polymer solubility at a given temperature:

$$\Gamma = C_p/C_v \qquad (13)$$

where C_v is the vapor concentration in the gas phase, while C_p is the concentration in the polymer. The partition coefficient measures the overall strength of interaction such that the larger the value of Γ, the stronger the interaction. The value of Γ can be obtained from gas-liquid chromatographic analysis. Selection criteria for coatings include: 1. stability over large ranges of time and temperature; 2. ability to discriminate between analytes; 3. ability to regenerate; 4. fast response.

To demonstrate a polymer-coating-based microcantilever sensor for the detection of volatile organic compounds, we

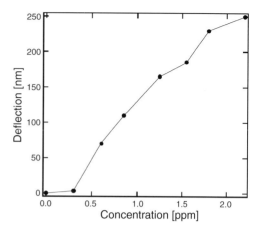

Fig. 6. Detection of benzene using polychloroprene coated cantilevers. The sensitivity can be increased by two orders of magnitude by using thinner cantilevers

modified a cantilever for the detection of benzene. Thin films of polychloroprene have been shown to have excellent sensitivity for benzene in SAW applications. We applied a thin layer of polychloroprene by dipping the cantilever in a solution of 0.3% polychloroprene by weight dissolved in a mixture of acetone and toluene. Since the silicon cantilever had a thin layer of gold on one side, the thin polychloroprene film was deposited mainly on one side. Figure 6 shows the bending of the polychloroprene-coated cantilever as a function of benzene concentration (Thundat T, Oak Ridge National Laboratory, unpublished results).

E. Cs^+ Ions

We have successfully detected Cs^+ ions in a flowing liquid solution using an ion-selective SAM-coated microcantilever capable of detecting cesium ions in the presence of high concentrations of potassium or sodium ions (Ji et al. 2000). The cesium recognition agent used in this work was 1,3-alternate

25,27-bis(11-mercapto-1-undecanoxy)-26, 28-calix[4]benzocrown-6, bound to a gold-coated microcantilever. The crown cavity of the 1,3-alternate conformation of calix[4]benzocrown-6 has been shown to be very suitable for accommodating cesium ions, with Cs/Na and Cs/K selectivity ratios in excess of 10^4 and 10^2, respectively, determined using a solvent extraction technique. Binding constant values of 10, 2×10^4, and 2.5×10^6 have also been determined by a fluorescence technique for Na^+, K^+, and Cs^+ in 1:1 methanol/methylene chloride solution. Based on the high selectivity exhibited by these compounds, the receptor molecule was designed and anchored onto the gold surface of the microcantilever by standard techniques.

Figure 7 shows the response of the SAM-coated cantilevers to Cs^+ concentration. The most dramatic response is exhibited when the concentration of cesium is in the range of 10^{-7} to 10^{-11} M. In contrast, the cantilever response to K^+ (the most prevalent interfering ion) in the same concentration range is very small at short response times (e.g., 20 nm deflection for 1×10^{-8} M solution

of K^+ compared to 330 nm deflection for Cs^+ at the same concentration). For Cs^+ concentrations larger than 10^{-6} M, the bending response at equilibrium reaches its maximum value at around 330 nm. A blank test performed on a gold-coated silicon nitride cantilever without the SAM revealed that even at high Cs^+ concentration (e.g., 10^{-3} M), the bending response was unaffected (with the bending response being the same as in de-ionized water).

The relationship between the cantilever displacement, h, and the molar concentration of Cs^+ can be obtained by manipulating Eq. (9). Since surface stress is directly proportional to ion absorption by the microcantilever, Eq. (9) can be rearranged to the form shown in Eq. (14),

$$h = B \left(\frac{3(1 - \nu)L^2}{d^2 E} \right) \left(\frac{b[M]R_0}{1 + K[M]} \right) \quad (14)$$

where B is a constant, b is the complexation constant between the ion receptor and the ions present in solution (1:1 stoichiometry), M is the concentration of ions in solution, and R_0 is the moles of ion

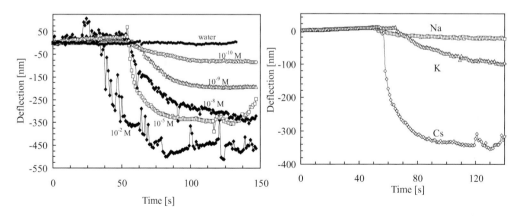

Fig. 7. Left: Bending response of three SAM coated microcantilevers as a function of time after exposure to different concentrations (10^{-2}, 10^{-5}, 10^{-8}, 10^{-9} and 10^{-10} M) of Cs^+ ions in solution. Right: Cantilever bending for Cs^+, K^+, and Na^+ at 10^{-5} M concentrations

receptor present on the cantilever before complexation.

Since the change in concentration of ions present in solution after absorption by the receptor is essentially unchanged in a continuous flow system, the value of M will be the same as the initial concentration of ions. The complexation constants determined from the plot of $1/h$ vs $1/[M]$ for cesium (b_{cs}) and potassium (b_k) using Eq. (14) are $2 \times 10^9 \, M^{-1}$ and $1.6 \times 10^7 \, M^{-1}$, respectively. These values are much higher (by three orders of magnitude) than the corresponding values observed for the derivatives of similar compounds in 1:1 methanol/methylene chloride solution, but the ratio b_{cs}/b_k is essentially the same. Similar enhancements in the association constant values for other SAM systems (relative to the free molecule) have also been reported. The large value of association constant observed for Cs^+ ions indicates that binding is essentially irreversible. This notion is further supported by our experimental results, which showed that cycling pure water through the system (to rinse the SAM-coated microcantilever containing Cs^+ ions) for several hours failed to regenerate the initial reading (water curve in Fig. 7).

The bending response of the SAM-coated microcantilever on Cs^+, K^+ and Na^+ complexation was also compared for the same concentration of each ion $(10^{-5} \, M)$ (see Fig. 7). The results indicated that the SAM-coated microcantilever was much more selective towards Cs^+ ions than towards K^+ and Na^+ ions. In fact, Na^+ ions have a minimal effect (if any) on the bending response, while K^+ ions exhibit enough sensitivity to interfere (as a perturbing ion when present) in the detection of cesium ions.

Our data show that the sensitivity of this cantilever-based sensor for in situ measurements is several orders of magnitude better than the sensitivities of currently available ion selective electrodes (ISE).

VI. Biological Sensing Using Single Microcantilever Sensors

The ability to operate a microcantilever sensor in liquid solution with extreme sensitivity makes it an ideal choice for the development of biosensors. Biosensing can be based on either mass adsorption or surface stress variation. In this section, we describe recent work on detecting specific biomolecular interactions such as DNA hybridization, antigen-antibody binding, and protein-ligand binding (Raiteri et al. 1999, Fritz et al. 2000 a, Hansen et al. 2001, Wu et al. 2001a,b).

A. Detection of DNA

Single-stranded DNA (ssDNA) can be immobilized using gold-thiol strong binding on one side of a cantilever by coating that side with gold and using a thiol linker at one end of ssDNA. ssDNA bound to the cantilever acts as the probe (or receptor) molecule for the target complementary strands. Figure 8 shows cantilever deflection as a function of time; in this case, the probe ssDNA was maintained at 20 nucleotides long (20 nt), whereas the target ssDNA varied in length, with lengths of 20 nt, 15 nt, 10 nt, and 9 nt. Note that single nucleotide length differences can be determined mechanically, while non-complementary ssDNA produces no mechanical signals. Wu et al. (2001a) investigated the origins of cantilever deflection due to biomolecular interactions and found that the deflection resulted from a change in free energy of one cantilever surface. The relative importance to enthalpic and entropic contributions to free energy determined the direction of cantilever motion. By controlling the entropy contribution to the free energy, the direction of motion could be controlled.

Hansen et al. (2001) have shown that for a 25 nt long probe ssDNA, single base pair mismatches in 25 nt long target ssDNA can be mechanically detected. In addition, if the target ssDNA is 10 nt long, single and double base pair mismatches produce cantilever deflections of opposite sign compared to those of 25 nt long target ssDNA. This observation has so far remained unexplained.

B. Detection of PSA

Antigen-antibody interactions are a class of highly specific protein–protein binding that play a critical role in molecular biology. When antibody molecules were immobilized to one surface of a cantilever, specific binding between antigens produced cantilever deflection. Figure 9 shows cantilever deflection as a function of time for quantitative detection of prostate-specific antigen (PSA) against a background of bovine serum albumin (BSA). Similar tests have been performed against backgrounds of human serum albumin (HSA) and human plasminogen, both of which are found abundantly in human sera. Figure 10 shows the steady-state cantilever deflections as a function of PSA concentration for three different cantilever lengths and thicknesses. It is clear that longer cantilevers are more sensitive. However, what is critical is that PSA concentrations can be detected below concentrations of 4 ng/ml, which is the clinical threshold for prostate cancer. Notably, one can detect concentrations down to 0.2 ng/ml. Since for the same PSA concentrations, cantilever deflections varied with their geometry, it is important to standardize these measurements by evaluating the surface stress. This can be done using Stoney's formula [Eq. (8)]. Indeed, all the data for different cantilever lengths collapse onto a single curve, suggesting that the surface stress is more fundamental than cantilever deflections (Wu et al. 2001b).

Since cantilever motion originates from the free energy change induced by specific biomolecular binding, this technique offers a common platform for high-throughput, label-free analysis of protein–protein binding, DNA hybridization, and DNA-protein interactions, as well as drug discovery. Hence, one can build an array of cantilever beams, each specialized to detect one specific molecule or a combination of molecules. One can design and fabricate an array of microcantilevers that will enable simultaneous experiments with hundreds of proteins and DNA sequences, thus providing a glimpse of the intricate molecular interactions that operate within a living cell. The nanomechanical assay described here needs no label and can be performed in a single reaction without additional reagents. The potential advantages of a label-free assay, which can measure multiple analytes in a single step without the addition of other reagents, are enormous and could ultimately translate to much lower cost per test.

VII. Conclusions

In general, the sensitivity and specificity of microcantilever sensors can be optimized by careful geometrical design of the cantilever and its coatings. For example, the mass sensitivity of a cantilever is proportional to $(\rho d)^{-1}$, where ρ is the density of the cantilever material and d is the thickness of the cantilever. Therefore, by reducing the thickness of the cantilever, mass sensitivity can be improved by several orders of magnitude. However, smaller thickness requires shorter cantilever length and increased resonance frequency. The sensitivity of detection can also be increased by judicious optimization of damping effects by choice of cantilever materials, operating media, and the geometry of the cantilever. The cantilever deflection approach requires longer cantilevers with smaller spring constants.

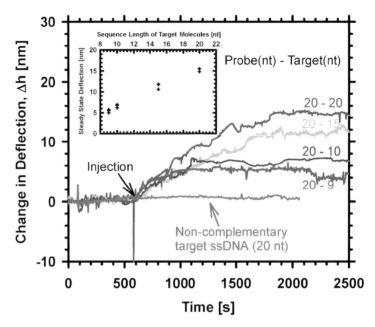

Fig. 8. Changes in Au–Si cantilever deflection due to hybridization of a probe ssDNA (sequence K-20 at 50 ng/μl or 8 μM concentration) in the distal end with complementary target ssDNA of different lengths — 20 nt, 15 nt, 10 nt, and 9 nt (40 ng/μl or 3–6 μM concentration). Also shown is the absence of cantilever deflection for a non-complementary target ssDNA (sequence NC20). The data clearly suggests that differences in nanomechanical motion due to one nucleotide difference in length can be observed

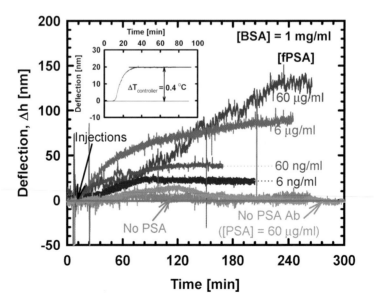

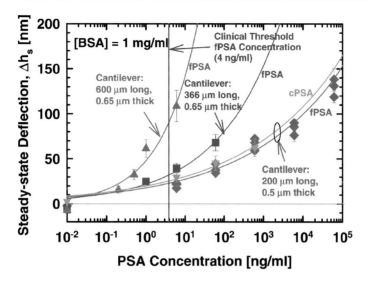

Fig. 10. Steady state cantilever deflections as a function of fPSA and cPSA concentrations for three different cantilever geometries. Note that longer cantilevers produce larger deflections for the same PSA concentration, thereby providing higher sensitivity. Using 600 µm long and 0.65 µm thick silicon nitride cantilevers, it was feasible to detect fPSA concentration of 0.2 ng/ml. Every data point on this plot represents an average of cantilever deflections obtained in multiple experiments done with different cantilevers, whereas the range of deflections obtained from these experiments is shown as the error bar. The only exception is the data for fPSA detection using 200 µm cantilevers, where the data (diamonds) from multiple experiments at a given concentration is shown as a cluster plot. The error bar in each of these data points represents the fluctuation of the cantilever during the particular measurement

The chemical selectivity of cantilevers depends on the selection of surface coating for chemical interactions.

The primary advantages of the microcantilever method are; 1) sensitivity based on the ability to detect cantilever motions with subnanometer resolution; 2) ability to fabricate in multi-element sensor arrays; 3) ability to work in air, vacuum, or liquids; 4) capability of physical, chemical and biological sensing by engineering cantilever design and functionalization. To our knowledge, no other sensing technology offers such versatility.

Acknowledgments

We would like to thank Drs. A. Passian, G. M. Brown, G. Y. Chen, S. Cherian, E. Finot, K. M. Hansen, Z. Hu, H. F. Ji, and A. Mehta, as well as R. Cote and R. Datar from the University of Southern California Medical School for their

Fig. 9. Cantilever deflection versus time for free PSA (fPSA) detection sensitivity against a background of 1 mg/ml of BSA using 200 µm long and 0.5 µm thick silicon nitride microcantilevers. fPSA detection was feasible over a concentration range 6 ng/ml to 60 µg/ml using this cantilever geometry. Note the lack of deflection in absence of both the ligand (anti-PSA antibody) and the ligate (fPSA). The inset plots cantilever deflection for a 0.4 K temperature change of the system and shows that the thermal stability is within the noise of the system

contributions and help with cantilever-based research. Financial support from DOE's Office of
Biological and Environmental Research (OBER),
EMSP, and the National Institutes of Health-
National Cancer Institute (NIH-NCI) is gratefully
acknowledged. Oak Ridge National Laboratory is
managed by UT-Battelle, LLC, for the U.S. Dept.
of Energy under contract DE-AC05-00OR22725.
The work of A. Majumdar was supported by DOE
and NCI.

References

Albrecht TR, Akamine S, Carver TE, Quate CF
  (1990) Microfabrication of cantilever styli for
  the atomic force microscope. J Vac Sci Technol
  A8: 3386–3396
Ballantine DS, White RM, Martin SJ, Ricco AJ,
  Frye GC, Zellers ET, Wohltjen H (1997)
  Acoustic Wave Sensors: Theory, Design, and
  Physicochemical Applications. Academic Press,
  San Diego
Baller MK, Lang HP, Fritz J, Gerber Ch,
  Gimzewski JK, Drechsler U, Rothuizen H,
  Despont M, Vettiger P, Battison FM, Ramseyer
  JP, Fornaro P, Mayer E, Guntherdot HJ (2000)
  A cantilever array based artificial nose. Ultra-
  microscopy 82: 1–9
Barnes JR, Stephenson RJ, Welland ME, Gerber
  Ch, Gimzewski JK (1994) Photothermal spec-
  troscopy with femtojoule sensitivity using a mi-
  cromechanical device. Nature 372: 79–81
Battiston FM, Ramseyer JP, Lang HP, Baller,
  MK, Gerber Ch, Gimzewski JK, Mayer E,
  Guntherdot HJ (2001) A chemical sensor based
  on a microfabricated cantilever array with simultaneous resonance-frequency and bending
  readout. Sens Actuator B 77: 122–131
Betts TA, Tipple CA, Sepaniak MJ, Datskos PG
  (2000) Selectivity of chemical sensors based on
  microcantilevers coated with thin polymer
  films. Anal Chim Acta 422: 89–99
Boisen A, Thaysen J, Jensenius H, Hansen O
  (2000) Environmental sensors based on micromachined cantilevers with integrated readout.
  Ultramicroscopy 82: 11–16
Butt HJ (1996) A sensitive method to measure
  changes in surface stress of solids. J Coll Inter
  Sci 180: 251–260
Chen GY, Thundat T, Wachter EA, Warmack RJ
  (1995) Adsorption-induced surface stress and
  its effects on resonance frequency of microcantilevers. J Appl Phys 77: 3618–3622

Cowburn RP, Moulin AM, Welland WE (1997)
  High-sensitivity measurement of magnetic
  fields using microcantilevers. Appl Phys Lett
  71: 2202–2204
Datskos PG, Oden PI, Thundat T, Wachter EA,
  Warmack RJ, Hunter RS (1996) Micromechanical temperature sensor. Appl Phys Lett 69:
  2986–2988
Finot E, Lesniewska E, Goudonnet JP, Thundat T
  (2001) Measuring magnetic susceptibilities
  of nanogram quantities of materials using
  microcantilevers.   Ultramicroscopy   86:
  175–180
Fritz J, Baller MK, Lang HP, Rothuizen H,
  Vettiger P, Meyer E, Guntherodt HJ, Gerber
  Ch, Gimzewski JK (2000 a) Translating biomolecular   recognition   into   nanomechanics.
  Science 288: 316–318
Fritz J, Baller MK, Lang HP, Strunz T, Meyer E,
  Guntherodt HJ, Delamarche HE, Gerber Ch,
  Gimzewski JK (2000 b) Stress at the solid–
  liquid interface of self-assembled monolayers
  on gold investigated with nanomechanical
  sensor. Langmuir 16: 9694
Gimzewski JK, Gerber CH, Mayer E, Schlitter RR
  (1994) Observations of a chemical reaction
  using micromechanical sensor. Chem Phys Lett
  217: 589–594
Grate JW, Martin SJ, White RM (1993) Acoustic
  wave microsensors. Anal Chem 65: 978A–996A
Hansen KM, Ji HF, Wu G, Datar R, Cote R,
  Majumdar A, Thundat T (2001) Cantilever-
  based optical deflection assay for discrimination of DNA single-nucleotide mismatches.
  Anal Chem 73: 1567–1571
Hu Z, Thundat T, Warmack RJ (2001) Investigations of adsorption and absorption-induced
  stresses using microcantilever sensors. J Appl
  Phys 90: 427–431
Jensenius H, Thaysen J, Rasmussen AA, Veje
  LH, Hansen O, Boisen A (2000) A microcantilever based alcohol vapor sensor — application
  and response model. App Phys Lett 76:
  2615–2617
Ji HF, Finot E, Thundat T, Dabestani R, Britt PF,
  Bonnesen PV, Brown GM (2000) A novel self-
  assembled monolayer (SAM) coated microcantilever for low level Cs ion detection. Chem
  Commun 6: 457–458
Ji HF, Hansen KM, Hu Z, Thundat T (2001)
  Detection of pH variation using modified
  microcantilever sensors. Sens Actuators B72:
  233–238
Lai J, Shi Z, Perazzo T, Majumdar A (1997) Optimization and performance of high-resolution

micro-optomechanical thermal sensors. Sens Actuator 58: 113–119

Manalis SR, Minne SC, Quate CF, Yaralioglu GG, Atalar A (1997) Two dimensional micro-mechanical bimorph arrays for detection of thermal radiation. Appl Phys Lett 70: 3311–3313

McGill RA, Abraham MH, Grate JW (1994). Choosing polymer coatings for chemical sensors. CHEMTECH 24: 27–37

Oden PI, Chen GY, Steele RA, Warmack RJ, Thundat T (1996 a) Viscous drag measurements utilizing microfabricated cantilevers. Appl Phys Lett 68: 1465–1469

Oden PI, Datskos PG, Thundat TG, Warmack RJ (1996 b) Uncooled thermal imaging using a piezoresistive microcantilever. Appl Phys Lett 99: 3277–3279

Perazzo T, Mao M, Kwon O, Majumdar A, Varesi J, Norton P (1999) Infrared vision using micro-optomechanical camera. Appl Phys Lett 74: 3567–3569

Raiteri R, Nelles G, Butt H-J, Knoll W, Skladal P (1999) Sensing biological substances based on the bending of microfabricated cantilevers. Sens Actuators B 61: 213–217

Salapaka MV, Bergh S, Lai J, Majumdar A, McFarland E (1997) Multimode noise analysis of cantilevers for scanning probe microscopy. J Appl Phys 81: 2480–2487

Sarid D (1994) Scanning Force Microscopy. Oxford University Press, New York

Stephan SC, Gaulden T, Brown A-D, Smith M, Miller LF, Thundat T (2002) Microcantilever charged particle detector. Rev Sci Instrum 73: 36–41

Stoney GG (1909) The tension of metallic film deposited by electrolysis. Proc Roy Soc (London) 82: 172–175

Thundat T, Warmack RJ, Chen GY, Allison DP (1994) Thermal and ambient-induced deflec-tions of scanning force microscope cantilevers. Appl Phys Lett 64: 2894–2896

Thundat T, Chen GY, Warmack RJ, Allison DP, Wachter EA (1995 a) Vapor detection using resonating microcantilevers. Anal Chem 67: 519–521

Thundat T, Wachter EA, Sharp SL, Warmack RJ (1995 b) Detection of mercury vapor using resonating cantilevers. Appl Phys Lett 66: 1695–1697

Thundat T, Oden PI, Warmack RJ (1997) Micro-cantilever sensors. Microscale Therm Eng 1: 185–199

Tufte ON, Stelzer EL (1993) Piezoresistive prop-erties of silicon diffused layers. J Appl Phys 34: 323–329

Varesi J, Lai J, Shi Z, Perazzo T, Majumdar A (1997) Photothermal measurements with picowatt resolution using micro-optomechani-cal sensors. Appl Phys Lett 71: 306–308

Wachter EA, Thundat T, Datskos PG, Oden PI, Sharp SL, Warmack RJ (1996) Remote optical detection using microcantilevers. Rev Sci In-strum 67: 3434–3439

Ward MD, Buttry DA (1990) In-situ interfacial mass detection with piezoelectric transducers. Science 249: 1000–1007

Wu G, Ji HF, Hansen K, Thundat T, Datar R, Cote R, Hagan MF, Chakraborty A, Majumdar A (2001 a) Origin of nanomechanical cantilever motion-generated from biomolecular interac-tions. Proc Natl Acad Sci 98: 1560–1564

Wu G, Datar R, Hansen K, Thundat T, Cote R, Majumdar A (2001b) Nanomechanical bioassay of prostate specific antigen (PSA). Nat Biotechnol 19: 856–860

Zhao Y, Mao M, Majumdar A (1999) Application of Fourier optics for detecting deflections of infrared-sensing cantilever arrays. Microscale Therm Eng 3: 245–251

The Embedding of Sensors

25. Embedded Mechanical Sensors in Artificial and Biological Systems

Paul Calvert

Abstract

Animals have many mechanical sensors to monitor the stresses on their structure and so allow control of motion. Machines also contain sensors which monitor their action, but only a few, with each sensor closely tied to the control of a particular parameter. There have been many studies of embedded stress sensors, particularly for damage detection in composite materials and for "health monitoring" of bridges. Wider use would be made of embedded mechanical sensing if there were simpler systems for incorporating them into composite and concrete structures and simpler readout methods.

I. Introduction: Synthetic and Biological Sensors

Animals live and move in a world where the environment is constantly changing. They have an abundance of sensors to monitor both body movements and the surroundings. Most machines are designed to function in a world that changes little. Thus, refrigerators expect a static environment, free of wind and rain and have little adaptability to environmental changes or to their level of usage. Similarly, automobiles do move but are designed for smooth roads without collisions and do not expect to be hunted. The vehicles that we build have only a few sensors to monitor the functioning of the engine and body components.

Many designers expect this to change. Equipment will be fitted with many more sensors and will have more ways to adapt their functioning in response. This chapter has a general discussion of mechanical sensing in biology and in the synthetic world, followed by a more detailed discussion of materials with embedded sensors, especially composite materials.

II. Animal Mechanical Sensing

Mechanical sensors of animals can be separated into exteroreceptors, such as tactile sensors that sense changes in the environment and proprioreceptors that allow control of the motion of the body. In mammals the proprioreceptors are embedded in the muscles, tendons and joints. The most numerous sensors are the muscle spindles. These organs are 4–10 mm long and are incorporated at from 50–500 per gram of muscle. The hand and neck have especially high densities of spindles. The human body as a whole has some millions of spindles.

Each spindle contains a set of muscle fibers within a capsule (Matthews 1981). These fibers are under tension and become unloaded when the main muscle (outside the capsule) contracts. The length change is detected by afferent (sensing) nerve fibers attached to the spindle muscle fibers. Both static length and rate of change can be detected. Since the system of muscle and tendon is elastic, rate of change of muscle length will have components reflecting both velocity and acceleration of the joint. In addition, the lengths of the fibers within the capsule are controlled by efferent (motor) nerve fibers. This is part of a feedback system that controls the length of the whole muscle by maintaining a given stress in the spindle fibers (Houk and Rymer 1981).

Muscle spindles are connected in parallel with the muscle and sense changes in length. In mammals, the Golgi tendon organs are located at the junction between the muscle and the tendon that connects it to bone. While the muscle spindles are sensitive to passive stretching of the muscle, the Golgi tendon organs are in series with the muscle and so sense the load on the muscle.

In addition, mammalian joints contain four types of mechanoreceptors, located in the joint capsule or in the ligaments. In the capsule are slowly adapting sensors that sense stress and rapidly-adapting sensors that measure changing stress. In the ligaments, there are sensors for the directions of movement and pain receptors.

The arthropod cuticular exoskeleton contains integrated mechanoreceptors. One group of these receptors is associated with sensing hairs and act as exteroreceptors to provide information about the environment (see chapters 9, 10 and 11, this volume). A second group is associated with holes in the external skeleton, the cuticle. These proprioreceptors allow control of the motion of the animal by providing information about strains in the skeleton. Mechanoreceptors of insects are reviewed by McIver (1985). One type of proprioreceptor is the campaniform sensillum, a dome-shaped structure less than 25 µm in diameter with a flexible membrane to which a sensory cell is attached. In spiders, the equivalent structures are in the form of elongated slits. In addition to cuticular receptors, insects do have sensors resembling muscle spindles associated with the internal organs and soft parts of the abdomen (McIver 1985).

Barth (1985, 2002) has described the slit sensilla of spiders. These slits are 8 to 200 µm long by 1 to 2 µm across and are arranged either singly or in groups, in particularly large numbers on the legs, with 100's of slits on each leg. The slit functions in parallel with the cuticle as a strain sensor. Strain perpendicular to the long axis of the slit causes an inward buckling of the soft membrane that covers the slit. A sensory cell is attached to the membrane. Many of the slits are arranged in parallel groups, the lyriform organs, where the sensitivity to strain depends on the length of the slit and position in the group. Different orientations of slits and parallel arrangements in "lyriform organs" give sensitivities to different strain directions. Measurements of cuticular strain and of the receptor response from spider legs subjected to applied loads gives a threshold sensitivity of about

50 µε (microstrain) at a threshold load of 8 mN (Blickhan and Barth 1985).

Most synthetic sensors have multiple sensitivities. A strain gauge, for example, shows a change of resistance with extension but resistance also changes with temperature. In addition, these gauges are frequently mounted on polymer films, which take up water and expand as humidity increases. These effects can largely be corrected for by using a control gauge that is not exposed to the strain, but does see changes in temperature and humidity. Arthropod cuticle is certainly temperature and humidity sensitive and inputs from multiple receptors with differing sensitivities must be required to separate these effects. Similarly, some chemoreceptors of insects show overlapping sensitivities to sugar, water and salt, but these can presumably be distinguished using inputs from several sensors (Morita and Shiaishi 1985).

Animals also have very large numbers of chemical sensors. A recent paper discussed why so many are useful (Derby and Steullet 2001). Advantages cited include better spatial resolution, a better signal-to-noise ratio, discrimination between different sources of stimulus, damage tolerance, and coupling to different processors to elicit different behaviors.

III. Synthetic Mechanical Sensors

Conventional stress sensing, or more strictly strain sensing, can be done with strain gauges, piezoelectric ceramics or polymers, or optical measurements of deflection. Each of these systems requires complex electronics to sense the change. In addition they cannot usually be incorporated into the structure but must be attached afterwards. For this reason, much of the work on damage detection in composites has focused on the use of optical fibers.

Fiber Bragg gratings can be formed by UV irradiation of germanium-doped silica single-mode fibers to introduce a series of planes of differing refractive index that are highly reflective for a selected wavelength. Strain or a change in temperature in the fiber causes a change in periodicity of the grating and so the reflected wavelength (Morey et al. 1989). A strain in the fiber of 10^{-3} (1000 microstrain) produces a wavelength shift of about 1 nm (Bocherens et al. 2000). Performance characteristics include a wide range of strain sensitivity from 10^{-9} to 10^{-2} with the lower limit being very dependent on the instrumentation. A typical sensitivity is 25 µε (Kersey et al. 1994). The lower limit depends on the monochromator used and the degree of temperature control. It is possible to multiplex 20–30 sensors on a single optical fiber, each operating at a different wavelength. Absolute strain sensing is possible without constant monitoring.

The need for a monochromator and laser light source does mean that the strain detection system may often be larger than the sample. This is a drawback for the use of fiber optics in any but very large structures. In the future, miniature integrated source and detection systems should remove this problem.

Various other approaches have been taken to detect stress or damage using *optical fibers*. Earlier systems aimed to detect bending of multimode fibers as the composite was stressed locally. The transmitted light intensity is sensitive both to macrobending, where the fiber is distorted by damage, and microbending, where local stresses lead to small changes in direction (Krohn 1991). These effects give rise to changes in transmission and to reflected beams. The changes in intensity are related to the extent of distortion but it is not possible to glean much detailed information about the nature of the damage. Using pulse methods, known as OTDR, optical time domain reflectometry, it is possible to determine the position of the damage along the fiber. This approach does not seem to offer enough information to be a useful tool for composites. Laser pulses of the order of 100 ps are needed to identify defects separated by a few centimeters (Bocherens et al. 2000). For large structures, such as bridges, the fibers can be coupled to mechanical devices in order to give a signal which is proportional to strain and OTDR can be used to determine the stress at each of an array of sensors (Luo et al. 1999). These damage detection systems may be more comparable with the pain sensors of the nervous system than with the proprioreceptors.

Metal strain gauges are the most common method of measuring strain. Metal lines are sputtered or printed onto a flexible support such as polyimide. The gauge is glued to the surface of a sample and changes in resistance due to strain are measured using a Wheatstone bridge circuit. Temperature compensation is achieved by using a second stress-free gauge. Strain gauges have a gauge factor, the ratio of fractional change in resistance to fractional strain, of 2–5. Strains of 10's of $\mu\epsilon$ (microstrain) can be measured. Silicon strain gauges can be made with a gauge factor of 30–120 but they are brittle, the resistivity-strain relationship is non-linear and temperature compensation is difficult. Better semiconductor gauges are emerging as a result of developments in MEMs technology.

Piezoelectric ceramics, usually lead zirconate titanate (PZT), are widely used as actuators where high frequency motion is needed, for example in ultrasonics, and where small controlled movements are needed, such as in atomic force microscopy. The piezoelectric effect arises from the titanium or zirconium ion being slightly off-center in the unit cell, resulting in a dipole. When the material is

cooled through the Curie point under high voltage the dipoles in each domain become parallel to one another and to the applied field (Moulson and Herbert 1990, Newnham and Ruschau 1991). A subsequent applied stress compresses the unit cell and so changes the length of the dipoles, giving rise to a voltage. As a stress sensor, the material is in the form of a thin, brittle, ceramic sheet, a fraction of a millimeter thick, with metallized surfaces.

Piezoelectric polymer, polyvinylidenefluoride, has been widely used as a strain sensor in the form of a film, usually about 20 μm thick, with metallized surfaces. The most effective polymer is a polyvinylidenefluoride copolymer (PVF2) that has been drawn and poled (Zhang and Scheinbeim 2001). The drawing orients the material and converts the normal inactive beta crystal form to the active alpha form (Kepler and Anderson 1992). Metal electrodes are evaporated onto each side and the film is then poled, normally by cooling from around 100°C under a large imposed field.

The piezoelectric coefficient of the resulting material is around 20 pC/N. This is much less than in piezoelectric ceramics. The polymers can be much more sensitive than the ceramics as microphones or hydrophones, where the low modulus allows a large strain to result from a given energy input, so giving rise to a large voltage. As embedded sensors, the polymer films are less liable to fracture and are less disruptive of the composite structure.

With a small area of stressed film, the charge developed under stress is small and rapidly lost by leakage. Hence, high frequency stresses can readily be sensed but static stresses need sensitive electronics with a high internal resistance and low capacitance connections. Using an oscilloscope, we found that an embedded film of 1 cm^2 of film gave a response of about 25 mV/kPa. We were also able to show that the piezoelectric response was retained during high temperature curing of a surrounding composite, up to about 180°C, the melting point of the polymer (Denham et al. 1997). This was unexpected as the piezoelectric response in free-standing films is normally lost at around 100°C. It is believed that the embedded film is constrained by the surrounding material and so cannot relax as an unconstrained film would.

There has been much interest in composite piezoelectric materials. Various porous ceramic structures can be back-filled with resin to reduce the overall modulus and improve the stress sensitivity (Safari and Danforth 1998). Several groups have studied dispersions of PZT powder in PVF2 or other polymers (Kowbel et al. 1999, Marra et al. 1999 a, b). Without substantial connectivity of the (high dielectric constant) ceramic phase, one would not expect a significant enhancement of the piezoelectric response over that of the base polymer, yet with good connectivity the material would be brittle. No one seems to have yet produced a very active and tough composite material. Efforts to use a piezoelectric paint have been described (Egusa and Iwasawa 1998).

Piezoelectric polymer films need to be drawn and poled and so must be embedded or attached as a unit; they cannot be printed or coated onto a material. The ideal material in this regard would be a printable piezoresistive polymer. There have been a number of claims for such properties in composites of carbon or metal powder in rubber or plastic but a reproducible, reversible piezoresistive material has yet to be fully developed.

The following section surveys recent work on embedding stress sensors into synthetic materials. A subsequent section will discuss the contrast between biological and synthetic stress sensing.

IV. Embedded Sensors in the Synthetic World

Many animal sensors monitor the mechanical and chemical functioning of the animal itself, rather than monitoring the environment. Two areas have emerged where there is intensive study of sensor systems in synthetic structures. One is in damage detection for composite materials in aircraft; the other is in health monitoring of large structures, especially bridges but also roads, buildings and ships. The first example is acute in that one is usually looking for a localized fault before it becomes critical. In the case of a bridge, that damage may be more delocalized and chronic over a long time. Both concrete and composites have the advantage that they undergo chemical curing, so it is relatively easy to embed sensors during the formation of the structure. With molded or machined parts it is only possible to attach sensors externally.

Much of the work to date on embedded sensors has focused on damage sensing of composites. Military aircraft have long had composite parts that are critical, in the sense that a failure may cause a crash. It has been found that composite laminates may suffer damage that cannot be easily detected from the outside, especially interlayer delamination but also cracking of the inner layers. This is in contrast to metal skins where damage can usually be seen from the outside.

Composite laminates are normally orthotropic, very stiff and strong in two directions but weak in response to stresses perpendicular to the plane. Efforts to produce 3-dimensional weaves have not yet been successful. This leads to a vulnerability to some types of loads but also means that engineers are unfamiliar with the response of the material to fatigue or local damage. As the aircraft industry moves to greater usage of composites in critical positions on civil aircraft, the problem of damage detection has become more important. This is both because of the large number of passengers at risk in the event of a failure and because the inspection may be less rigorous than for military aircraft.

In the case of bridges, there is a desire to save on the costs of regular inspection and a reluctance to use innovative materials or designs without some way of following changes in the structure over long times. There is also increasing use of composites in civil engineering structures, both for the primary structure and for repair after earthquake or other damage. Where the composites are formed on site, there are serious concerns with maintaining the quality of the material, especially in controlling the cure process. A "health" monitoring system would alleviate these concerns.

A. Optical Fibers in Composites

Fiber optic sensors have been tested in composite boat hulls (Wang G et al. 2001). Patches containing fiber Bragg gratings were attached at various points to the hull of a glass-fiber patrol boat and were used to monitor stresses, when weights were dropped on the deck and at sea. The sensor information was used to compare the actual stresses with the anticipated stresses used in the design of the ship. In larger composite structures, such as ships, a time-delay reflectometry system can be used to determine the site of breaks in fibers attached to the structure (Kageyama et al. 1998).

In contrast to these sensors, which are essentially external, there have been a number of papers describing test panels or trial systems with embedded fiber-optic sensors (Kuang et al. 2001, Lee et al. 2001). Generally these are fitted to sandwich panels within the foam core (Bocherens et al. 2000) or are directly embedded in the composite.

When they are embedded, Bragg gratings will respond to stress generated by the differential shrinkage on cooling. The system can then be used to determine temperature, although special methods are needed to distinguish between thermal stresses and external stresses on the fiber (Li et al. 2001). Fiber Bragg gratings have also been incorporated into pultruded rods during manufacture (Kalamkarov et al. 1999, 2000). Pultrusion is a process for forming composite rods or tubes by pulling a combination of reinforcing fibers and matrix through a die so that the fibers become highly aligned with the axis of the rod.

The relationship between the strain in the fiber and that in the composite is not simple. The composite will experience different strains parallel and perpendicular to the fibers within any layer. There will also be significant shear strains generated by the mismatch of moduli between layers with different orientations. The sensing fibers are comparable in diameter to the thickness of a single composite layer. In addition, there may be debonding at the fiber-coating or coating-resin interfaces that will reduce the strain in the fiber (Tang et al. 1998). Polyimide-coated fibers are chosen to provide good bonding to the surrounding resin matrix of the composite.

Most fiber optic sensors are "universal sensors" in that it may be difficult to separate the effects of stress, temperature, bending or other variables. The same is true of most types of sensors. Temperature changes in Bragg gratings produce a similar effect to strain. This can be resolved by using paired fibers with differing temperature and strain sensitivities or using different sensing structures in one fiber (Jin et al. 1998, Cavaleiro et al. 1999). There is also interest in using glass-ceramic materials with a zero thermal expansion coefficient (Beall 2000). Strain gradients will produce inhomogeneous changes in the fiber grating and broadening of the reflected peak. This can provide information about the local stress distribution, close to a crack for instance (Huang et al. 1996, Peters et al. 2001).

A serious concern in composite materials is that impact damage can lead to local delamination that later propagates to failure. During stressing of composites, fracture or debonding of individual fibers leads to acoustic emission, which can be detected by attached piezoelectric sensors. These events correlate with transient changes in the optical signal (Rippert et al. 2000). Thus, transient changes in fiber strain or microbending could be used to detect damage as it occurs, rather than after the event. In addition to detecting strain, optical fibers can be used for damage sensing in that a broken fiber will transmit only low levels of light.

In large structures, one would want to incorporate many sensors in an efficient manner. Arrays of Bragg grating sensors can be used with wavelength selection to allow individual gratings to be interrogated, which is usually done using a Fabry-Perot interferometer (Measures 1993). The approach has been used to monitor crack growth in a cracked aluminum plate that was repaired with a bonded composite patch with the sensors mounted on the surface of the patch (McKenzie et al. 2000). The practical use of this system would depend on the availability of a small, optoelectronic interface for collection of the information.

There are concerns about loss of strength in a composite when embedding optical fibers with a diameter over 100 μm in a composite containing carbon fibers of about 8 μm. Where the optical fibers run parallel to the reinforcing fibers, it is found that there is little change in the composite structure. Where the optical fiber runs at an angle to the reinforcing fibers, there may be resin-rich zones that could initiate failure of the composite, especially in compression (Case and Carman 1994, Carman and Sendeckyj 1995). In many cases, the chief design criterion for composite structures is the stiffness rather than strength and the small volume fraction of optical fibers has little adverse effect on stiffness. However, special attention must also be paid to avoid stress concentrations at the point where the optical fiber enters the composite. There is also evidence for adverse effects on fatigue (Surgeon and Wevers 2001).

Glass fibers fracture at a strain of about 1%. This is adequate for monitoring carbon fiber composites, since carbon fibers fracture at about 0.25%, but it limits the usefulness of optical fiber sensors in more elastic structures. This has been discussed in the context of fiber sensors embedded in polyester mooring ropes. The glass fibers would need to be wound in a spiral pattern to reduce the strain in the glass compared to that in the polyester (Rebel et al. 2000). There is also a concern that the fibers will be prone to damage during impact and so will fail before the composite. This can be partly avoided by careful choice of the position of the sensors (Jarlas and Levin 1999).

B. Resistivity Monitoring

It is an attractive idea to monitor changes in a carbon fiber composite by changes in conductivity. For continuous fiber composites, this would occur through fiber breakage or simply through changes in fiber resistivity with strain. Given the very high conductivity of carbon fibers, the strain dependence of resistance tends to depend on local flaws and so is unpredictable. More predictable results should be achievable with semiconducting fibers such as silicon carbide, where the strain dependence of resistivity is much higher. In short fiber composites, the conductivity of the composite is dependent on percolation effects between highly conducting fibers and a highly insulating matrix. The actual response of a given system will be simple for a simple volume expansion or compression but shape changes with no volume change would not be expected to lead to a resistance change in an isotropic system.

Carbon fibers in composites are piezoresistive but bare fibers are not, suggesting that the effect arises from rearrangements of fiber packing, not simply from strain in the fibers (Wang X et al. 1999). The response of an embedded sensor to damage depends on the host material, with damage resulting in a reduced signal in composites but an increased signal in concrete (Wang C et al. 2001). Both carbon fibers and silicon carbide fibers in composites can be used directly as damage detectors, based on an increase in DC resistance as the fibers break (Abry et al. 2001, Fankhanel et al. 2001). AC measurements can give information about other processes, such as delamination and debonding, which do not involve fiber breakage. Given that the resistance change can be detected, a method is still needed to determine the site and extent of damage; this can be done by electrical impedance tomography (Schueler et al. 2001). A carbon fiber bundle can be used as a sensor in a glass-fiber composite. The strain-to-break of carbon fibers is less than that for glass, so attention needs to be paid to the stress directions and to calibration (Tamiatto et al. 1998).

C. Piezoelectric Stress Sensors

Piezoelectric ceramics have been widely used attached to the surface of materials as strain sensors. Surface-attached piezoelectric sensors could be used to detect the location and size of delamination under a patch applied to a damaged composite panel (Koh et al. 2001). They have

been embedded in resins to sense water uptake by means of a reduction in the velocity of sound and an increase in attenuation (Zellouf et al. 1996). The effects on composite properties of embedded PZT ceramic sensors have been studied and are not serious (Mall and Coleman 1998). The sensors were embedded in a window cut through two center plies where the fibers were running across the major stress direction.

We have embedded PVF2 films in composite sandwich panels between the outer skin and the honeycomb or foam core (Denham et al. 1997, Calvert et al. 1999). These samples were subjected to an instrumented impact test and the response of the embedded sensor compared with that of the impactor. As illustrated in Fig. 1, the impactor tends to show a smooth loading and unloading while the embedded sensor shows a loading peak, followed by a constant stress as the core crushes and then a reflected stress wave of opposite sign. This approach could be used to distribute many sensors within a composite panel connected by wiring printed onto the composite sheets. Damage detection would require either constant monitoring or a periodic assessment of changes in response to a known external stress pattern.

One approach to detecting stress is simply to monitor the sensors at all times. Alternatively, embedded or attached piezoelectrics may be used to generate a stress wave and detect its reflection from damage in the structure (Beard and Chang 1997). The mechanical impedance between a piezoelectric source and sensor can be used to detect damage but the system is very sensitive to changes close to the sensors (Lopes et al. 2000). Neural networks are proposed to assess the nature of the damage. Given that the new generation of passenger aircraft have large critical composite parts, such as tailfins, some of which have failed, this

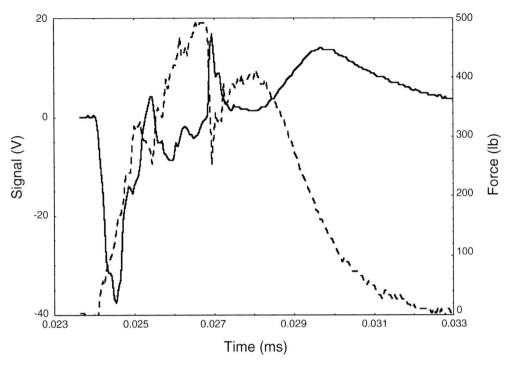

Fig. 1. An embedded PVF2 piezoelectric sensor is used to detect damage in a composite sandwich panel with a foam core. The output from the embedded sensor (solid line) is compared with the output from a stress sensor inside the weight of a dropped weight impact tester (dotted line)

ability to detect delamination may be crucial to the continued use of composites.

D. Other Sensing Methods

The obvious method to detect stress in a composite would be to use an embedded strain gauge (Salzano et al. 1992). One disadvantage with this method is that the resistance of the embedded leads will also change. MEMs have been embedded in composites as sensors (Hautamaki et al. 1999) but there are still questions about how to embed these sensors without damaging them (Javidinejad and Joshi 2001).

Fracture and debonding in composites can lead to light emission that can be picked up by embedded optical fibers. If the fiber is photoluminescent, the re-emitted light can be efficiently collected and transmitted by the fiber (Sage and Bourhill 2001, Sage et al. 2001).

E. Health Monitoring in Concrete and Roads

Beyond detection of localized damage, there is the possibility of using embedded stress sensors to monitor the "health" of the material in the sense of detecting any progressive loss of stiffness or other property changes. This has been discussed particularly in the context of concrete buildings or bridges. Concrete structures are prone to cracking, which can lead to water penetration and corrosion of reinforcing bars, followed by failure. It is attractive to embed optical fibers or other monitoring systems in bridges, roads and buildings in order to provide health monitoring. The sensors need to have a lifetime comparable to that expected for the structure, typically 100 years. They need to be spaced closely enough that cracks can be detected well before they are large enough to threaten the integrity of the structure.

A review paper describes test systems for monitoring bridges with embedded sensors (Aktan et al. 1998). One suggestion is to coat a short carbon fiber composite onto the concrete and monitor the strain through piezoresistivity (Wang X and Chung 1998). The system could detect strains up to 0.1%, before being damaged itself. Given the very low strain to failure for concrete, this is adequate. Fiber Bragg grating sensors have also been embedded into bridges. They are attached to prestressed composite tendons, which are used to reinforce the concrete (Ohno et al. 2001). Fiber Bragg gratings have also been bonded to steel reinforcing bars and embedded in concrete test systems (Idriss et al. 1998). For large structures, optical time delay systems become attractive.

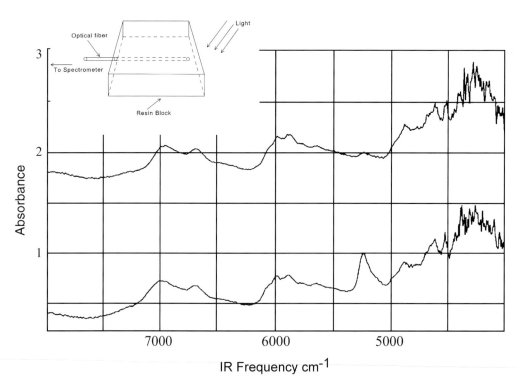

Fig. 2. A fiber optic sensor can be used to detect extent of cure and water content of epoxy matrix material for a composite. Insert: In this simple sensor geometry, ambient light on the sample is collected by a fiber end embedded below the resin surface. As the light passes through the resin, the absorption peaks due to the water and resin are superimposed on the incoming spectrum. Main figure: Near infrared spectra from an optical fiber embedded in an epoxy resin (MY721/DDS) block, dry and after exposure to high humidity (over water at 60°C for 24 hours). The water peak is at 5200 cm^{-1}

F. Cure Monitoring of Composites

An issue of great concern in composites is that of cure sensing. Most composites are formed with a thermosetting matrix, often an epoxy. The materials are often "pre-pregged" in that the liquid resin components are forced into the fiber layers and the soft mat is covered with release paper and shipped cold. If the prepreg becomes hot during shipping, it will partially cure and the final curing cycle will be changed. This is both due to the effect of partial reaction on the subsequent path and due to changes in the reaction temperature if the curing exotherm is reduced by pre-reaction. Consequently, there are severe quality control problems with composites. One solution is to embed a sensor into the panel so that the extent of reaction can be monitored and the curing cycle modified to allow for precure. A similar argument might be applied to any other processing step that involves chemical reaction, the setting of concrete for example.

An early and successful method for monitoring cure used an embedded field effect transistor to follow changes in dielectric constant during curing (Senturia and Sheppard 1986, Senturia et al. 1983). This could be used on test samples but the probes were left embedded in the composite so that neither they nor the part could be used further. Recent efforts to reduce the cost of composites have focused on vacuum assisted resin-transfer molding (VARTM) where liquid resin is injected over a preformed fibrous mesh in a mold. The large sizes and complex shapes have raised a need for good methods to monitor the pressure and temperature changes during the process (Mathur et al. 2001, Vaidya et al. 2001). One approach has been to monitor conductivity changes as the resin front moves through the fiber perform (Fink et al. 1995).

A variety of fiber optic methods can be used (Doyle et al. 1998). One approach is simply to use the fiber to monitor temperature since the curing reaction of the resin is exothermic (Murukeshan et al. 2000). It is also possible to follow refractive index (Cusanoa et al. 2000) or changes in the near-IR spectrum. We embedded optical fibers in epoxy (MY721/DDS) blocks and followed the curing reaction by near-IR spectroscopy using external illumination as the source (Fig. 2). A loss of the amine peak at $6690\,\mathrm{cm}^{-1}$ and an increase in the hydroxyl peak at $6950\,\mathrm{cm}^{-1}$ could be related to the extent of cure (Calvert et al. 1996). The same system could also be used to follow water uptake as the cured resin was exposed to a humid atmosphere. This is probably the simplest of the many methods for sensing vibrational spectra from a composite, since it uses ambient light and a fiber that terminates in the resin. However, the near-IR detector is large.

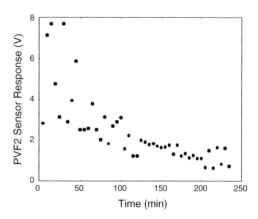

Fig. 3. An embedded sensor can be used to follow the change in elastic modulus of a composite as the resin cures. Output of a PVF2 sensor embedded in an epoxy block subjected to a small impact from a small dropped weight. As the block cures and hardens at room temperature the sensor output decreases

We have also embedded PVF2 sensing films into epoxy resin as cure monitors (Denham et al. 1996, 1997). The response of the sensor to a small external impact was followed as the sample was cured. With a room temperature cure, the signal decreased as cure took place. This was interpreted in terms of a parallel model for stress distribution in which the matrix carries more of the load, as it becomes stiffer than the sensor material. With high temperature curing, there is a problem that the sensor changes reversibly with temperature up to about 80°C and then irreversibly loses sensitivity as the structure depoles at higher temperatures.

These changes make it difficult to obtain information about the curing without, at least, returning the sample to room temperature. PVF2 melts at 170°C, so we did expect complete loss of piezoelectric response after heating to this temperature. However, this does not occur if the film is constrained under pressure between the layers of a composite, since these prevent relaxation of the oriented structure. As a result, the film does retain activity to 220°C (Denham et al. 1996). Figure 3 shows the response of an embedded sensor to repeat small impacts on the sample during room temperature cure and Fig. 4 shows the responses during heating of an

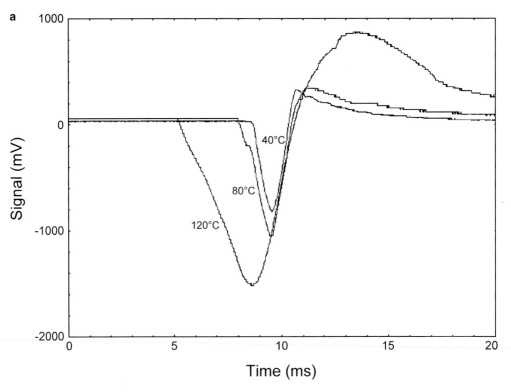

Fig. 4. Response of an embedded sensor to a dropped weight as in Fig. 3, as an epoxy sample is (a) heated and cured and (b) cooled after curing. The initial impact stress and the later reflected stress wave are both seen. Both the sensor and resin have temperature-dependent properties, which affect the sensor response in addition to changes due to the curing reaction

b

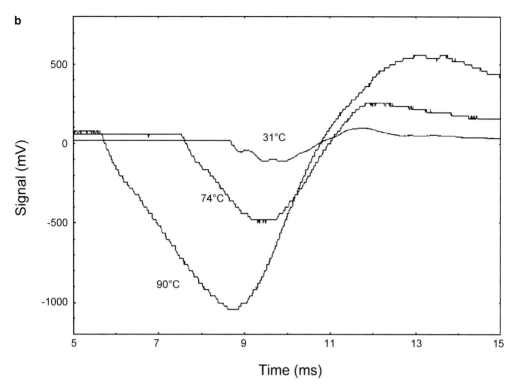

Fig. 4 (continued)

V. Self-Repair

uncured sample to induce curing and then subsequent cooling.

The ability to detect damage, in a composite or concrete structure, has led to the discussion of self-repair mechanisms. These are usually based on the release of reactive resin components at the site of damage. There have been demonstrations of self-healing by release of resin and curing agent at a crack. This has been demonstrated in resins using an epoxy-amine system (Dry 1996) and using micro-encapsulated dicyclopentadiene, which is released and contacts a Ruthenium metathesis catalyst dispersed in the epoxy matrix (White et al. 2001). A similar sys-

tem has been applied to healing cracks in concrete (Dry and McMillan 1996, Li et al. 1998). This approach is still in its infancy. So far, research has demonstrated a retention of strength in cracked samples but it is not clear whether they could be viable in any real situations. Biological healing relies on some network to transport fluid to the damage site and it is not clear whether we would want to embed such an extensive structure into a bridge, for instance.

Another approach has been to embed TiNi shape memory alloy wires in a composite and then use their contraction to close up cracks induced by tension (Jang et al. 2001). It has also been suggested that an array of piezoelectric actuators could both detect damage and damp new vibration modes that might lead to increased damage (Ray et al. 2000).

VI. Comparison of Synthetic and Biological Stress Sensing: Future Needs for Synthetic Sensors

In animals some proprioreceptors provide information that allows the control of motion. Most current industrial robots are robot arms, driven by hydraulics or motors with either no or very few force sensors. A current description of "Monroe", an advanced robot, includes "attitude sensors mounted on the body and grounding force sensors mounted on each sole. The attitude sensing system has composite gyros and an inclinometer. The floor reaction force sensor system consists of six force sensors mounted on each sole. Four force sensors are arranged in the toe and two are located at the heel, in order to detect the center of gravity and whether the heel contacts with floor or not." (Monroe 2002) . This could be contrasted with the innumerable sensors associated with the human foot.

There is increasing military interest in autonomous flying, swimming and walking vehicles. For many years, NASA has been working on autonomous vehicles for planetary exploration. It seems likely that all such vehicles would benefit from sensory information on the functioning of their bodies in addition to carrying long-range sensors such as cameras. Driven by concepts such as virtual reality, there is much interest in the design of arrays of stress sensors that would provide information similar to that from our sense of touch (DeRossi et al. 1987, Domenici and deRossi 1992). There is a continuing interest in "smart" structures, which will contain both sensors and actuators, principally in order to provide vibration control (Fanucci 1991). There is also much work on building man-machine interfaces where force sensors and actuators deliver a virtual reality (Fletcher 1996) or where sensing fabrics sense and respond to the motion of the wearer

(De Rossi et al. 1999; see also chapter 26, this volume).

A number of advances in technology are needed if we are to provide a robot vehicle with a sensory system comparable to that of an animal. We need small sensors, simple ways to embed and connect them, miniature transducers to provide a convenient electrical output and control systems that can use the input from large numbers of sensors.

One obvious gap between the synthetic and biological world is in the scale of sensors. Most synthetic sensors are several millimeters in their largest dimension where biological sensors are 10–100 times smaller in each direction. It is also necessary to connect the sensors to a data processing system. While we can connect electronic systems on a board with very fine wiring, there is no obvious way to build the equivalent of a nervous system running through a three dimensional functioning structure. Given large sensors and connections, there is no obvious way to duplicate the large numbers of sensors found in biology, except in a large structure, such as a bridge. Fiber optic sensors, with a diameter of around 200 µm, are one answer to this problem but are still too constraining for a true nervous system analogue. There remains a need for small embeddable electrical stress sensors for composites.

The ideal system would be a simple miniature piezoresistor. Strain gauges are characterized by a gauge factor, the ratio of fractional change in resistance to fractional strain, with a typical range up to 20. When attached to rigid structures operating at strains below 0.1%, strain gauges require sensitive circuitry to detect stress. The gauges themselves are fairly inflexible and so cannot be used on soft materials. In contrast, the proprioreceptors of insect and spider cuticle are apparently connected through a pore to a flexible membrane to magnify the strain. The shape of the pore helps to define the direction of the detected stress component. Simple, soft, large strain piezoresistors do not exist yet.

Similar arguments can be applied to chemical sensors. There are certainly many structures where distributed sensors for humidity, oxygen or pH would be a desirable part of a health monitoring system. Like wood, composites are very sensitive to humidity. Changes in water content lead to anisotropic swelling and to large changes in stiffness and strength.

Printable, resistive chemical sensing materials are readily available. Shown in Fig. 5 is a prototype chemical sensor array formed by extrusion freeform fabrication. Silver-epoxy paste connections are embedded in the resin and carbon-polymer resistors are exposed at the edge. These change resistance on swelling with a solvent such as ethanol. Each individual sensor responds to changing levels of alcohol vapor in air (Fig. 6). The whole array is also sensitive to the direction of a

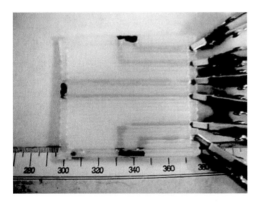

Fig. 5. Embedded chemical sensors within a solid object. Conducting lines of silver ink are embedded in an epoxy block formed using extrusion freeform fabrication. At three points on the surface these lines connect to carbon-polymer sensors which detect solvents, as shown in Fig. 6. The clips show where connections are made to the resistance measurement system

source of vapor in an open room, and to whether the room air conditioning is turned on. A more complex system could thus be used to seek a source of vapor.

New processing methods will be needed to incorporate large numbers of sensors in monolithic parts. One approach for laminated composites would be to add a sheet of stress sensors and connections to a laminate. Such a flexible sheet would need to carry the sensors and printed wiring and to be about the thickness of a single layer, about 100–200 μm. Since many composites have 0°/90° layup patterns, the sheet will not disrupt properties significantly if it matches the modulus and strength of the composite across the fiber direction. Clearly, the sensors need to withstand the processing conditions of the composite.

A second approach is to use layerwise processing methods, such as freeform fabrication to embed sensors as the part is built up. This has been used for demonstrations in metal (Golnas and Prinz 1998, Li et al. 2001) and epoxy (Denham et al. 1996) parts. The connections to the sensors are no longer restricted to a plane but can be fully three-dimensional. If the individual connections are also fully embedded, they will be less vulnerable to fracture as the part is subjected to cyclic loads.

Progress in integration of electronic and optoelectronic systems onto chips means that transducers can more readily be sited near to groups of sensors. When combined with new technologies for wireless communication between devices, it should become possible to embed many sensors in a large structure and analyze the information centrally without large amounts of extra wiring or optical fiber.

Robotic control tends to be based upon using a single sensor to control a single actuator. It is not clear how we should use simultaneous inputs from a large number of sensors. Hasan and Stuart (1988)

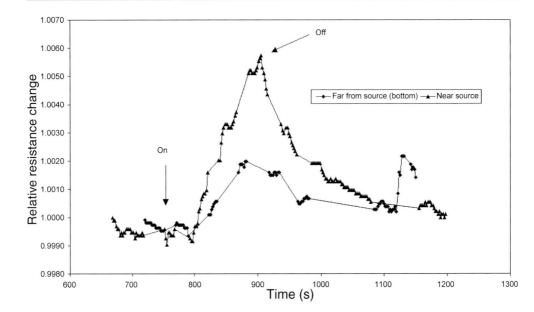

Fig. 6. Response by change of resistance of the composite sensors, as shown in Fig. 5, to airborne alcohol vapor. Upstream and downstream sensors, on either side of the sample, show differing responses. In principle, such a device could hunt upstream to the source of the alcohol

discuss the control strategies associated with mammalian propioreception. One suggestion is that animals are designed for the maximum maneuverability that the nervous system can control as opposed to robot designs, which will aim for maximum stability combined with enough maneuverability required for performance. Loeb (2001) has reviewed control of limb motion in vertebrates.

VII. Conclusions

The wider use of composites in aircraft, bridges and buildings will increase the need for good monitoring methods. The processing methods for these materials do lend themselves to embedding of sensors, and fiber Bragg gratings have been shown to be effective at detecting damage. However, the whole system is still too complex to be practical for general use. New, finer sensor systems based on

resistance or potential changes would make embedded sensors in small parts more feasible. For large structures, the need is really for reliable technologies to distribute sensors throughout a material and to couple them to monitoring equipment.

Acknowledgments

I would like to thank Friedrich Barth for advice on the animal sensing section of the manuscript.

References

Abry J, Choi Y, Chateauminois A, Dalloz B, Giraud G, Salvia M (2001) In-situ monitoring of damage in CFRP laminates by means of AC and DC measurements. Compos Sci Technol 61: 855–864

Aktan A, Helmicki A, Hunt V (1998) Issues in health monitoring intelligent infrastructure. Smart Mater Struct 7: 674–692

Barth FG (1985) Slit sensilla and the measurement of cuticular strains. In: Barth FG (ed)

Neurobiology of Arachnids. Springer Verlag Berlin pp 162–188

Barth FG (2002) A Spider's World: Senses and Behavior. Springer Verlag, Berlin

Beall G (2000) Glass-ceramics for photonic applications. Glass Science and Technology-Glastechnische Berichte 73 Suppl C1: 3–11

Beard S, Chang F (1997) Active damage detection in filament wound composite tubes using built-in sensors and actuators. J Intell Mater Syst Struct 8: 891–897

Blickhan R, Barth FG (1985) Strains in the exoskeleton of spiders. J Comp Physiol A 157: 115–147

Bocherens E, Bourasseau S, Dewynter-Marty V, Py S, Dupon M, Ferdinand P, Berenger H (2000) Damage detection in a radome sandwich material with embedded fiber optic sensors. Smart Mater Struct 9: 310–315

Calvert P, George G, Rintoul L (1996) Monitoring of cure and water uptake in a freeformed epoxy resin by an embedded optical fiber. Chem Mater 8: 1298–1301

Calvert PD, Denham HB, Anderson TA (1999) Free-form fabrication of composites with embedded sensors. Proc SPIE 3670: 128–133

Carman G, Sendeckyj G (1995) Review of the mechanics of embedded optical sensors. J Compos Technol Res 17: 183–193

Case S, Carman G (1994) Compression strength of composites containing embedded sensors or actuators. J Intell Mater Syst Struct 5: 4–11

Cavaleiro P, Araujo F, Ferreira L, Santos J, Farahi F (1999) Simultaneous measurement of strain and temperature using Bragg gratings written in germanosilicate and boron-codoped germanosilicate fibers. IEEE Photonic Tech L 11: 1635–1637

Cusanoa A, Breglioa G, Giordano M, Calabrò A, Cutoloa A, Nicolais L (2000) An optoelectronic sensor for cure monitoring in thermoset-based composites. Sens Actuator A-Phys 84: 270–275

DeRossi D, Nannina A, Domenici C (1987) Biomimetic tactile sensors with stress-component discrimination capability. J Mol Electronics 3: 173–181

De Rossi D, Della Santa A, Mazzoldi A (1999) Dressware: Wearable hardware. Materials Science & Engineering C-Biomimetic and Supramolecular Systems 7: 31–35

Denham H, George G, Rintoul L, Calvert P (1996) Fabrication of polymers and composites containing embedded sensors. Proc SPIE 2779: 742–747

Denham HB, Anderson TA, Madenci E, Calvert P (1997) Embedded pvf2 sensors for smart composites. Proc SPIE 3040: 138–147

Derby CD, Steullet P (2001) Why do animals have so many receptors? The role of multiple chemosensors in animal perception. Biol Bull 200: 211–215

Domenici C, DeRossi D (1992) A stress component-selective tactile sensor array. Sensors and Actuators A: Physics 31: 97–100

Doyle C, Martin A, Liu T, Wu M, Hayes S, Crosby P, Powell G, Brooks D, Fernando G (1998) In-situ process and condition monitoring of advanced fibre-reinforced composite materials using optical fibre sensors. Smart Mater Struct 7: 145–158

Dry C (1996) Procedures developed for self-repair of polymer matrix composite materials. Compos Struct 35: 263–269

Dry C, McMillan W (1996) Three-part methylmethacrylate adhesive system as an internal delivery system for smart responsive concrete. Smart Mater Struct 5: 297–300

Egusa S, Iwasawa N (1998) Piezoelectric paints as one approach to smart structural materials with health-monitoring capabilities. Smart Mater Struct 7: 438–445

Fankhanel B, Muller E, Mosler U, Siegel W, Beier W (2001) Electrical properties and damage monitoring of SiC-fibre-reinforced glasses. Compos Sci Technol 61: 825–830

Fanucci JP (1991) Smart structures. In: Lee SM (ed) International Encyclopedia of Composites. VCH Publishers, Weinheim, Germany 5: 155–168

Fink BK, Walsh SM, DeSchepper DC, Gillespie JW, McCullough RL, Don RC, Waibel BJ (1995) Advances in resin transfer molding flow monitoring using smart weave sensors. Proc ASME Materials Division 69: 999–1015

Fletcher R (1996) Force transduction materials for human-technology interfaces. IBM Systems J 35: 630–638

Golnas T, Prinz F (1998) Thin film thermomechanical sensors embedded in metallic structures. Materials Science Forum 287-2: 201–204

Hasan Z, Stuart DG (1988) Animal solutions to problems of movement control: The role of proprioceptors. In: Cowan WM, Shooter EM, Stevens CF, Thompson RF (eds) Annual Reviews in Neuroscience. Annual Reviews Inc., Palo Alto CA 11: 199–224

Hautamaki C, Zurn S, Mantell S, Polla D (1999) Experimental evaluation of MEMs strain sensors embedded in composites. J Micro-electromech Syst 8: 272–279

Houk JC, Rymer WZ (1981) Neural control of muscle length and tension. In: Brooks VB (ed) Handbook of Physiology, Section 1 The Nervous System. Amer Physiol Soc, Bethesda MD, Vol II Pt 1: 257–323

Huang S, Ohn M, Measures R (1996) Phase-based Bragg intragrating distributed strain sensor. Appl Optics 35: 1135–1142

Idriss R, Kodindouma M, Kersey A, Davis M (1998) Multiplexed Bragg grating optical fiber sensors for damage evaluation in highway bridges. Smart Mater Struct 7: 209–216

Jang B, Toyama N, Koo J, Kishi T (2001) Mechanical properties and manufacturing of TiNi/GFRP smart composites. Int J Mater Prod Technol 16: 117–124

Jarlas R, Levin K (1999) Location of embedded fiber optic sensors for minimized impact vulnerability. J Intell Mater Syst Struct 10: 187–194

Javidinejad A, Joshi S (2001) Autoclave reliability of MEMs pressure and temperature sensors embedded in carbon fiber composites. J Electron Packaging 123: 79–82

Jin X, Sirkis J, Chung J, Venkat V (1998) Embedded in-line fiber etalon/Bragg grating hybrid sensor to measure strain and temperature in a composite beam. J Intell Mater Syst Struct 9: 171–181

Kageyama K, Kimpara I, Suzuki T, Ohsawa I, Murayama H, Ito K (1998) Smart marine structures: An approach to the monitoring of ship structures with fiber-optic sensors. Smart Mater Struct 7: 472–478

Kalamkarov A, Fitzgerald S, MacDonald D, Georgiades A (1999) On the processing and evaluation of pultruded smart composites. Compos Pt B-Eng 30: 753–763

Kalamkarov A, Fitzgerald S, MacDonald D, Georgiades A (2000) The mechanical performance of pultruded composite rods with embedded fiber-optic sensors. Compos Sci Technol 60: 1161–1169

Kepler R, Anderson R (1992) Ferroelectric polymers. Adv Phys 41: 1–57

Kersey AD, Koo KP, Davis M (1994) Fiber optic Bragg grating laser sensors. SPIE Proceedings 2292: 102–112

Koh Y, Chiu W, Marshall I, Rajic N, Galea S (2001) Detection of disbonding in a repair patch by means of an array of lead zirconate titanate and polyvinylidene fluoride sensors and actuators. Smart Mater Struct 10: 946–962

Kowbel W, Xia X, Champion W, Withers JC, Wada BK (1999) Pzt/polymer flexible composites for embedded actuator and sensor applications. Proc SPIE 3675: 32–42

Krohn DA (1991) Fiber Optic Sensors. Research Triangle Park, NC, Instrument Society of America

Kuang K, Kenny R, Whelan M, Cantwell W, Chalker P (2001) Residual strain measurement and impact response of optical fibre Bragg grating sensors in fibre metal laminates. Smart Mater Struct 10: 338–346

Lee D, Lee J, Kwon I, Seo D (2001) Monitoring of fatigue damage of composite structures by using embedded intensity-based optical fiber sensors. Smart Mater Struct 10: 285–292

Li V, Lim Y, Chan Y (1998) Feasibility study of a passive smart self-healing cementitious composite. Compos Pt B-Eng 29: 819–827

Li X, Prinz F, Seim J (2001) Thermal behavior of a metal embedded fiber Bragg grating sensor. Smart Mater Struct 10: 575–579

Loeb G (2001) Learning from the spinal chord. J Physiol 533: 111–117

Lopes V, Park G, Cudney H, Inman D (2000) Impedance-based structural health monitoring with artificial neural networks. J Intell Mater Syst Struct 11: 206–214

Luo F, Liu J, Ma N, Morse T (1999) A fiber optic microbend sensor for distributed sensing application in the structural strain monitoring. Sens Actuator A-Phys 75: 41–44

Mall S, Coleman J (1998) Monotonic and fatigue loading behavior of quasi-isotropic graphite/epoxy laminate embedded with piezoelectric sensor. Smart Mater Struct 7: 822–832

Marra S, Ramesh K, Douglas A (1999 a) The mechanical and electromechanical properties of calcium-modified lead titanate/poly(vinylidene lidene fluoride-trifluoroethylene) 0-3 composites. Smart Mater Struct 8: 57–63

Marra S, Ramesh K, Douglas A (1999 b) The mechanical properties of lead-titanate/polymer 0-3 composites. Compos Sci Technol 59: 2163–2173

Mathur R, Heider D, Hoffmann C, Gillespie J, Advani S, Fink B (2001) Flow front measurements and model validation in the vacuum assisted resin transfer molding process. Polym Compos 22: 477–490

Matthews P (1981) Muscle spindles. In: Brooks VB (ed) Handbook of Physiology. Section 1: The Nervous System. Amer Physiol Soc Bethesda MD, Vol II Pt 1: 189–228

McIver SB (1985) Mechanoreception. In: Kerkut GA, Gilbert LI (eds) Comprehensive Insect Physiology, Biochemistry, and Pharmacology, Vol. 6 Nervous System: Sensory. Pergamon Press, Oxford, pp 71–132

McKenzie I, Jones R, Marshall I, Galea S (2000) Optical fibre sensors for health monitoring of bonded repair systems. Compos Struct 50: 405–416

Measures R (1993) Fiber optic sensing for composite smart structures. Composites Engineering 3: 715–750

"Monroe" (2002) Website Http://www. Mechatronics.Mech.Tohoku.Ac.Jp/ ∼ kumagai /research/monroe/biped_e.Html

Morey WW, Meltz G, Glenn WH (1989) Fiberoptic Bragg grating sensor. Proc SPIE 1169: 98–107

Morita H, Shiaishi A (1985) Chemoreception physiology. In: Kerkut GA, Gilbert LI (eds) Comprehensive Insect Physiology, Biochemistry, and Pharmacology, Vol. 6 Nervous System: Sensory. Pergamon Press, Oxford pp 133–170

Moulson AJ, Herbert JM (1990) Electroceramics : Materials, Properties, and Applications. Chapman and Hall, London

Murukeshan V, Chan P, Ong L, Seah L (2000) Cure monitoring of smart composites using fiber Bragg grating based embedded sensors. Sens Actuator A-Phys 79: 153–161

Newnham RE, Ruschau GR (1991) Smart electroceramics. J Am Ceramics Soc 74: 463–480

Ohno H, Naruse H, Kihara M, Shimada A (2001) Industrial applications of the BOTDR optical fiber strain sensor. Opt Fiber Technol 7: 45–64

Peters K, Studer M, Botsis J, Iocco A, Limberger H, Salathe R (2001) Embedded optical fiber Bragg grating sensor in a nonuniform strain field: Measurements and simulations. Exp Mech 41: 19–28

Ray L, Koh B, Tian L (2000) Damage detection and vibration control in smart plates: Towards multifunctional smart structures. J Intell Mater Syst Struct 11: 725–739

Rebel G, Chaplin C, Groves-Kirkby C, Ridge I (2000) Condition monitoring techniques for fibre mooring ropes. Insight 42: 384–390

Rippert L, Weavers M, Van Huffel S (2000) Optical and acoustic damage detection in laminated CFRP composite materials. Compos Sci Technol 60: 2713–2724

Safari A, Danforth SC (1998) Development of novel piezoelectric ceramics and composites for transducer applications. Proc SPIE 3341: 184–195

Sage I, Bourhill G (2001) Triboluminescent materials for structural damage monitoring. J Mater Chem 11: 231–245

Sage I, Humberstone L, Oswald I, Lloyd P, Bourhill G (2001) Getting light through black composites: Embedded triboluminescent structural damage sensors. Smart Mater Struct 10: 332–337

Salzano T, Calder C, Dehart D (1992) Embedded-strain-sensor development for composite smart structures. Exp Mech 32: 225–229

Schueler R, Joshi S, Schulte K (2001) Damage detection in CFRP by electrical conductivity mapping. Compos Sci Technol 61: 921–930

Senturia S, Sheppard N (1986) Dielectric analysis of thermoset cure. Adv Polymer Science 80: 1–47

Senturia S, Sheppard N, Lee H, Marshall S (1983) Cure monitoring and control with combined dielectric temperature probes. SAMPE Journal 19: 22–26

Surgeon M, Wevers M (2001) The influence of embedded optical fibres on the fatigue damage progress in quasi-isotropic CFRP laminates. J Compos Mater 35: 931–940

Tamiatto C, Krawczak P, Pabiot J, Laurent F (1998) Integrated sensors for in-service health monitoring of glass resin composites. J Adv Mater 30: 32–37

Tang L, Tao X, Du W, Choy C (1998) Reliability of fiber Bragg grating sensors embedded in textile composites. Compos Interfaces 5: 421–435

Vaidya U, Abraham A, Bhide S (2001) Affordable processing of thick section and integral multifunctional composites. Compos Pt A-Appl Sci Manuf 32: 1133–1142

Wang C, Wu F, Chang F (2001) Structural health monitoring from fiber-reinforced composites to steel-reinforced concrete. Smart Mater Struct 10: 548–552

Wang G, Pran K, Sagvolden G, Havsgard G, Jensen A, Johnson G, Vohra S (2001) Ship hull structure monitoring using fibre optic sensors. Smart Mater Struct 10: 472–478

Wang X, Chung DDL (1998) Short carbon fiber reinforced epoxy coating as a piezoresistive strain sensor for cement mortar. Sensors and Actuators A: Physical 71: 208–212

Wang X, Fu X, Chung D (1999) Strain sensing using carbon fiber. J Mater Res 14: 790–802

White S, Sottos N, Geubelle P, Moore J, Kessler M, Sriram S, Brown E, Viswanathan S (2001) Autonomic healing of polymer composites. Nature 409: 794–797

Zellouf D, Saint-Pierre N, Jayet Y, Tatibouet J (1996) Piezoelectric implants: Monitoring the water degradation in polymer-based composites. Proc SPIE 2779: 158–163

Zhang Q, Scheinbeim J (2001) Electric EAP. In: Bar-Cohen Y (ed) Electroactive Polymer Actuators as Artificial Muscles. SPIE Press, Bellingham, WA, pp 89–122

26. Active Dressware: Wearable Kinesthetic Systems

Danilo De Rossi, Federico Lorussi, Alberto Mazzoldi, Piero Orsini, and Enzo P. Scilingo

Abstract

Artificial sensory motor systems granting the power to reach out and interact with illusory objects and granting the objects the power to resist movement or to manifest their presence are now under development in a truly wearable form using an innovative technology based on electroactive polymers. The integration of electroactive polymeric materials into wearable garments endows them with strain sensing and mechanical actuation properties. Woven active electronic components and energy storage devices now under investigation would also potentially provide all essential instrumental functions (sensor, actuator, processor, power supply) in materials and forms which could be incorporated into garments. The methodology underlying the design of haptic garments has necessarily to rely on knowledge of biological perceptual processes which is, however, scattered and fragmented. Integration of afferent and efferent neuromuscular responses and commands to build up complex functions such as kinesthesia, stereognosis and haptics is far out of reach of our present understanding. Nonetheless, use of new polymeric electroactive materials in the form of fibers and fabrics, combined with emerging biomimetic concepts in sensor data analysis, pseudomuscular actuator control and biomechanic design, may not only provide new avenues toward the realization of truly wearable kinesthetic and haptic interfaces, but also clues and instruments to better comprehend human manipulative and gestural functions. In this chapter, the biological bases which characterize sensory-motor functions in humans are summarized, focusing on their perceptual features. Biological muscle action and control are also outlined, with the purpose of providing essential information needed to analyze and design pseudomuscular actuation systems. Electroactive polymer actuators, which we are currently investigating, are then discussed with emphasis given to their unique capabilities in the phenomenological mimicking of skeletal muscle actuation. Finally, the conception, early stage implementation and preliminary testing of a fabric-based wearable interface endowed with spatially redundant strain sensing and distributed actuation are illustrated with reference to a wearable upper limb artificial kinesthesia system.

I. Kinesthesia: Body Self Awareness

Kinesthesia, taken literally, means a sense of movement, although current usage of the term often includes the sense of static limb position. Today the term kinesthesia is used in the broadest sense to include the awareness of the positions and movements of the limbs (and other body parts), whether self-generated or externally imposed (Clark and Horch 1986). Our kinesthetic sense arises from activity in specialized sensory receptors which provide information about the angles of the joints, the lengths of the muscles and the tension they produce. They also code for the rates at which these values change. Relative bone motion can occur by sliding, spinning and rolling, as a function of the torque generated in the joints by muscles.

The central nervous system (CNS) controls position and movements hierarchically and it processes sensory motor information at different levels (decision, process, executive) in local and global closed loops to minimize the time of data treatment. The CNS obtains information about position and movement from receptors, which respond to mechanical events caused by joint displacements, and constitute the base of the kinesthetic sensation. Such mechanoreceptors include joint and skin receptors, muscle spindles and tendon organs. They can be grouped into two functionally different classes: slowly adapting (SA) and rapidly adapting (RA) receptors (Burgess et al. 1982). The first type would represent a natural candidate for coding on either passively or actively maintained static joint angles, while the second one would code for rate phenomena. The vast majority of cutaneous mechanoreceptors is of the rapidly adapting type and responds to the velocity of skin displacement during the joint movement, but it does not appear to provide an essential input for awareness of static joint position. However, a great deal of mechanoreceptor activity was shown only near extreme positions of the joint with a lack of activity over a wide range of angles. Similarly, mechanoreceptors in the ligaments respond only at the very extreme positions of the joint. Most important of all, orthopedic surgeons have consistently reported that entirely replacing human diseased joints (and their receptor apparatus) with prostheses leaves the patient's kinesthetic

sense intact. Joint receptors appear to be load rather than angle detectors, possibly serving a protective function when excessive muscle effort tend to force a joint beyond its physiological range of motion. Muscle receptors alone remain as candidates for encoding joint position. Since in a joint the musculo-skeletal geometry defined by the local anatomical arrangement is invariant, knowledge of the length of the muscles around the joint permits determination of joint position and movement.

Muscles contain two types of slowly adapting mechanoreceptors: the Golgi tendon organs and the muscle spindles. Tendon organs lie in series with the main muscle fibers, well situated to measure tension in the muscle, showing no response to position. Muscle spindles lie parallel with the main muscle fibers, well situated to measure muscle length and rate of change of length. Each spindle receives two types of innervation: sensory and motor. Motor activity alters both the length and the rate response of spindle afferents independently. In case of multi-joint muscles, ambiguity can be resolved only if the CNS refers also to signals simultaneously delivered by other muscles which cross only some of these joints. Therefore, a key feature of spindle receptors resides in the efferent control of its afferent activity. The spindle responds not only to changes in length, but also to changes in its efferent control in the absence of any muscle length change. This could preclude the spindle from providing information on absolute muscle length and the CNS from extracting similar information. There is evidence that central (cortical) feedback pathways provide information (namely, a copy of the efference) that is used to decode muscle afferent peripheral signals. According to the equilibrium point hypothesis (Feldman 1980), central commands might define a reference point and scaling of the afferent activity from peripheral receptors reflecting joint torque which is related to both central commands and external load conditions. The CNS can monitor its own motor commands to the muscles (and to the spindles) sending back a copy of them, originally defined as "corollary discharge" or "efference copy", to the perceptual regions of the brain. This constitutes the second mechanism utilized by the CNS to obtain kinesthetic information. In other words, the perception of motor commands is able to give rise to sensations of movement, but not in the total absence of afferent inflow.

A model in which positional information derives from superimposition in the brain sensory structures of the efferent copy and afferent signals reflecting muscle torque has been proposed (Feldman and Latash 1982). In this view, motor control develops a joint perception, establishing a system of coordinates for evaluating torque afferent signals. In fact, in the so-called spring model of the muscle behavior, a torque-angle characteristic represents a single-valued function of the angle for a given motor command. A family of such muscle invariant characteristics represents the mechanical effect of the range of the motor central command in the context of motor performance execution (equilibrium-point hypothesis). Muscle torque is quantified by the difference, depending on the task requirements and external conditions, between the actual joint angle and the measure of the corresponding motor commands, both expressed in terms of muscle length or joint angle. Afferent signal from the muscle spindle interface can generate similar signals reflecting the difference between externally and internally imposed change in the intrafusal fiber state and in this way can provide information on muscle length. In essence, the control and perception of joint angle are closely interrelated mechanisms.

II. From Biological to Artificial Sensing

Biological kinesthesia and haptics internally reconstruct mechanical events by relying largely on the activity of the subject. Their inherent bidirectional nature implies sensing performed by peripheral receptors, but the full perception is not confined to them. Artificial human-like kinesthetic systems should be able to adequately embed artificial signals referable to the joints possibly in a structured map of local inputs from a number of individual joints. Presently, however, human movement tracking systems generate only sparse real-time data (Mulder 1994). Improvements in accuracy, spatial resolution and parallel processing will lead to devices suitable for tracking fine manipulations and complex gestural recognition. Trajectory tracking of a joint should not depend on sensor technology and location. In an available glove usually sensors are located in a finger to detect a planar movement for each degree of freedom; in several cases a single sensor on the back of each joint should unambiguously signal joint position and movement. However, joint surface geometry dictates some degree of associated rotation, unless an intended and undetected counter-rotation cancels it. On the other hand, from a functional point of view, such rotation is not influential in the sense that it is a part of the only permitted movement. In this case, the sensor displays a one to one correspondence between the trajectories parameterized in a certain curvilinear frame (which includes rotations) to a set of values, which exactly reconstructs the position. A major problem is the embedding of this manifold into Cartesian space (i.e. the definition of the geometry, including rotation in the Cartesian frame). What we propose in section VI is in an identification procedure that permits the control system to test a redundant sensor set and reconstructs a global input-output function. This step could correspond to a learning phase in which the CNS also acquires knowledge of the muscular-skeletal system it has to control.

III. Wearable Sensing System

In biological systems, the intrinsic noisy, sloppy and poorly selective characteristics of individual mechanoreceptors are masterfully compensated by redundant allocation, powerful peripheral processing and efficient and continuous calibration through supervised and unsupervised learning and training. A truly biomimetic sensing system should possess these features to some extent, not just as a mimicking exercise, but as a result of solid engineering reasoning. Guided by these arguments we are investigating strain sensing elastic fibers and fabrics to realize adherent wearable systems with excellent mechanical matching with body tissues. Fabrics we use have been knitted using two different threads (De Rossi et al. 2001a):

- polypyrrole (PPY) coated cotton-Lycra threads
- carbon loaded rubber (CLR) coated cotton-Lycra threads

PPY based threads have been prepared as described in De Rossi et al. (1999), while CLR based threads have been obtained as an experimental sample (SMARTEX Srl, Prato, Italy). Fabrics have been characterized in terms of their electromechanical transduction properties, which very much depend upon their knitting topology. In Fig. 1, quasi-static responses in terms of percent change in electrical resistance versus strain are reported for the two fabrics. The calculated gauge factors ($GF = \Delta R/\varepsilon R_0$, where $\varepsilon = \Delta L/L_0$ is the strain) have been experimentally quantified as -12 for the PPY/lycra-cotton and $+2.5$ for CLR/lycra-cotton. These are excellent figures of merit for strain gauges. The dynamical electromechanical transduction properties of the two fabrics have also been investigated under sinusoidal and stepwise stretching.

PPY Lycra-cotton fabrics under stepwise stretching show (Fig. 2) the ability of coding for rapid onset of stretching and relaxing as proved by Scilingo et al. (2002). They are not, however, suitable for reading slowly varying strain levels. Phenomenologically, their behavior is analogous to RA mechanoreceptors in terms of temporal adaptation. Conversely, CLR Lycra-cotton fabrics (Fig. 2) properly code for steady and slowly varying strain levels, their bandwidth being from DC to 8 Hz. Phenomenologically, their behavior is analogous to SA mechanoreceptors in terms of adaptation. The location and

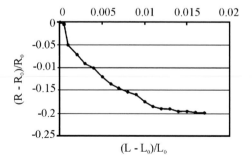
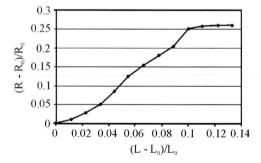

Fig. 1. Quasi-static responses in terms of change in electrical resistance versus strain for PPY/lycra-cotton (left) and CLR/lycra-cotton (right) fabrics

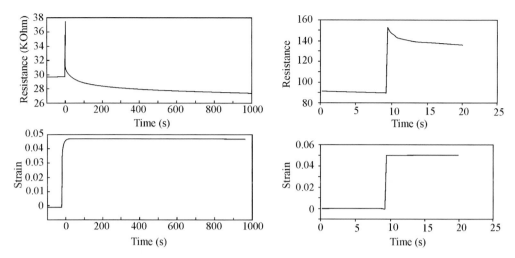

Fig. 2. Response in terms of change in electrical resistance under stepwise stretching, for PPY/lycra-cotton (left) and CLR/lycra-cotton (right) fabrics

density of strain fabric sensors are factors in the hands of the system designer through appropriate textile techniques (embroidery, jacquard). Selecting appropriate spatial distribution of sensing elements, however, is a complex process since the root of the problem resides in solving an inverse problem which is (mathematically) ill-posed, not presenting a unique and stable solution. Heuristic and commonsense reasoning can help in setting sensor allocation and assessing the level of redundancy required to cope with subject anatomical variability and relative displacement between the worn system and the body, inevitably occurring when the subject moves. We address these issues and the need for a simple and effective global calibration of the sensing system in Section VI.

IV. From Biological to Artificial Muscles

The CNS generates neural stimuli which set the status of body dynamics through the action of muscles. Phenomenological muscle models are needed to enable the formulation of hierarchical control theories. According to the equilibrium point theory, the force (F) exerted by a muscle depends on its intrinsic compliance (α), its effective length (x), and its rest length (λ), which is the unstrained length, as:

$$F = f(\alpha, x - \lambda)$$

This model can be specialized into two submodels, depending on the choice of the parameter controlled by the CNS. If the CNS, in the activation phase, generates stimuli which modify the instantaneous viscoelastic characteristics of the muscle itself (α hypothesis), the formula written above becomes:

$$F = \begin{cases} \alpha(x - \lambda_0) & x > \lambda_0 \\ 0 & x \leq \lambda_0 \end{cases}$$

where α is the parameter controlled by the CNS and $\lambda = \lambda_0$ is the maximum length which the muscle assumes without generating tensile force, constant for each value of α and depending only on the physical properties of the muscle. This equation gives origin to a set of linear characteristics, which differ only by their slope. Conversely, in the λ hypothesis, the parameter $k = \alpha$ is assumed constant and the CNS controls the variable λ. The force F is translation-invariant and generates a set (spanned by λ) of functions which depend on x. The equilibrium of the kinematic chain results from balancing the action of two muscular groups (agonist and antagonist) and the external load. Under the λ hypothesis, it is easy to prove that, if the function F is non linear, setting a couple (or more) of λ, one for each muscle, permits the control not only of the position of the kinematic chain, but also of its stiffness. In next section we will explain how, using a quadratic characteristic of the form

$$F = \begin{cases} k(x - \lambda)^2 & x > \lambda \\ 0 & x \leq \lambda \end{cases}$$

a pseudo-muscular actuation system can be controlled to obtain stiffness control.

Usually, in biological systems, the set of muscles that drive a kinematic chain is larger than the set of degrees of freedom of the same chain, and the concept of tunable stiffness can be used to solve the problem of the redundancy of controls. It is also easy to prove that a linear choice for F does not allow one to set any value for the compliance of the system. If the values of the parameter λ are set by the CNS and they are held constant in time, the system reaches its equilibrium position following a particular path determined by the external load. If velocity and stiffness during the movement are defined a priori, the CNS must control the parameters during the movement, instantaneously modifying their values. This operation happens, obviously, at a superior hierarchical level. From a physiological point of view, a muscle is structured in a set of active elements (motor units) each one with its rest length. These units are activated by progressive recruitment. The precise order of recruitment reduces the number of degrees of freedom that the CNS otherwise should control, while the progressive activation of fibers give rise to the quadratic viscoelastic characteristic of the entire muscle. Substantially, the macroscopic characteristic is obtained from a composition of parallel elementary units. These concepts are crucial in developing artificial muscles based on active elements having strongly different physicochemical characteristics from biological muscle, but having similar phenomenological behavior.

The basic matter of which biological systems are made is a sort of jelly-like composite material swollen in a multi-component aqueous solution. From a mechanical standpoint, it is a poor material with which it is impossible to design precise mechanisms. However, it is versatile and can be used for transduction, conduction, computation and actuation. Human muscles, for example, are very unconventional actuators, from an engineering point of view. They are neither pure force generators (like DC electric motors) nor pure motion generators (like stepper motors); rather they behave like springs, with tunable elastic parameters. This property turns out to be a key element for confronting the complexity of sensory motor problems, such as motor redundancy, trajectory formation, negotiation of impacts and motor learning (Rack and Westbury 1969, Honk and Rymer 1999). Attempting to mimic both muscular and neural aspects in a machine has practical relevance in a number of significant cases, particularly for those applications in which conventional control techniques compare poorly with human performance. This is the case, for example, in the handling of natural objects in a natural environment. The built-in compliance of muscles is a winning feature for achieving versatility and robustness, although at the cost of precision. In a natural environment, all sort of uncertainties of interacting objects (on shape, location, surface and bulk properties) make pure mechanical precision unnecessary because what is of functional significance is a global characterization of the integrated sensory-motor system. Starting from the early fifties, water swollen polyelectrolyte gels have been largely investigated, as chemomechanical energy converters (De Rossi et al. 1990). Only recently, however, has the potential usefulness of electrically driven polymeric actuators been fully realized and substantial research efforts devoted to their implementation.

V. Wearable Actuating System

To design and realize an artificial muscle, three classes of active materials are under study in our laboratory: electron conducting polymers (CP), electrostrictive polymers (EP) and carbon nanotubes (CN). Carbon nanotube mats and fibers are the most recent entry in this class of actuation materials (Baughman et al. 1999). Despite their potential to surpass all existing materials in some aspects of their performance (generated stress, low drive voltage, high energy density), their development is still at a very early stage and major fabrication issues need to be solved to make them useful in actuation. Conducting polymers may lead to useful realization because of:

- sizeable active strain (of the order of 1–3%);
- large active stress (up to tens of MPa; more than 30 times larger than human skeletal muscles);
- low electrical potential difference required to elicit their response (a few volt);
- built-in tunable compliance;
- relatively easy processability in film, fiber and bundle forms.

On the other hand, limited cycle lifetime and long time of response still negatively affect CP actuation systems, both factors being determined by the need for an electrochemical driving force. Attempts are under way to realize CP fiber actuators with the intent to overcome these limits and to integrate them into active fabrics. Continuum and lumped parameter models for CP fiber actuators have also been formulated and validated (Mazzoldi et al. 2000), providing a necessary tool to implement biomimetic control strategies and algorithms. Electrostrictive polymers and dielectric elastomers possess excellent figures of merit in several respects: linear actuation strain up to 60%, fast response time (down to tens of milliseconds) and sizable generated stresses (of the order of a MPa) (Bar-Cohen 2001). The price for achieving these performances is the very high electric fields required (up to 100 MV/m). By assuming that deformation in EP fibers occurs under isovolumic conditions and what changes is the rest length and not the Young's modulus, the relation between force and length in a fiber of this material can be expressed as:

$$f_b = EV_0 \left(\frac{1}{\mu} - \frac{1}{x} \right) u(x - \mu)$$

where x is the actual length of the fiber, μ corresponds to λ and it is the equivalent of the muscle rest length and $u(t)$ represents the Heaviside function, f is the (elastic) force performed by a fibers, E the Young's modulus, V_0 the volume. This relation can be obtained by integrating Cauchy's elasticity equations with the conditions $V = V_0$ (constant).

A possible way to implement an artificial muscle using these materials is to arrange them in a configuration that increases the strain performed by the actuators due to geometric factors. This solution allows larger strains in the case of CP muscle and reduced operation voltages in EP. An interesting device to amplify the strain is inspired to the

McKibben pneumatic muscle (Chou et al. 1996). In this device, a cylindrical bundle of CP or EP fibers is covered by a braid mesh (with flexible but not extensible threads) fixed at one end. If the force applied to the free end of the mesh or the radial strain of the bundle is changed, the system changes shape. The diameter increases, the length decreases and the angle α between the axes of the cylinder and the threads changes. It has been shown (De Rossi et al. 2001b) that if the initial value of α is larger than $\pi/4$, then the radial expansion is transduced into a linear contraction with an amplification factor larger than 1, as expressed by the relation:

$$(\Delta L_m/L_m) \approx -tg^2(\alpha) \, (\Delta R_m/R_m)$$
$$\text{for } (\Delta R_m/R_m) \ll 1,$$

where R_m, L_m are the initial radius and length of the artificial muscle. This approach proves useful to solve the problems of high voltage for EP or small strain for CP.

We intend to realize a parallel bundle fiber configuration and to implement an artificial recruitment paradigm controlled by a local device in order to speed up computation and central processor overload, using the λ model. Two of these actuators, in an agonist-antagonist configuration, are controlled by a couple of parameters (eventually time-varying) which impose position and stiffness of the mechanical system, as a function of the muscular characteristics implemented by the local controller. Obviously, to implement the λ model, the local control unit needs to know the effective length of the muscle. This problem can be solved by using strain gauge sensors. Significantly, in the last section we have noted that the possibility to obtain a control of the stiffness (or compliance) of a kinematic chain is due to the intrinsic nonlinearity of the variable compliance of

the muscle. In particular, the quadratic model we have introduced allows one to control this parameter. In order to replicate the trend of a musculo-skeletal system (including compliance) it is necessary that the fiber bundle performance replicate the viscoelastic muscular characteristic with a certain precision in terms of force and slope of the force/length characteristic. Let us consider the relation between strain and force introduced above.

When many collinear fibres – a set I – are grouped in a bundle B, the resultant force generated by the bundle is

$$F_B = \sum_{i \in I^*} f_{b_i} = \sum_{i \in I^*} EV_0 \left(\frac{1}{\mu_i} - \frac{1}{x} \right) u(x - \mu_i)$$

where I^* is the set of active fibres, that is $x > \mu_i$. It is worth noting that the global behavior of the bundle can be modified by selecting a suitable activation order for the fibres I^*. Hence it is also possible to approximate Feldman's muscle model by using a bundle. In other terms, it is possible, for a chosen tolerance ε, to define an activation order I^* that satisfies the following relation

$$\|F - F_B\|_{C^1} = \sup_{x \in [0,l]} |F - F_B| +$$

$$+ \sup_{x \in [0,l]} \left| \frac{\partial F}{\partial x} - \frac{\partial F_B}{\partial x} \right| = \sup_{x \in [0,l]} \left| k(x - \lambda)^2 - \right.$$

$$- \sum_{i \in I^*} EV_0 \left(\frac{1}{\mu_i} - \frac{1}{x} \right) u(x - \mu_i) \Big| +$$

$$+ \sup_{x \in [0,l]} \left| 2k(x - \lambda) - \sum_{i \in I^*} EV_0 \left(\frac{1}{x^2} \right) \times \right.$$

$$\times u(x - \mu_i) \Big| < \varepsilon$$

$(x > \mu_i, I \in I^*)$ which ensures the possibility of obtaining a position and stiffness control. Although in this realization, the parameter λ has lost analogy with the biological system, the relation written above uniquely determines it. Furthermore, if λ is time varying and the response time of the fiber actuator is short enough, the system follows exactly a trajectory of equilibrium points. The importance of this last solution resides in the fact that the velocity of the trajectory can be controlled during movement, and stiffness regulation during movement permits to control the effects of external perturbations.

VI. Upper Limb Artificial Kinesthesia

From an application point of view, research on artificial kinesthetic and haptic interfaces spurs the development of numerous fields of use supporting a great number of activities in different environments. Historically, research interest in

this area has mainly been motivated by work in the domain of teleoperation. We are now approaching the stage at which immersive, multisensory displays and human machine interfaces, as well as the increased availability of transmission bandwidth, computing power and digital resources will enable reproduction and simulation of "being in motion" and "being in touch" to be created with unprecedented perceived quality. The long-term goal of our research is to develop a family of wearable, bidirectional (sensing and display) haptic interfaces to be used in surgery and rehabilitation (Fig. 3). To achieve this distant goal, several methodologies and techniques need to be developed in terms of sensing (tactile and kinesthetic), actuation and control. Skin-like tactile sensors both for fine-form

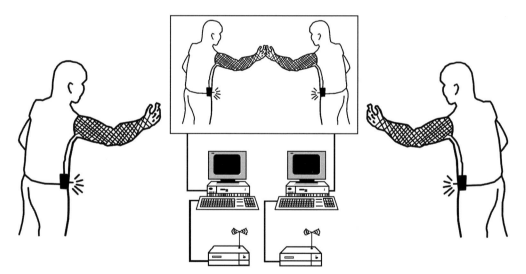

Fig. 3. Scheme of telerehabilitation, where a bilateral active haptic interface is worn by the patient and it is telemetrically controlled and monitored by a medical specialist from a remote position. Another domain of interest is telesurgery, where a haptic interface could be worn by the surgeon in a master-slave system to provide better maneuverability, dexterity, and an ergonomic coupling such that the surgeon can maneuver the system as though he were directly manipulating the remote object itself. Finally, a bilateral haptic interface could be used as a wearable active orthos is for a paralyzed arm. In this case the impaired subject could perform by himself the physical therapy and the haptic interface could also provide assistance in arm movements

discrimination (Caiti et al. 1995) and for incipient object slippage detection (Canepa et al. 1998) and tactile displays for haptic discrimination of softness (Bicchi et al. 2000) have been investigated in our laboratory. In their original implementation, however, they are not usable as wearable, non-obtrusive devices. Although work is in progress in the direction of realizing fabric-based tactile sensor arrays using piezoelectric polyvinylidene fluoride yarns, more work is needed before being worth reporting. On the other hand, the technology described in this chapter may enable the realization of redundant systems of sensors and actuators in the form of wearable kinesthetic (not yet haptic) exoskeleton.

A few prototypes already realized by us (De Rossi et al. 2001b) have shown reasonable capabilities to detect and monitor body segment position by reading the mutual angles between the bones. In particular, gleno-humeral joint, elbow joints, and the joints of the hand have been investigated. We have attributed three degrees of freedom to the shoulder (flexion-extension, adduction- abduction, rotation), two to the elbow (flexion-extension and pronation-supination), one degree of freedom for each interphalangeal joint of the hand, two to each metacarpo-phalangeal joint and two degrees of freedom to the trapezium-metacarpal joint. Moreover, relative movements between metacarpal bones have been considered. In these early prototypes, sensors have been intuitively located in correspondence to each joint in a number equal to the degrees of freedom. In the new generation of prototypes, the strategy of redundant allocation is adopted and a large set of sensors is distributed over the garment.

The adopted philosophy has been in a certain sense functional, i.e., the final aim of our work is to know which gesture a subject makes, and not which individual sensor has modified its status. From this point of view, redundant sets of sensing fabric patches linked in different topological networks can be regarded as a spatially distributed sensing field. Existing interconnections allow one to drastically reduce the number of channels of the data acquisition system and, by simultaneously comparing the sensing field with the value of the joint variables in the identification phase, to reconstruct the posture in the data acquisition phase. By comparing data obtained in the identification phase with set of movements (known), it is possible to determine whether the adopted allocation of sensors is appropriate. The emphasis of the method is, anyway, the observation of the global status of the system, comparing all the sensors simultaneously. The identification phase has been executed by having a subject wearing the sensing device and pointing by a handheld laser at a lattice of markers fixed in a measurement environment. An entire set of sensor data has been recorded for each pointed position. This data have then been interpolated by a piecewise linear function.

We have tested the posture detection system over a second "target" lattice in the space of the positions, where data obtained by the sensors have not been used to interpolate the piecewise linear function, but only to check the device accuracy. In Fig. 4, as an example, the reconstruction of a target related to the position of an arm is reported. Abduction-adduction angle and flexion-extension angle of the gleno-humeral joint are reported respectively in the abscissa and the ordinate. Each piecewise line is the solution of the equations holding for significant sensors (responses from sensors which are not influential for the detection of this particular posture degenerate into the entire plane), projected from the entire space of joints variables into the plane of the coordinates of the shoulder. The estimated position is given by the

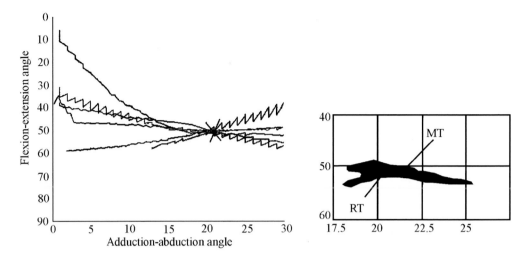

Fig. 4. Traces of the values held by different sensors corresponding to a fixed position. The estimated position is given by the intersection of the traces. Right: the high-density zone of probability for the estimated position is represented. The measured target (MT) position differs from the real target position (RT) by less than 4%

intersection of these solutions. In Fig. 4 (right), the intersection zone is enlarged. Due to the redundant allocation of the sensors, the solution is calculated in the "least square sense", i.e., by considering the entire zone with high density of possible solutions. The position of the target is calculated by the average value of all the points contained on this zone. The distance between the calculated position of the target and the real one is of about 10 cm in a range of about 2.5 m, i.e., an error less than 4%. In other cases we have obtained larger errors, always less than 8%.

At this stage of our work, biomimetic actuation has only been investigated in relation to an active glove. Two different muscle systems act on the hand: muscles that lie in the hand (intrinsic) or in the forearm (extrinsic). The extrinsic muscles realize flexion and extension of the fingers, while the intrinsic muscles generate adduction/abduction of the fingers, stabilization and part of the opposition of the thumb. Muscles that lie in the forearm are connected to finger bones by long tendons. In order to emulate the functionality of the hand, we can choose to replicate the extrinsic system either by placing a certain number of actuators over the forearm or by covering the fingers with active fibers lying in the direction of tendons. In fact, the stress generated by the artificial muscles, considered in the last section, may permit generation of the same force as human muscle with a considerable reduction in volume. Practically, a set of three groups of fibers for each finger (replicating agonist and antagonist muscle to control position and stiffness) lies in correspondence with the flexor and extensor tendons to realize flexion and extension of the finger. The same results hold for the flexion-extension of the thumb.

The technique we employ is an application of that described in Section V. Because of the redundancy of the actuators in respect to the degrees of freedom to control, we have chosen to introduce additional state variables (called generalized compliances). Let us consider a more general case of a kinematic chain with n degrees of freedom collected in a geometric state vector $\mathbf{q}$ (length $[\mathbf{q}] = n$). The chain is actuated by a functional group controlled by a vector λ (length $[\lambda] = m > n$). By introducing suitable functions $y_i(\mathbf{q})$ of the state $\mathbf{q}$, it is possible to define a 'supplementary' state vector $\mathbf{c}$ or $\mathbf{s}$ (length $[\mathbf{c}] = p$) of generalized compliances C_i (or stiffnesses S_i)

$$C_i = \left\| \frac{\partial y_i(\mathbf{q})}{\partial P} \right\| \left(S_i = \left\| \left(\frac{\partial y_i(\mathbf{q})}{\partial P} \right)^{-1} \right\| \right)$$

$$i = 1, \ldots, p$$

so that $p + n = m$, where P is a perturbation (in term of force) to the system. Hence, the redundancy of controls can be solved by defining a global state vector $\mathbf{z}$, obtained by appending the supplementary vector $\mathbf{c}$ (or $\mathbf{s}$) to the geometric vector $\mathbf{q}$ with, i.e. $\mathbf{z} = (\mathbf{q}|\mathbf{c})$ (respectively $\mathbf{z} = (\mathbf{q}|\mathbf{s})$). The choice of the functions y_i is strictly related to the entity of the physical compliance we consider. In particular, if $p < n$, it is not possible to control the compliance of each single joint, and only a distributed compliance over certain segments of the chain can be controlled. A simple application of this concept is given by the actuation of a long finger. Because of the movement of distal phalanx is bounded by a functional synergy to the movement of the medial phalanx, the flexion-extension of all the joints of a long finger can be controlled by only three actuators. If we neglect the distal phalanx (due to the cited synergy), only a generalized compliance can be defined, and we have chosen a function of both the angles involved.

In general, by completing the set of state variables, i.e. $\mathbf{z} = (\mathbf{q}|\mathbf{c})$, we have obtained a full rank local function $\mathbf{Z} = h(\Lambda)$ from controls space Λ to state space $\mathbf{Z}$. This function, that is just a change of coordinates, stands at the basis of the feed-forward control of the kinematic chain linked to the actuators. Unfortunately, dynamic terms that complete the equilibrium equations influence remarkably the evolution of the system. In this case direct commands from a central control unit lead to random motion. On the other hand, computer simulations have proved that the calculation expressing the instantaneous control explodes for chains having complex structure. To overcome this problem, the well-known concept of habitual movements, i.e., the movements that a subject performs without keeping particular attention to what he is doing, may be employed. Without going into the specific details of this argument, we cite the most accredited theories (Flash and Hogan 1985, Uno et al. 1989), which demand the choice of the trajectories according to the minimization of variational functionals, such as the minimum jerk and the minimum torque change hypotheses respectively

$$J = \int_{t_0}^{t} \sum_{k+1}^{n} \left(\frac{d^3 q_k}{dt^3} \right)^2 dt,$$

$$T = \int_{t_0}^{t} \sum_{k=1}^{n} \left(\frac{d^3 \tau_k}{dt^3} \right)^2 dt$$

where q_k are the geometrical state variables and τ_k the torques at the joints. The utilization of one of these two criteria allows the problem to be broken down into simpler parts. In fact, if a central control unit (CCU) has to coordinate the movement of several functional groups, it can demand the role of trajectory computing to several semi-peripheral units (SPU), one for each group. These devices receive the final value of the state variables from the CCU and the actual state values from the peripheries (the actuators). Then SPUs compute, when possible, the path and the velocity of the movement simultaneously and pass them to the peripheral dynamical control (PC) and to the fibers bundle of the kinematic segment. When they cannot perform the computation, SPUs send a feedback to the CCU that divides the path in more elementary steps to be interpreted by the SPUs. In this way, the CCU has enough resources for controlling more functional groups, ensuring the possibility of carrying out complex movements of the whole kinematic chain. Anyway, it is possible to exclude the SPUs and to execute a non-habitual movement directly under the control of the CCU. Computer simulations obtained with this strategy have given appreciable results in terms of coherence of movements and delay times.

Acknowledgement

The authors acknowledge the generous financial support of DARPA and ONR-IFO through NICOP grant N00014-01-1-0280, Pr N 01PR04487-00.

References

Bar-Cohen Y (2001) Electroactive polymer (EAP) actuators as artificial muscles – Reality, potential and challenges. SPIE Press, Billingham, USA

Baughman RH, Cui C, Zakhidov AA, Iqbal Z, Barisci JN, Spinks GM, Wallace GG, Mazzoldi A, De Rossi D, Rinzler AG, Jaschinski O, Roth S, Kertesz M (1999) Carbon nanotube actuators. Science 284: 1340–1344

Bicchi A, Scilingo EP, De Rossi D (2000) Haptic discrimination of softness in teleoperation: the role of the contact area spread rate. IEEE Trans Robotics Automation 16: 496–504

Burgess PR, Wei JY, Clark FJ, Simon J (1982) Signaling of kinesthetic information by peripheral sensory receptors. Ann Rev Neurosci 5: 171–187

Caiti A, Canepa G, De Rossi D, Germagnoli F, Magenes G, Parisini T (1995) Towards the realization of an artificial tactile system – fine-form discrimination by a tensorial tactile sensor array and neural inversion algorithms. IEEE Trans Syst Man Cybern 25: 533–546

Canepa G, Petrigliano R, Campanella M, De Rossi D (1998) Detection of incipient object slippage by skin-like sensing and neural network processing. IEEE Trans Syst Man Cybern B 20: 348–356

Chou CP, Hannaford B (1996) Measurements and modeling of McKibben pneumatic artificial muscles. IEEE Trans Robotics Automation 12: 90–102

Clark FJ, Horch KW (1986) Kinesthesia. In: Boff KR, Kaufman L, Thomas JP (eds): Handbook of Perception and Human Performance. Wiley, New York, pp 1–62

De Rossi D et al. (2001a) Sensing threads and fabrics for monitoring body kinematic and vital signs. Proceedings of fibers and textiles for the future, Tampere, Finland

De Rossi D, Della Santa A, Mazzoldi A (1999) Dressware: wearable hardware. Mat Sci Eng C 7: 31–35

De Rossi D, Kajiwara K, Yamauchi A, Osada Y (1990) Polymers Gels – Fundamentals and Biomedical Applications. Plenum Press, London

De Rossi D, Lorussi F, Mazzoldi A, Scilingo EP, Rocchia W (2001b) Strain amplified electroactive conducting polymer actuator. Proceedings of SPIE [4329-07]

Feldman AG (1980) Superposition of motor programs I & II, Neurosci 5: 81–90, 91–95

Feldman AG, Latash ML (1982) Interaction of afferent and efferent signals underlying joint position sense: Empirical and theoretical approaches. J Motor Behav 14: 174–93

Flash T, Hogan N (1985) The co-ordination of arm movements: An experimentally confirmed mathematical model. J Neurosci 5: 1688–1703

Honk JC, Rymer WZ (1999) Neural control of muscle length and tension. The nervous system. In: Brooks VB (ed) Handbook of Physiology, Vol II, Part 2. Am Physiol Soc, Baltimore, USA

Mazzoldi A, Della Santa A, De Rossi D (2000) Conducting polymers actuators: properties and modeling. In: Osada Y, De Rossi D (eds) Polymers, Sensors and Actuators. Springer, Berlin, pp 207–244

Mulder A (1994) Human movements tracking technology. Hand Centered Studies of Human Movement Project. Tech Rep, Simon Fraser University, School of Kinesiology

Rack PM, Westbury DR (1969) The effects of length and stimulus rate on tension in the isometric cat soleus muscle. J Physiol 204: 443–460

Scilingo EP, Lorussi F, Mazzoldi A, De Rossi D (2002) Strain sensing fabrics for wearable kinaesthetic systems, IEEE Sensors J, in press

Uno Y, Kawato M, Suzuki R (1989) Formation and control of optimal trajectory in human multijoint arm movement -minimum torque-change model. Biol Cybernetics 61: 89–101

Subject Index*

A

absolute sensitivity   6, 54, 61, 130, 138, 145, 152, 159, 166, 176
acceleration detectors   175, 176
acoustic – electrical analog of ear function   **64**
– inputs, cricket ear   **43**
– map   260
– wave devices   340
action potentials   5, 154, 166
active oscillations   73
– dressware   379, 381
– glove   389
– materials   385
– oscillation   231
– perception   232
adaptation   134, 191, 279
– mechanism   73
adaptive responses   **192**
adsorption cantilever   340
aerial mini-robot   **227**
– robots   230
aerodynamics   325
aerodynamic isolation   333
– sampling   324
– sniffer   **332**, 334
air exchange rate   214
airflow, canine olfaction   326
airjets   331, 329
alar fold   **328**
Ambystoma tigrinum   307
"animaloid" robots   119
array, tactile sensors   116
arthropod mechanosensory hair   **161**, **163**
arthropods, tactile sensor   159
artificial hair arrays   139
– kinesthesia   387
– microsensor, hair flow-sensitive   139
– muscle   383, 385
– nose   305

B

babybot   252, **253**
background noise   10, 156, 179, 183
bacteria, movement   295, 299
bats   37, 60, **61**
bee   9, 238
beetles   **9**
behavior   8
Békésy von   103
bending, tactile hair   **164**
– moment   164, **165**
– response, cantilever   **349**
benzene   348
best frequency, cochlea   **104**
binocular disparity   255
biocomponents   27
biologist's view   3
biomimetic actuation   389
– systems, flow-sensitive hair   139, 143
biomimicry   330
bionics   12
biorobotics   118, 224, 245, 253, 259
biosensors   **28**
bird   9
blood flow   187, 190, 198
Boltzmann constant   54, 344
boundary layer   150, 162, 174, 333
– layer thickness   132
Bragg gratings   368
– grating sensors   370
Brownian motion   145, 154, 292, 294
bulk micromachining   112
bush crickets   45

– olfactory system   **314**, **315**, **318**
– sensing, kinesthetic system   381
– sensors, flow-sensitive hair   129, 130, 140
attributed relational graph   276, **281**
auxiliary structures   4
aviation security   333
axial ratio, gradient following   295
– stresses, arthropod hair   165, **166**, **167**

* Bold numbers refer to pages with figures.

SpringerMedicine

Santiago Ramón y Cajal

Texture of the Nervous System of Man and the Vertebrates

An annotated and edited translation of the original Spanish text
with the additions of the French version by Pedro Pasik and Tauba Pasik

Volume I
1999. XL, 631 pages. 270 figures, partly in colour.
Hardcover **EUR 161,–***. ISBN 3-211-83057-X

Volume II
2000. XIII, 667 pages. 361 figures, partly in colour.
Hardcover **EUR 161,–***. ISBN 3-211-83201-7

Volume III
2002. XII, 663 pages. 377 figures, partly in colour.
Hardcover **EUR 161,–***. ISBN 3-211-83202-5

Set: 3 Volumes
2002. 2026 pages. 1008 figures, partly in colour.
Set-price: Hardcover **EUR 363,–*** (approx. 25 % off the regular price)
ISBN 3-211-83056-1

* Recommended retail prices.
Net-prices subjet to local VAT.

The three volumes contain the works and thoughts of Santiago Ramón y Cajal, originally published in Spanish as the Texture of the Nervous System of Man and the Vertebrates (1899–1904), and later translated into French (1909–1911). These non-English versions have been quoted close to 1000 times over the last five years, and are therefore beyond mere historical importance.

Springer Wien New York

A-1201 Wien, Sachsenplatz 4–6, P.O. Box 89, Fax +43.1.330 24 26, e-mail: books@springer.at, Internet: **www.springer.at**
D-69126 Heidelberg, Haberstraße 7, Fax +49.6221.345-229, e-mail: orders@springer.de
USA, Secaucus, NJ 07096-2485, P.O. Box 2485, Fax +1.201.348-4505, e-mail: orders@springer-ny.com
Eastern Book Service, Japan, Tokyo 113, 3–13, Hongo 3-chome, Bunkyo-ku, Fax +81.3.38 18 08 64, e-mail: orders@svt-ebs.co.jp

SpringerLifeSciences

Facundo Valverde

Golgi Atlas of the Postnatal Mouse Brain

1998. XII, 146 pages. 50 figures and 2 plates.
Hardcover EUR 106,–
(Recommended retail price)
Net-price subject to local VAT.
ISBN 3-211-83063-4

The Atlas provides a complete overview of all major structures of the mouse brain that can be identified in Golgi preparations. The most important feature is its three-dimensional integrity since all structures and nerve tracts can be followed from one section to the next one with uninterrupted continuity.

The Golgi Atlas presents a series of camera lucida drawings of the entire telencephalon and upper brain stem of the young postnatal mouse in 24 transverse, 11 sagittal and 15 horizontal planes. The drawings were prepared from selected brains stained in toto with the Golgi method, that have been serially sectioned in the three orthogonal planes.

The text includes an introduction of the material and methods used for the construction of this Atlas and a survey with a complete bibliography on the previous studies made with the Golgi method in Rodents. In this account, a number of issues concerning particular anatomical details are considered in relation to the interpretations obtained by other students. Reference is made to some relevant reviews and key articles.

SpringerWienNewYork

A-1201 Wien, Sachsenplatz 4–6, P.O. Box 89, Fax +43.1.330 24 26, e-mail: books@springer.at, Internet: **www.springer.at**
D-69126 Heidelberg, Haberstraße 7, Fax +49.6221.345-229, e-mail: orders@springer.de
USA, Secaucus, NJ 07096-2485, P.O. Box 2485, Fax +1.201.348-4505, e-mail: orders@springer-ny.com
Eastern Book Service, Japan, Tokyo 113, 3–13, Hongo 3-chome, Bunkyo-ku, Fax +81.3.38 18 08 64, e-mail: orders@svt-ebs.co.jp

Springer-Verlag
and the Environment

WE AT SPRINGER-VERLAG FIRMLY BELIEVE THAT AN international science publisher has a special obligation to the environment, and our corporate policies consistently reflect this conviction.

WE ALSO EXPECT OUR BUSINESS PARTNERS – PRINTERS, paper mills, packaging manufacturers, etc. – to commit themselves to using environmentally friendly materials and production processes.

THE PAPER IN THIS BOOK IS MADE FROM NO-CHLORINE pulp and is acid free, in conformance with international standards for paper permanency.